CISM International Centre for Mechanical Sciences

Courses and Lectures

Volume 581

Series editors

The Rectors
Elisabeth Guazzelli, Marseille, France
Franz G. Rammerstorfer, Vienna, Austria
Wolfgang. A. Wall, Munich, Germany

The Secretary General
Bernhard Schrefler, Padua, Italy

Executive Editor
Paolo Serafini, Udine, Italy

The series presents lecture notes, monographs, edited works and proceedings in the field of Mechanics, Engineering, Computer Science and Applied Mathematics. Purpose of the series is to make known in the international scientific and technical community results obtained in some of the activities organized by CISM, the International Centre for Mechanical Sciences.

More information about this series at http://www.springer.com/series/76

Jörg Schröder · Doru C. Lupascu
Editors

Ferroic Functional Materials

Experiment, Modeling and Simulation

Springer

Editors
Jörg Schröder
Institute of Mechanics
University of Duisburg-Essen
Essen
Germany

Doru C. Lupascu
Instiute for Materials Science
University of Duisburg-Essen
Essen
Germany

ISSN 0254-1971 ISSN 2309-3706 (electronic)
CISM International Centre for Mechanical Sciences
ISBN 978-3-319-68881-7 ISBN 978-3-319-68883-1 (eBook)
https://doi.org/10.1007/978-3-319-68883-1

Library of Congress Control Number: 2017955266

Printed on acid-free paper

This Springer imprint is published by Springer Nature
The registered company is Springer International Publishing AG
The registered company address is: Gewerbestrasse 11, 6330 Cham, Switzerland

Preface

Functional materials play a key role in many modern technical devices ranging from consumer market items to applications in high-end equipment for automotive, aircraft and spacecraft, medical, and information technologies. Among functional materials, smart materials represent a class that transforms one basic physical property into another. The development of devices utilizing smart materials, as well as their testing, is generally very expensive. Therefore, considerable effort has been made to develop modeling tools that allow bypassing many of the experimental steps previously required in design. Important smart materials are ferroelectrics (coupling between electric polarization and strain), ferromagnets (coupling between magnetization and strain), shape-memory alloys (coupling between temperature and strain), and magnetoelectric multiferroics (coupling between electric polarization and magnetization). The latter ones combine the mutual controllability of magnetic and electric state variables. They are of great interest in the development of multifunctional devices. In single-phase multiferroics, the magnetoelectric interaction is generally very weak and mostly occurs at cryogenic temperatures. Therefore, the experimental preparation and characterization of composite materials, as well as their constitutive description based on homogenization strategies, are key challenges for the optimization of such magnetoelectric composites. The development of such composites made from two different ferroics is based on a comprehensive understanding of both the experimental and theoretical details of these materials. Thus, this CISM course covers the modeling of ferroelectric materials, ferromagnetic materials and shape-memory alloys, the formation of ferroic microstructures and their continuum-mechanical modeling, the experimental preparation and characterization of magnetoelectric multiferroics, computational homogenization, and the algorithmic treatment in the framework of numerical solution strategies.

The CISM course on "Ferroic Functional Materials: Experiment, Modeling, and Simulation," held in Udine from September 8 to 12, 2014, was addressing doctoral students and postdoctoral researchers in civil and mechanical engineering, materials science, physics and applied mathematics, and industrial researchers who wished to broaden their knowledge in experiments and theory of ferroic materials. The main focus was on the state-of-the-art experimental methods and advanced modeling techniques, which are essential to qualify young scientists for high-quality research, and the development of innovative products and applications.

It is our pleasure to thank the lecturers of the CISM course Kaushik Bhattacharya (Pasadena, USA), Manfred Fiebig (Zurich, Switzerland), John Huber (Oxford, UK), Christopher Lynch (Los Angeles, USA), Marc-André Keip (Stuttgart, Germany) as well as the additional contributors to these CISM lecture notes Dorinamaria Carka (Old Westbury, USA), Irina Anusca (Essen, Germany), Morad Etier (Essen, Germany), Yanling Gao (Essen, Germany), Gerhard Lackner (Essen, Germany), Ahmadshah Nazrabi (Essen, Germany), Mehmet Sanlialp (Essen, Germany), Harshkumar Trivedi (Essen, Germany), Matthias Labusch (Essen, Germany), and Naveed Ul-Haq (Essen, Germany). We furthermore thank the participants who made the course a success. Finally, we extend our thanks to the Rectors, the Board, and the staff of CISM for the excellent support and kind help.

Essen, Germany

Jörg Schröder
Doru C. Lupascu

Contents

Fundamentals of Magneto-Electro-Mechanical Couplings: Continuum Formulations and Invariant Requirements

Jörg Schröder

Abstract Couplings of magnetic and electric fields in materials could allow for promising applications in medical and information technology. In this contribution, we recapitulate well-known aspects of magneto-electro-mechanical properties and their couplings. At first, we echo basic aspects of electricity and magnetism and Maxwell's equations. Secondly, we summarize the governing equations for electrostatics and magnetostatics, point out the properties of physical fields across internal surfaces, and discuss the work-energy theorem of electrodynamics, the so-called Poynting's theorem. Thirdly, we will discuss some fundamental concepts of magneto-electro-mechanical couplings in matter. Here, we will formulate thermodynamic potentials depending on different basic variables in order to be flexible with a view to different modeling aspects. Afterwards, we discuss aspects of form-invariance of physical laws under coordinate transformations: Lorentz invariance, Galilean transformation and time reversal. Here, we focus on piezoelectric as well as on magnetic symmetry groups and give remarks on classical invariant theory suitable for coordinate-invariant modeling of thermodynamical potentials.

Parts of work presented in this contribution are taken from common works together with my former co-workers Holger Romanowski, Ingo Kurzhöfer, Marc-André Keip, and Matthias Labusch. The author greatly appreciates the "Deutsche Forschungsgemeinschaft" (DFG) for the financial support under the research grant SCHR 570/12-1 within the research group FOR 1509 on "Ferroische Funktionsmaterialien—Mehrskalige Modellierung und experimentelle Charakterisierung".

J. Schröder (✉)
Faculty of Engineering, Department Civil Engineering, Institute of Mechanics,
University of Duisburg-Essen, Essen, Germany
e-mail: j.schroeder@uni-due.de

J. Schröder and Doru C. Lupascu (eds.), *Ferroic Functional Materials*,
CISM International Centre for Mechanical Sciences 581,
https://doi.org/10.1007/978-3-319-68883-1_1

1 Introduction

In this course, we are interested in the continuum modeling of ferroic functional materials. Of particular interest are materials which allow for couplings between different physical quantities as for instance coupling between electric and mechanical (ferroelectrics), mechanical and magnetic (ferromagnetics), or magnetic and electric fields (multiferroics), which are defined as

- ferroelectric materials have a spontaneous electric polarization, where the internal electric dipoles are coupled to the material lattice, that can be switched by an applied electric field,
- ferromagnetic materials, governed by their crystalline structure and microstructure, have a spontaneous magnetic polarization that can be reversed by a magnetic field, and
- multiferroic materials exhibit two or more ferroic properties, like ferroelectricity, ferromagnetism, and ferroelasticity, in the same phase.

However, the coexistence of magnetic and electric orderings in the same phase is rather difficult. Based on theoretical studies, it has been shown that usual atomic-level mechanisms which are driving ferromagnetism and ferroelectricity are mutually exclusive. For example, ferromagnetism requires partially filled orbitals ("d"-shells) and ferroelectricity empty orbitals in the atomic structure. In general, the properties of functional materials emerge on different scales. Some exist on the atomic scale, as for example the magnetization. Others, as e.g. the electric polarization, are present on the unit cell level of a crystal. Furthermore, some materials obtain their functional properties only when the above quantities couple over a larger length scale, as for instance in case of multiferroic composites.

In this course, we will recapitulate well known basic properties, balance equations and some formulations of thermodynamics. Furthermore, we will focus on some transformation properties: Physical laws, valid in a frame of reference, satisfy specific symmetry conditions. For example, the symmetry which reflects the form-invariance of physical laws under coordinate transformations. Invariance means, that some physical quantities remain unchanged under specific transformations. In order to get a deeper insight into the mathematical modeling of electro-mechanically coupled materials, we discuss the properties of the associated physical quantities under the coordinate-transformations: rotations, spatial reflections, and time-reversal.

An overview of electromagnetic theories is given in Landau and Lifschitz (1985), Fabrizio and Morro (2003), Tipler (1999), Jackson (2002), Griffiths (2008), Bobbio (2000), Fließbach (1999), Fließbach (2000), Grehn and Krause (2007), and Zohdi (2012). For a detailed discussion of Maxwell Equations see Maugin et al. (1991), Eringen and Maugin (1989), Eringen and Maugin (1990) or Weile et al. (2014).

2 Foundations of Magneto-Electro Couplings

The field equations of classical electrodynamics, also denoted as classical electromagnetism, are associated with the scientists Michael Faraday[1] and James Clerk Maxwell.[2] In this field, we study the interactions between electric charges and currents. This theory provides a set of equations which allow for a suitable description of electromagnetic phenomena on relevant (large enough) length scales and field strengths, i.e. when quantum mechanical effects can be neglected. The four underlying partial differential equations of electromagnetism, the so-called Maxwell equations, are the

- Gauß's law for electric fields
- Gauß's law for magnetic fields
- Faraday's law of induction
- (extended) Ampère's law

have firstly presented in their complete form in Maxwell's textbooks *"A Treatise on Electricity and Magnetism"* in 1873, see Maxwell (1873). Another important contribution of him is Maxwell (1865) in which he shows that electric and magnetic fields travel through space as waves at the speed of light. The achievements of Maxwell in the field of electromagnetism have been called the

"second great unification in physics",

see Nahin (1992).

In the following sections, we discuss the basic characteristics of electric and magnetic fields, introduce the notion of charge, conductor, insulator, equipotential surface, electric field, electric potential, polarization, magnetic force, current and give some remarks concerning the distinctive features of electrodynamics in free space and in solids. Then, we summarize the set of Maxwell's equations, Ampère's law, Faraday's law, Gauß's law for electric field, and Gauß's law for magnetic fields, in integral and differential form. Furthermore, we emphasize the jump conditions of electric and magnetic quantities across singular surfaces. Finally, we derive the electromagnetic wave equations in vacuum.

---

[1]Michael Faraday, 1791–1867, English scientist and excellent experimentalist, electromagnetism: electromagnetic induction, diamagnetism.

[2]James Clerk Maxwell, 1831–1879, Scottish scientist, electrodynamics: formulation of the complete set of equations that describe electricity, magnetism, and optics.

2.1 Preliminaries, Definitions and Units

Notation: Mechanical, Electrical, Magnetical Quantities, SI-units. In Table 1, we summarize the basic physical units necessary for the description of mechanical, electrical and magnetical quantities.[3]

The International System of units, denoted by SI (french: Le Système international d'unités), are based on the meter-kilogram-second (MKS) system. Another, often used framework, especially in electrodynamics, is based on the centimeter-gram-second (CGS) system, which unfortunately has several variants. The conversion factors c_f are needed to compute the Gaussian quantities in terms of the SI-Units. Here, the parameter α is a parameter chosen as a factor for the speed of light c:

$$c = \alpha \cdot 10^8 \, \frac{\text{m}}{\text{s}} \quad \text{with} \quad \alpha = 2.99792458 \approx 3.0 \,. \tag{1}$$

Table 1 Physical quantities, SI-Units, Gaussian-Units

Quantity	Symbol	SI-Unit; abbr. (MKS-system)	Gaussian-Unit (CGS-system)	Factor (c_f)
length	l	meter; m	centimeter	10^2
mass	m	kilogram; kg	gram	10^3
time	t	second; s	second	1
force	$\boldsymbol{F}$	newton; N	dyne	10^5
energy	$\mathcal{E}$	joule; J = Ws = CV	erg	10^7
power	$\mathcal{P}$	watt; W	erg/second	10^7
charge	q	coulomb; C = As	esu	$\alpha \cdot 10^9$
current	$\boldsymbol{I}$	ampere; A = C/s	esu/second	$\alpha \cdot 10^9$
electric–				
- field	$\boldsymbol{E}$	$\frac{\text{volt}}{\text{meter}}$; V/m	$\frac{\text{statvolt}}{\text{centimeter}}$	$\alpha^{-1} \cdot 10^{-4}$
- potential	ϕ^e	volt; V = J/C	statvolt	$1/300$
- displacement	$\boldsymbol{D}$	$\frac{\text{coulomb}}{\text{meter}^2}$; C/m^2	$\frac{\text{statcoulomb}}{\text{centimeter}^2}$	$\alpha \, 4\pi \cdot 10^5$
capacitance	C_p	farad; F = C/V	centimeter	$\alpha^2 \cdot 10^{11}$
magnetic–				
- field	$\boldsymbol{B}$	tesla; T	gauss	10^4
- auxiliary field	$\boldsymbol{H}$	$\frac{\text{ampere}}{\text{meter}}$; A/m	oersted	$4\pi \cdot 10^{-3}$

[3]The auxiliary field $\boldsymbol{H}$ has no proper name. For a discussion of the babel about this, see A. Sommerfeld, Electrodynamics, NY, Academic press, 1952, page 45.

The magnetic field $\boldsymbol{B}$ is sometimes also denoted as *magnetic induction* or *flux density*, in this context see the criticism of Griffiths (2008), page 271.

Two often used constants are the *permittivity constant* of free space

$$\epsilon_0 = 8.85 \cdot 10^{-12} \, \frac{\mathrm{C}^2}{\mathrm{N\,m}^2} \tag{2}$$

and the *permeability* of free space

$$\mu_0 = 4\,\pi \cdot 10^{-7} \, \frac{\mathrm{N}}{\mathrm{A}^2} \, . \tag{3}$$

The latter two constants satisfy the fundamental relation

$$\epsilon_0 \, \mu_0 = \frac{1}{c^2} \, . \tag{4}$$

In the following, we denote the SI-Unit of a physical quantity $(\bullet)$ by $[(\bullet)]_{\mathrm{SI}}$, for example: $[\phi^e]_{\mathrm{SI}}$ means

$$[\phi^e]_{\mathrm{SI}} = \frac{\mathrm{J}}{\mathrm{C}} =: \mathrm{V} \, . \tag{5}$$

A fundamental exercise in the field of electromagnetics is the computation of forces acting on electric charges, as well as the trajectories of the charges. The applied principle of superposition facilitates this problem. Firstly, we focus on the fundamentals in electrostatics. Secondly, we discuss some phenomenon in magnetism. Finally, we summarize the Maxwell's equations.

2.2 A Primer in Electrostatics

Charge, Conductor, Insulator, Equipotential Surface. The electric charge is not a continuous field; in fact it is quantized. The charge is given by multiples ($\pm 1, \pm 2, \ldots$) of the *elementary charge*

$$e = 1.602177 \cdot 10^{-19} \, \mathrm{C} \, . \tag{6}$$

Another fundamental law of nature is the

law of conservation of charge,

which was firstly discussed by Benjamin Franklin.[4] It states, that the net charge of an isolated system cannot change. Many materials allow for the free movement of electrons, e.g. metals. This kind of materials are called conductors. On contrary, if the electrons cannot move freely within the materials, these materials are insulators. Adjacent points with identical potential values form an equipotential surface. Excess

[4]Benjamin Franklin, 1706–1790, American scientist and statesman.

charges on an isolated conductor will distribute themselves on its surface in such a way that the whole conductor has a constant potential. Therefore, the free movement of charges in conducting media induces that every conducting surface is an equipotential surface, see Fig. 1.

Michael Faraday introduced the idea of electric field lines, which extend away from positive charge towards negative charge. The tangents to the electric field lines define the direction of the vectorial electric field $\boldsymbol{E}$. In fact, we infer that the electric field $\boldsymbol{E}$ (outside the conductor) is perpendicular to the equipotential surface, see Fig. 2:

$$\boldsymbol{E} \perp \boldsymbol{t} \quad \text{equivalently} \quad \boldsymbol{E} \cdot \boldsymbol{t} = 0 \,, \tag{7}$$

where $\boldsymbol{t}$ characterizes the tangent vectors with respect to the equipotential surfaces.

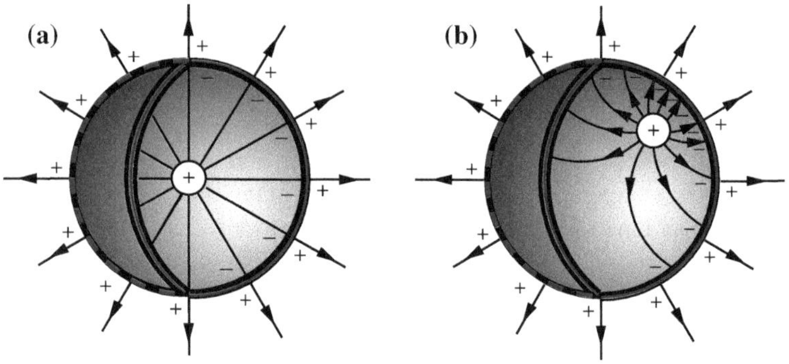

Fig. 1 A point charge in a conduction spherical shell. The electric fields outside the sphere are identical in both cases: **a** centered point charge, **b** non-centered point charge. Conducting surfaces are equipotential surfaces

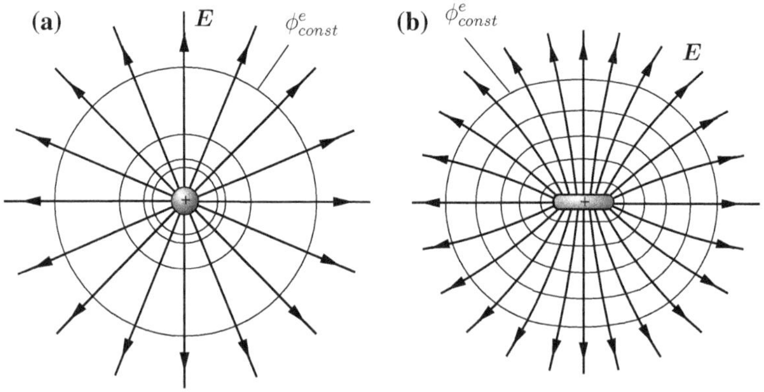

Fig. 2 Equipotential surfaces ϕ^e = constant (ϕ^e_{const}) and electric field lines $\boldsymbol{E}$ outside uniformly charged conductors

Electric field, electric potential. The amplitude of the electrostatic force F_C between two point charges q_1 and q_2 which are separated by the distance r is given by *Coulomb's law*

$$F_C = k\frac{|q_1|\,|q_2|}{r^2}\ . \tag{8}$$

This is a repulsion force for equal charged particles and an attraction force for differently charges particles. The extension of Coulomb's law to charge distribution in space, on surfaces and lines yield to integral expressions for the electric field, which constitute that the curl of the electric field vanishes

$$\operatorname{curl}\boldsymbol{E} = \boldsymbol{0}\ . \tag{9}$$

The *electrostatic constant* k, defined in terms of the *permittivity constant* of free space ϵ_0, has the value

$$k = \frac{1}{4\,\pi\,\epsilon_0} = 8.99\cdot 10^9\frac{\mathrm{N\,m^2}}{\mathrm{C^2}}\ . \tag{10}$$

The electrostatic force is conservative, thus we can assign an *electric potential energy* ψ^{el}. Therefore, the work done by the electrostatic force is path independent. Thus, we can calculate the electric work W^{el} by the change of the electric potential between two different states of interest:

$$-\,W^{el} = \Delta\psi^{el} = \psi^{el}_{final} - \psi^{el}_{initial}\ . \tag{11}$$

Without any further limitation, we set the reference potential energy to zero, i.e. $\psi^{el}_{initial} := 0$. The electric potential per unit charge is defined by

$$\phi^e = -W^{el}/q \quad \text{in} \quad [\phi^e]_{\mathrm{SI}} = \frac{\mathrm{J}}{\mathrm{C}} =: \mathrm{V}\ . \tag{12}$$

In order to compute the electric potential ϕ^e from the electric field $\boldsymbol{E}$, we consider the work W

$$W^{el} = \int \mathrm{d}W \quad \text{with} \quad \mathrm{d}W = \boldsymbol{F}_C\cdot\mathrm{d}\boldsymbol{s}\ . \tag{13}$$

Substituting $\boldsymbol{F}_C = q\boldsymbol{E}$ yields

$$W^{el} = \int q\;\boldsymbol{E}\cdot\mathrm{d}\boldsymbol{s} = \int \boldsymbol{E}\cdot\mathrm{d}\boldsymbol{D} =: -q\;\phi^e\ , \tag{14}$$

where (d)$\boldsymbol{D}$ denotes the (differential) electric displacement. A positively (negatively) charged particle produces a positive (negative) electric potential. Obviously, we can compute the electric field by the negative gradient of the electric potential:

$$E = -\mathrm{grad}\phi^e \ . \tag{15}$$

The SI-Units of the electric field are

$$[E]_{\mathrm{SI}} = \frac{\mathrm{N}}{\mathrm{C}} = \frac{\mathrm{N}}{\mathrm{C}} \left(\mathrm{V}\frac{\mathrm{C}}{\mathrm{J}}\right) \frac{\mathrm{J}}{\mathrm{Nm}} = \frac{\mathrm{V}}{\mathrm{m}} \ . \tag{16}$$

Note: From (14), we compute the difference of the potential between the initial and final adjacent states by

$$\int_{initial}^{final} E \cdot \mathrm{d}s = -\left(\phi^e(s_{final}) - \phi^e(s_{initial})\right) \ . \tag{17}$$

Considering an equipotential surface $\phi^e(s_{f,final}) = \phi^e(s_{i,initial})$ induces that the electric field in this direction vanishes. Vice versa, for a vanishing electric field, we obtain a constant electric potential.

The electric displacement D in free space as a function of the electric field E is given by the constitutive relation

$$D = \epsilon_0 E \quad \text{with} \quad [D]_{\mathrm{SI}} = \frac{\mathrm{C}}{\mathrm{m}^2} \ . \tag{18}$$

Therefore, the energy density, see (11), is given by

$$\psi^{el} = \int E \cdot \mathrm{d}D = \int \epsilon_0 E \cdot \mathrm{d}E = \frac{1}{2}\epsilon_0 E \cdot E = \frac{1}{2}\epsilon_0 \| E \|^2 \ . \tag{19}$$

The Millikan experiment. The elementary charge can be determined within the Millikan[5] experiment. The test facility of Millikan and Fletcher[6] consists of two parallel horizontal metal plates of distance d, a schematic illustration is given in Fig. 3. Applying an electric potential difference $\Delta\phi^e$ on the parallel plates introduces a constant electric field

$$E = \Delta\phi^e / d \ . \tag{20}$$

Taking into account the acting forces on an oil droplet, the drag force F_d, the weight force F_g, the lifting force F_l, and the Coulomb-force in an electric field

$$F_C = q_e E \ , \tag{21}$$

[5]Robert Andrews Millikan, 1868–1953, American experimental physicist, Nobel laureate in chemistry (1923).

[6]Harvey Fletcher, 1884–1981, American physicist, dissertation on methods to determine the charge of an electron.

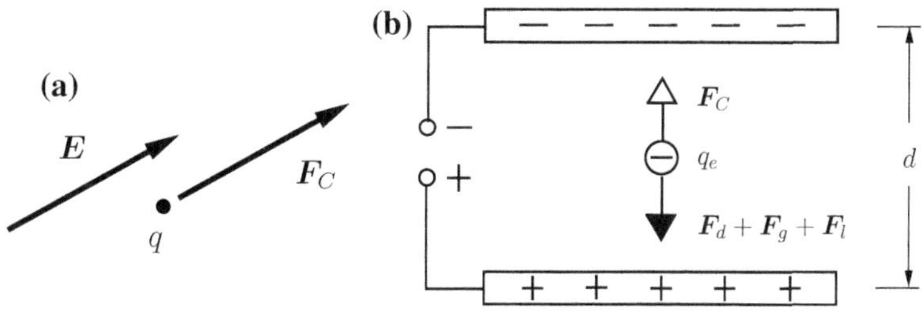

Fig. 3 Forces on a point charge q: **a** Coulomb force $\boldsymbol{F}_C$ in an electric field $\boldsymbol{E}$. **b** Millikan experiment for the computation of elementary charge

allow by evaluation the equilibrium condition, $\boldsymbol{F}_C + \boldsymbol{F}_d + \boldsymbol{F}_g + \boldsymbol{F}_l = \boldsymbol{0}$, for the computation of the electric charge q_e.

Law of conservation of charges. The continuity equation states, that the density of charges in a domain $\mathcal{B}$ can only change if there is a net flux of charges across the surface $\partial\mathcal{B}$ of the domain:

$$\int_{\partial\mathcal{B}} \boldsymbol{J} \cdot d\boldsymbol{a} = -\int_{\mathcal{B}} \dot{\rho}_e \, dv \,, \quad \text{local form:} \;\; \text{div}\boldsymbol{J} = -\frac{\partial \rho_e}{\partial t} \,, \tag{22}$$

where $\frac{\partial \rho_e}{\partial t}$ denotes the time derivative of the charge density ρ_e.

Gauß's law for electric fields. The flux of the electric field $\boldsymbol{E}$ through a surface $\partial\mathcal{B}$ of the body $\mathcal{B}$ is defined by

$$\int_{\partial\mathcal{B}} \boldsymbol{E} \cdot \mathrm{d}\boldsymbol{a} = \int_{\partial\mathcal{B}} \boldsymbol{E} \cdot \boldsymbol{n} \, \mathrm{d}a \,, \tag{23}$$

with the vectorial area element $\mathrm{d}\boldsymbol{a} = \boldsymbol{n}\,\mathrm{d}a$ and the outward uni normal $\boldsymbol{n}$. Thus, the flux through a closed surface is the measure of the net charge of the volume inside the surface:

$$\oint_{\partial\mathcal{B}} \boldsymbol{E} \cdot \mathrm{d}\boldsymbol{a} = \frac{Q_V}{\epsilon_0} \,, \tag{24}$$

where Q_V denotes the total charge of the volume V enclosed by $\partial\mathcal{B}$. It should be noted, that the surface-independency of Gauß's law is a result of the structure of Coulomb's law, in this context see e.g. Griffiths (2008). Applying the divergence theorem and substituting the charge density ρ_e,

$$\oint_{\partial\mathcal{B}} \boldsymbol{E} \cdot \mathrm{d}\boldsymbol{a} = \int_{\mathcal{B}} \text{div}\boldsymbol{E} \, \mathrm{d}v \quad \text{and} \quad \frac{Q_V}{\epsilon_0} = \int_{\mathcal{B}} \frac{\rho_e}{\epsilon_0} \, \mathrm{d}v \,, \tag{25}$$

respectively, yields

$$\int_{\mathcal{B}} \operatorname{div} \boldsymbol{E} \, \mathrm{d}v - \int_{\mathcal{B}} \frac{\rho_e}{\epsilon_0} \, \mathrm{d}v = 0 \quad \rightarrow \quad \operatorname{div} \boldsymbol{E} = \frac{\rho_e}{\epsilon_0} , \tag{26}$$

where $(26)_2$ is the local form of Gauß's law.

Electric fields in matter. The above considerations have been carried out for free space. Here, we make some extensions for electric fields in matter, where we assume that the fields are averaged over a suitable small domain in order to get rid of undesirable fluctuations. If we fill the free space with a *dielectric* material, we have to take into account the *dielectric constant* ϵ_r of this insulating material. As an example, we consider the capacitance $C_{p-plate}$ of a parallel-plate capacitor: plates with area A, distance between the plates d, charge of top plate $+q$, charge of bottom plate $-q$. $C_{p-plate}$ is the proportionality constant between the charge an der potential difference, i.e.

$$q = C_{p-plate} \, \Delta\phi^e \quad \text{with} \quad [C_{p-plate}]_{\mathrm{SI}} = \frac{\mathrm{C}}{\mathrm{V}} =: \mathrm{F} \, . \tag{27}$$

The SI unit for the capacitance is *farad* (F). Faraday has discovered, that the capacitance of a capacitor with a dielectric material $C^{dielectric}_{p-plate}$ is κ-times the value of the capacitor with air between the plates $C^{air}_{p-plate}$:

$$C^{dielectric}_{p-plate} = \epsilon_r \, C^{air}_{p-plate} \quad \text{and} \quad C^{air}_{p-plate} = \varepsilon_0 \, A/d \, . \tag{28}$$

Here, we have replaced the permittivity constant of free space ε_0 with $\epsilon_r\varepsilon_0$ in the regions completely filled with a dielectric material exhibiting the dielectric constant $\epsilon_r \geq 1$. In (homogeneous, linear) anisotropic media the scalar-valued permittivity constant, also denoted as the dielectric constant, has to be replaced by the second-order tensor, the dielectric moduli $\boldsymbol{\epsilon}$:

$$\boldsymbol{D} = \boldsymbol{\epsilon} \cdot \boldsymbol{E} \quad \text{with} \quad \boldsymbol{\epsilon} := \epsilon_0 \boldsymbol{\epsilon}_r \, . \tag{29}$$

An electric field acting on a dielectric material causes a dipole-moment in field direction. A measure for this is the polarization vector $\boldsymbol{P}$, characterizing the dipole-moment per unit-volume. This polarization influences the effect of the charge density described in $(26)_2$. In order to differentiate between bounded and free (effective) charges, we introduce the density of bounded charges ρ_b and the density of free (effective) charges ρ_f. Both are related to the total charge density by

$$\rho_e = \rho_b + \rho_f \, . \tag{30}$$

Replacing ρ_e by $(\rho_f + \rho_b)$ in $(26)_2$ yields the modified expression

$$\operatorname{div} \boldsymbol{E} = \frac{\rho_f + \rho_b}{\epsilon_0} \, . \tag{31}$$

The bounded charge density ρ_b is associated to the polarization $\boldsymbol{P}$ of the material by

$$\operatorname{div}\boldsymbol{P} = -\rho_b \ . \tag{32}$$

Therefore, we reformulate the expression for the free charge density:

$$\rho_f = \rho_e - \rho_b = \rho_e + \operatorname{div}\boldsymbol{P} \ . \tag{33}$$

From Gauß's law (31), we conclude

$$\operatorname{div}[\epsilon_0 \boldsymbol{E} + \boldsymbol{P}] = \rho_f \ . \tag{34}$$

Substituting the electric displacement $\boldsymbol{D}$, in the latter equation, with

$$\boldsymbol{D} = \epsilon_0 \boldsymbol{E} + \boldsymbol{P} \ . \tag{35}$$

leads to the following differential form of Gauß's law

$$\operatorname{div}\boldsymbol{D} = \rho_f \ . \tag{36}$$

Computing the curl of (35) and taking (9) into account, we obtain

$$\operatorname{curl}\boldsymbol{D} = \operatorname{curl}\boldsymbol{P} \ . \tag{37}$$

Thus, we conclude, that the curl of the electric displacement is identical to the curl of the polarization.

2.3 A Primer in Magnetostatics

Biot-Savart law. Steady currents I produce magnetic fields $\boldsymbol{B}$, which are constant in time; this defines the field of magnetostatics. The Biot-Savart law (Fig. 4) yields for a steady line current I the infinitesimal magnetic field

$$\mathrm{d}\boldsymbol{B} = \frac{\mu_0}{4\pi}\,\frac{I\ \mathrm{d}\boldsymbol{s} \times \boldsymbol{r}}{r^2} \ . \tag{38}$$

Here, the electric current $\{I, [I]_{\mathrm{SI}} = \mathrm{A} = \mathrm{C/s}\}$ is defined by the rate of flow of the electric charge through a cross sectional area of a wire. The direction of the current is identical with the direction of the flow of positive charge.The term $\mathrm{d}\boldsymbol{s}$ denotes the infinitesimal length of a segment of the wire. Replacing the term $I\ d\boldsymbol{s}$ in (38) by $q\ \boldsymbol{v}$, returns an expression of the magnetic field of a moving point charge

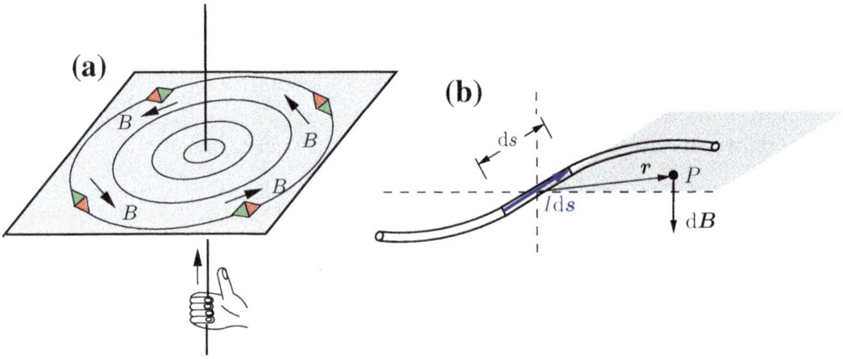

Fig. 4 **a** Around a straight wire carrying an electric current, the magnetic field strength $\boldsymbol{B}$ is organized in concentric circles. **b** Biot-Savart law: magnetic field $\mathrm{d}\boldsymbol{B}$ at point P due to current element $I\ \mathrm{d}\boldsymbol{s}$

$$\boldsymbol{B} = \frac{\mu_0}{4\pi} \frac{q\ \boldsymbol{v} \times \boldsymbol{r}}{r^2}\ , \tag{39}$$

with the velocity $\boldsymbol{v}$. However, the latter statement is imprecise,[7] it only holds for non-relativistic charges retardation-less conditions (Fig. 4).

The SI unit of the magnetic field is called tesla (T):

$$[\boldsymbol{B}]_{\mathrm{SI}} = \mathrm{T} = \frac{\mathrm{N}}{\mathrm{C\ m/s}} = \frac{\mathrm{N}}{\mathrm{A\ m}} = \frac{\mathrm{V\ s}}{\mathrm{m}^2}\ . \tag{40}$$

Magnetic force, currents. In the previous considerations, we have seen that electric charges q set up electric fields $\boldsymbol{E}$. However, the existence of magnetic counterparts within the meaning of charges have not been confirmed. Magnetic fields are set up (i) by moving electric charges and (ii) by elementary particles with intrinsic magnetic fields.

In order to define a magnetic field $\boldsymbol{B}$, we analyze the force $\boldsymbol{F}_L$, the so called Lorentz[8] force, acting on a moving electrically charged particle. The Lorentz force is given by the cross products of the vectors $q\boldsymbol{v}$ and $\boldsymbol{B}$:

$$\boldsymbol{F}_L = q\ \boldsymbol{v} \times \boldsymbol{B}\ , \tag{41}$$

where $\boldsymbol{v}$ is the speed of the moving particle. Obviously, the magnetic force $\boldsymbol{F}_L$ is always perpendicular to the velocity $\boldsymbol{v}$ and the magnetic field $\boldsymbol{B}$, see Fig. 5. The infinitesimal work done by the magnetic force vanishes, i.e.

$$\boldsymbol{F}_L \cdot (\mathrm{d}\boldsymbol{s}) = \boldsymbol{F}_L \cdot (\boldsymbol{v}\ \mathrm{d}t) = (q\ \boldsymbol{v} \times \boldsymbol{B}) \cdot (\boldsymbol{v}\ \mathrm{d}t) = 0\,, \tag{42}$$

[7] In this context see the discussion in Griffiths (2008), page 219.

[8] Hendrik Antoon Lorentz, 1853–1928, Dutch physicist, Nobel laureate in physics (1902).

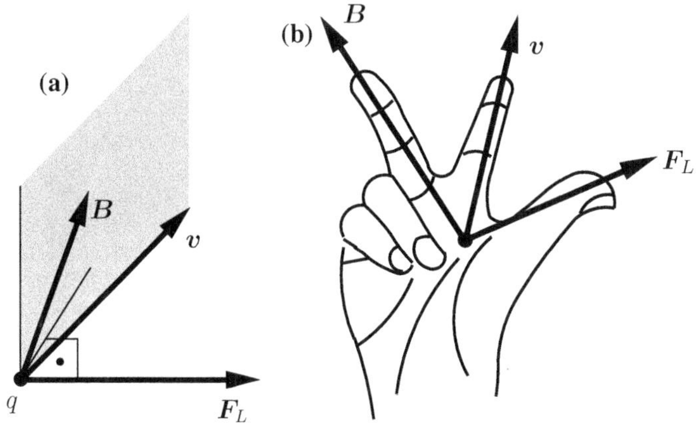

Fig. 5 Force on a point charge q: **a** Lorentz force $\boldsymbol{F}_L$ in a magnetic field $\boldsymbol{B}$. The magnetic field $\boldsymbol{B}$ is defined by its action on moving charges. **b** Right-hand-rule: index finger in the moving direction of the current ($\boldsymbol{v}$), middle finger in direction of magnetic field, thumb (perpendicular w.r.t. index and middle fingers) points in direction of force

because $\boldsymbol{v}$ is perpendicular to $\boldsymbol{v} \times \boldsymbol{B}$. Therefore, the Lorentz force can only alter the direction of the velocity and not its amplitude.

Gauß's law for magnetism. Magnetic fields are source-free, i.e. they are solenoidal fields, because there exist no magnetic monopoles. Therefore, all magnetic field lines are closed. In analogy to (24), with vanishing right hand side, we can define a magnetic flux through a surface. Following the procedure used for the derivations of (25) and (26), we obtain

$$\oint_{\partial\mathcal{B}} \boldsymbol{B} \cdot \mathrm{d}\boldsymbol{a} = 0\,, \qquad \int_{\mathcal{B}} \mathrm{div}\boldsymbol{B}\, \mathrm{d}v = 0 \quad \text{and} \quad \mathrm{div}\boldsymbol{B} = 0\,. \tag{43}$$

Equation $(43)_3$ can be interpreted as Gauß's law of magnetism, compare $(26)_2$. In analogy to $\boldsymbol{D}$ in electrostatics, we introduce an auxiliary field $\boldsymbol{H}$

$$\boldsymbol{H} = \frac{1}{\mu_0}\boldsymbol{B} \quad \text{with} \quad [\boldsymbol{H}]_{\mathrm{SI}} = \frac{\mathrm{A}}{\mathrm{m}}\,, \tag{44}$$

with the *permeability* of free space μ_0.

Electric current. The current measures the *charge per unit time* passing a given point. Let $\boldsymbol{J}$ be the volume current density, which is defined in terms of the mobile volume charge density by

$$\boldsymbol{J} = \rho_f\, \boldsymbol{v} \quad \text{with} \quad [\boldsymbol{J}]_{\mathrm{SI}} = \frac{\mathrm{C}}{\mathrm{s}} = \mathrm{A}\,. \tag{45}$$

This definition yields the following expression for the magnetic force on a body $\mathcal{B}$, compare (41):

$$\int_{\mathcal{B}} \rho_f \; \boldsymbol{v} \times \boldsymbol{B} \; \mathrm{d}v = \int_{\mathcal{B}} \boldsymbol{J} \times \boldsymbol{B} \; \mathrm{d}v \, . \tag{46}$$

The total charge, leaving the volume of $\mathcal{B}$, is

$$\oint_{\partial\mathcal{B}} \boldsymbol{J} \cdot \mathrm{d}\boldsymbol{a} = \int_{\mathcal{B}} \mathrm{div}\boldsymbol{J} \; \mathrm{d}v \, . \tag{47}$$

The global form of conservation of charge is

$$\oint_{\partial\mathcal{B}} \boldsymbol{J} \cdot \mathrm{d}\boldsymbol{a} = -\frac{\mathrm{d}}{\mathrm{d}t} \int_{\mathcal{B}} \rho_f \; \mathrm{d}v \, . \tag{48}$$

Substituting (47) into (48) results in

$$\int_{\mathcal{B}} \mathrm{div}\boldsymbol{J} \; \mathrm{d}v = -\frac{\mathrm{d}}{\mathrm{d}t} \int_{\mathcal{B}} \rho_f \; \mathrm{d}v \quad \rightarrow \quad \mathrm{div}\boldsymbol{J} = -\frac{\partial \rho_f}{\partial t} \, , \tag{49}$$

where $(49)_2$ is the local form of the continuity equation.

Magnetic fields in matter. For the description of the magnetic fields in matter, we always consider fields which are averaged over a suitable ensemble of atoms in order to smooth out highly fluctuating fields.

Under some technical assumptions, we can assume (this is not allowed for all materials, like ferromagnetic materials) a linear relation between the magnetic field $\boldsymbol{B}$ and $\boldsymbol{H}$:

$$\boldsymbol{B} = \boldsymbol{\mu} \cdot \boldsymbol{H} \quad \text{with} \quad \boldsymbol{\mu} := \mu_0 \boldsymbol{\mu}_r \quad \text{and} \quad \boldsymbol{\mu}_r = \mathbf{1} + \chi_M \, , \tag{50}$$

with the second-order tensor of the material permeability $\boldsymbol{\mu}_r$ and the magnetic susceptibility χ_M.

If we split the current $\boldsymbol{J}$ into bound currents $\boldsymbol{J}_b$ and free currents $\boldsymbol{J}_f$, we obtain

$$\boldsymbol{J} = \boldsymbol{J}_b + \boldsymbol{J}_f \quad \text{with} \quad \boldsymbol{J}_b = \mathrm{curl}\boldsymbol{M} \quad \text{and} \quad \mathrm{div}\boldsymbol{J}_b = 0 \, , \tag{51}$$

where $\boldsymbol{M}$ denotes the vector of magnetization. Substituting this, results in Ampère's law, i.e.

$$\frac{1}{\mu_0}\mathrm{curl}\boldsymbol{B} = \boldsymbol{J}_b + \boldsymbol{J}_f = \mathrm{curl}\boldsymbol{M} + \boldsymbol{J}_f \tag{52}$$

and yields

$$\mathrm{curl}\left(\frac{1}{\mu_0}\boldsymbol{B} - \boldsymbol{M}\right) = \boldsymbol{J}_f \, . \tag{53}$$

The term in parenthesis is denoted as the auxiliary field $\boldsymbol{H}$:

$$\boldsymbol{H} = \frac{1}{\mu_0}\boldsymbol{B} - \boldsymbol{M} \,. \tag{54}$$

Using this definition, we reformulate Ampère's law as

$$\operatorname{curl}\boldsymbol{H} = \boldsymbol{J}_f \,. \tag{55}$$

In the following, we skip the index f.

Ohm's law: In electric conductors (metals, semiconductors, electrolyte), charges are carried by electrons or ions. *Ohm's law* is known as the linear relation between the electric current and the electric field

$$\boldsymbol{J} = \sigma_c \, \boldsymbol{E} \,, \tag{56}$$

with the conductivity σ_c. Its resistivity is the reciprocal, i.e. σ_c^{-1}.

2.4 Maxwell's Equations

In this section, we summarize the well known Maxwell equations. This set of equations describe the electromagnetism within a unified theory. They contain the Gauß's laws for the electric and magnetic fields, Faraday's and Ampère's law. Furthermore, they describe the dynamics (forces and trajectories) of electric charges; see the summary in Table 2 with the geometrical quantities defined in Fig. 6. In 1865, Maxwell showed that the classical form of Ampère's law

$$\oint_{\partial\mathcal{B}_S} \boldsymbol{H}\cdot d\boldsymbol{s} = \int_{A_S} \boldsymbol{J}\cdot d\boldsymbol{a} \,, \quad \text{local form:} \quad \operatorname{curl}\boldsymbol{H} = \boldsymbol{J} \,, \tag{57}$$

Table 2 Summary of the Maxwell equations

Maxwell Equations	Integral form	Differential form
Ampère's law	$\oint_{\partial\mathcal{B}_S} \boldsymbol{H}\cdot d\boldsymbol{s} = \int_{A_S} (\boldsymbol{J}+\dot{\boldsymbol{D}})\cdot d\boldsymbol{a}$	$\operatorname{curl}\boldsymbol{H} = \boldsymbol{J}+\dot{\boldsymbol{D}}$
Faraday's law	$\oint_{\partial\mathcal{B}_S} \boldsymbol{E}\cdot d\boldsymbol{s} = -\int_{A_S} \dot{\boldsymbol{B}}\cdot d\boldsymbol{a}$	$\operatorname{curl}\boldsymbol{E} = -\dot{\boldsymbol{B}}$
Gauß's law (electric)	$\int_{\partial\mathcal{B}} \boldsymbol{D}\cdot d\boldsymbol{a} = \int_{\mathcal{B}} \rho_f \, dv$	$\operatorname{div}\boldsymbol{D} = \rho_f$
Gauß's law (magnet.)	$\int_{\partial\mathcal{B}} \boldsymbol{B}\cdot d\boldsymbol{a} = 0$	$\operatorname{div}\boldsymbol{B} = 0$

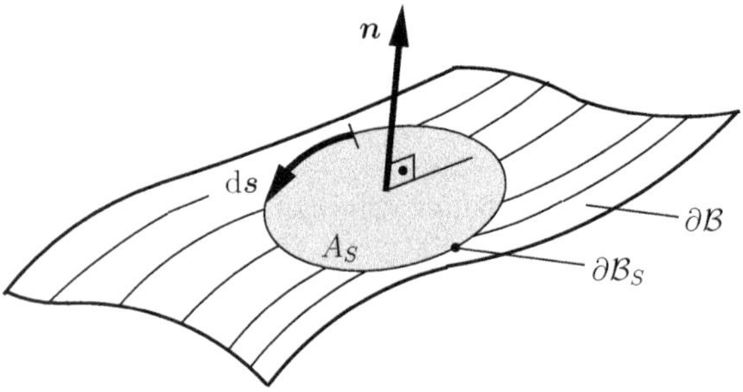

Fig. 6 Geometry: $\partial\mathcal{B}_S$ denotes a closed line at the surface $\partial\mathcal{B}$ of $\mathcal{B}$, A_S is the area inside this closed line. The orientation of the line segments d$\boldsymbol{s}$ of $\partial\mathcal{B}_S$ build with the outward unit normal $\boldsymbol{n}$ a right handed system

violates the conservation of charges (22). Computing the divergence of $\mathrm{curl}\boldsymbol{H} = \boldsymbol{J}$ yields

$$\mathrm{div}\boldsymbol{J} = 0\,, \quad \text{because} \quad \mathrm{divcurl}\boldsymbol{H} = 0\,, \tag{58}$$

which obviously contradicts (22).

Adding the displacement current[9]

$$\frac{\partial \boldsymbol{D}}{\partial t} =: \dot{\boldsymbol{D}} \tag{59}$$

to (57), leads to the modified expression

$$\oint_{\partial\mathcal{B}_S} \boldsymbol{H} \cdot \mathrm{d}\boldsymbol{s} = \int_{A_S} (\boldsymbol{J} + \dot{\boldsymbol{D}}) \cdot \mathrm{d}\boldsymbol{a}\,, \quad \text{local form:} \quad \mathrm{curl}\boldsymbol{H} = \boldsymbol{J} + \dot{\boldsymbol{D}}\,, \tag{60}$$

which satisfies the continuity equation. Ampère's law, especially the *displacement current*, relates changing electric fields $\boldsymbol{E}$ to magnetic fields $\boldsymbol{B}$: *a changing electric field induces a magnetic field.*

Faraday's law of induction: Faraday explored the questions whether magnetic fields are accompanied by electric fields. He observed, that a time varying magnetic field induces an electric current, see Fig. 7.

The electromagnetic induction relates the electric field $\boldsymbol{E}$ with the change of the magnetic field $\boldsymbol{B}$:

$$\oint_{\partial\mathcal{B}_S} \boldsymbol{E} \cdot \mathrm{d}\boldsymbol{s} = -\int_{A_S} \dot{\boldsymbol{B}} \cdot \mathrm{d}\boldsymbol{a}\,, \quad \text{local form:} \quad \mathrm{curl}\boldsymbol{E} = -\dot{\boldsymbol{B}}\,. \tag{61}$$

The 3rd Maxwell equation appears as

$$\int_{\partial\mathcal{B}} \boldsymbol{D} \cdot \mathrm{d}\boldsymbol{a} = \int_{\mathcal{B}} \rho_f\, dv = Q\,, \quad \text{local form:} \quad \mathrm{div}\boldsymbol{D} = \rho_f\,. \tag{62}$$

[9]In matter $(\dot{\bullet})$ is associated to the material time derivative.
$(\dot{\bullet}) = \dfrac{\partial(\bullet)}{\partial t} + \mathrm{grad}(\bullet) \cdot \dot{\boldsymbol{x}}$.

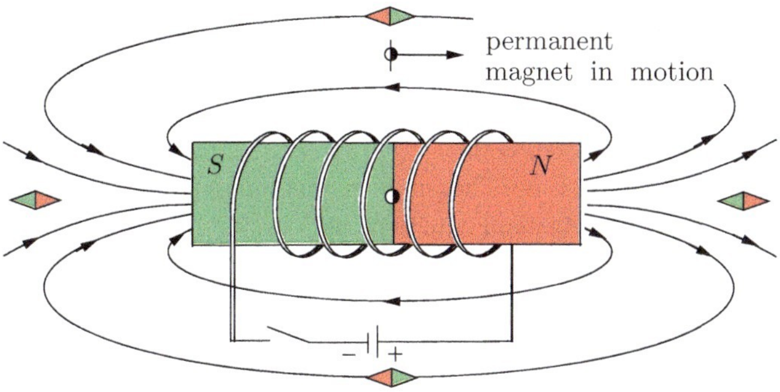

Fig. 7 Electromagnetic induction

Finally, the 4th Maxwell equation analyzed the magnetic field $\boldsymbol{B}$

$$\int_{\partial\mathcal{B}} \boldsymbol{B} \cdot \mathrm{d}\boldsymbol{a} = 0\,, \quad \text{local form:} \quad \operatorname{div}\boldsymbol{B} = 0\,. \tag{63}$$

2.5 *Special Cases*

Some special cases simplify the set of equations:

Quasi-stationary charges. In this case, the amplitude of the displacement current is much smaller than the current, i.e. $|\dot{\boldsymbol{D}}| \ll |\boldsymbol{J}|$. Thus, we can reduce the expression for Ampère's law to

$$\oint_{\partial\mathcal{B}_S} \boldsymbol{H} \cdot \mathrm{d}\boldsymbol{s} \approx \int_{A_S} \boldsymbol{J} \cdot \mathrm{d}\boldsymbol{a}\,. \tag{64}$$

Electrostatics. Stationary charges produce electric fields $\boldsymbol{E}$ which are constant in time. Ampère's law and Gauß's law for the magnetic field are not used. Thus, we only have to take into account Gauß's law for electric fields

$$\operatorname{div}\boldsymbol{D} = \rho_f \tag{65}$$

and the modified Faraday's law

$$\operatorname{curl}\boldsymbol{E} = \boldsymbol{0} \quad \text{implying} \quad \boldsymbol{E} = -\operatorname{grad}\phi^e\,. \tag{66}$$

Hence, we can compute the electric field by the gradient of an electric potential ϕ^e. The minus sign guarantees that the electric field points in the direction of the steepest descent of the potential.

Magnetostatics are associated with *steady currents*, which means that we consider magnetic fields constant in time. This implies, that the *induction term* $\frac{\partial \boldsymbol{B}}{\partial t}$ vanishes. Therefore, the remaining Maxwell equations are

$$\operatorname{div}\boldsymbol{B} = 0 \quad \text{and} \quad \operatorname{curl}\boldsymbol{H} = \boldsymbol{J} \ . \tag{67}$$

2.6 Electromagnetic Waves in Vacuum

In vacuum, there are no charge densities and no currents. As a result, the Maxwell Equations reduce to a set of coupled, first-order partial differential equations for the electric and the magnetic field:

$$\left.\begin{aligned} \operatorname{curl}\boldsymbol{H} &= +\frac{\partial \boldsymbol{D}}{\partial t} \\ \operatorname{curl}\boldsymbol{E} &= -\frac{\partial \boldsymbol{B}}{\partial t} \\ \operatorname{div}\boldsymbol{D} &= 0 \\ \operatorname{div}\boldsymbol{B} &= 0 \end{aligned}\right\} . \tag{68}$$

In order to decouple the electric and the magnetic field, we calculate the curl of $(68)_2$,

$$\operatorname{curl}\left[\operatorname{curl}\boldsymbol{E}\right] = -\operatorname{curl}\left[\frac{\partial \boldsymbol{B}}{\partial t}\right] , \tag{69}$$

with $\operatorname{curl}\left[\operatorname{curl}\boldsymbol{E}\right] = \nabla \times \left[\nabla \times \boldsymbol{E}\right]$ and $-\operatorname{curl}\left[\frac{\partial \boldsymbol{B}}{\partial t}\right] = -\nabla \times \left[\frac{\partial \boldsymbol{B}}{\partial t}\right]$. Evaluating the "double"-curl operator shows

$$\nabla \times \left[\nabla \times \boldsymbol{E}\right] = \nabla\left[\nabla \cdot \boldsymbol{E}\right] - \nabla^2 \boldsymbol{E} \ , \tag{70}$$

where $\nabla\left[\nabla \cdot \boldsymbol{E}\right] = \operatorname{grad}\left[\operatorname{div}\boldsymbol{E}\right]$ and $\nabla^2 \boldsymbol{E} = \Delta \boldsymbol{E}$. The symbol Δ denotes the Laplace operator. Exploiting the linear relation (18) in $(68)_3$ results in $\operatorname{div}\boldsymbol{E} = 0$. Therefore, we conclude from (69), using (70) and the linear constitutive relation (42), the reduced equation

$$-\,\Delta \boldsymbol{E} = -\mu_0 \frac{\partial \left[\operatorname{curl}\boldsymbol{H}\right]}{\partial t} \ . \tag{71}$$

Substituting now $\operatorname{curl}\boldsymbol{H}$ by $(68)_1$ and again using the linear constitutive relation for $\boldsymbol{D}$ as a function of $\boldsymbol{E}$ (18), yields

$$- \Delta \boldsymbol{E} + \mu_0 \epsilon_0 \frac{\partial^2 \boldsymbol{E}}{\partial t^2} = \boldsymbol{0} . \tag{72}$$

Taking the fundamental relation (4) into account, we conclude

$$\Box \boldsymbol{E} = \boldsymbol{0} , \tag{73}$$

where we have introduced the D'Alembert operator

$$\Box (\bullet) := \frac{1}{c^2} \frac{\partial^2 (\bullet)}{\partial t^2} - \Delta(\bullet) . \tag{74}$$

In analogy, we can derive the second wave equation

$$\Box \boldsymbol{B} = \boldsymbol{0} . \tag{75}$$

The presentations (73) and (75) are the wave equations of second order, separated for the electric and magnetic fields.

2.7 *Jump Conditions Across Interfaces*

In order to complete the set of differential Maxwell equations, we have to introduce an additional jump condition across singular surfaces and boundary conditions. We first consider the material time derivative of a field $\phi(\boldsymbol{x})$ over a material volume, i.e.

$$\frac{\mathrm{d}}{\mathrm{d}t} \int_{\mathcal{B}_t} \phi \, dv = \int_{\mathcal{B}_t} \frac{\partial \phi}{\partial t} \, dv + \int_{\partial \mathcal{B}_t} \phi \, \boldsymbol{v} \cdot \boldsymbol{n} \, da , \tag{76}$$

where $\boldsymbol{n}$ denotes the outward unit normal on $\partial \mathcal{B}_t$.

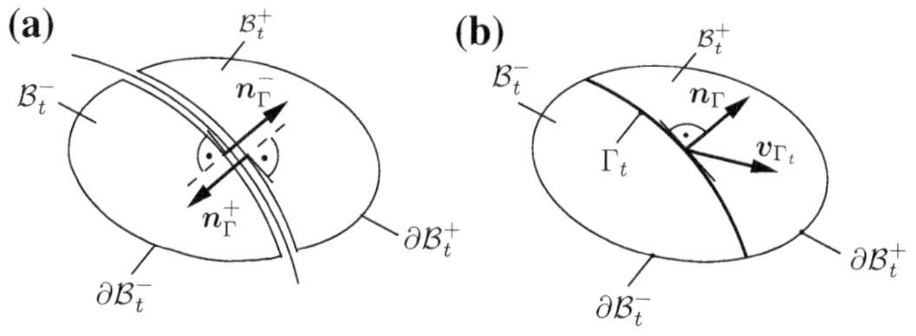

Fig. 8 Intersection of the body $\mathcal{B}_t$ by a moving discontinuity Γ_t into two parts, $\mathcal{B}_t^+$ and $\mathcal{B}_t^-$. The velocity of the discontinuity is $\boldsymbol{v}_\Gamma$

Let us consider a material body $\mathcal{B}_t$ that is divided by a discontinuity surface Γ_t into $\mathcal{B}_t^+$ and $\mathcal{B}_t^-$, see Fig. 8. If the physical quantities of this surface experience a discontinuity, it is termed a singular surface with the velocity $\boldsymbol{v}_{\Gamma_t}$.

Furthermore, the surfaces of the subdivided bodies $\mathcal{B}_t^-$ and $\mathcal{B}_t^+$ are

$$\partial\mathcal{B}_t^- \bigcup \Gamma_t \quad \text{and} \quad \partial\mathcal{B}_t^+ \bigcup \Gamma_t \,, \tag{77}$$

respectively. Unit normal vectors on Γ_t that point from $\mathcal{B}_t^-$ to $\mathcal{B}_t^+$ are denoted by $\boldsymbol{n}_\Gamma^-$; the opposite unit normal vectors are denoted by $\boldsymbol{n}_\Gamma^+$ as

$$\boldsymbol{n}_\Gamma = \boldsymbol{n}_\Gamma^- = -\boldsymbol{n}_\Gamma^+ \quad \text{on} \quad \Gamma_t \,, \tag{78}$$

where the tangential to the surface is denoted by $\boldsymbol{t}$.

Let ϕ be a physical quantity of interest, then ϕ^+ and ϕ^- are defined as

$$\phi^+ = \lim_{\boldsymbol{x}\in\mathcal{B}_t^+\to\boldsymbol{x}\in\Gamma_t} \phi(\boldsymbol{x}) \quad \text{and} \quad \phi^- = \lim_{\boldsymbol{x}\in\mathcal{B}_t^-\to\boldsymbol{x}\in\Gamma_t} \phi(\boldsymbol{x}) \,. \tag{79}$$

For further applications, it is comfortable to introduce the abbreviation for the jump of a physical quantity

$$[\![\phi]\!] = \phi^+ - \phi^- \quad \text{across} \quad \Gamma_t \,. \tag{80}$$

Applying the master balance law (76) to both parts of the subdivided body and adding both contributions, yields

$$\frac{D}{Dt}\int_{\mathcal{B}_t\backslash\Gamma_t} \phi \, dv = \int_{\mathcal{B}_t\backslash\Gamma_t} \frac{\partial\phi}{\partial t}\, dv + \int_{\partial\mathcal{B}_t} \phi\boldsymbol{v}\cdot\boldsymbol{n}\, da - \int_{\Gamma_t} [\![\phi\boldsymbol{v}_\Gamma]\!]\cdot\boldsymbol{n}_\Gamma \, da \,. \tag{81}$$

The localization theorem states that a balance law must be valid for any part of the body. Therefore, we conclude that

$$\frac{\partial\phi}{\partial t} + \operatorname{div}(\phi\boldsymbol{v}) = 0 \quad \text{in} \quad \mathcal{B}_t\backslash\Gamma_t \tag{82}$$

and

$$[\![\phi(\boldsymbol{v}-\boldsymbol{v}_\Gamma)]\!]\cdot\boldsymbol{n}_\Gamma = 0 \quad \text{on} \quad \Gamma_t \tag{83}$$

must be fulfilled separately. For a detailed discussion of this topic, we refer to Eringen and Maugin (1990), and Hutter and Jöhnk (2004).

Jump conditions for electrical quantities. The set of Maxwell's equations of elastostatics in its differential form have to completed with boundary and jump conditions. In order to describe the physical behavior at a material interphase, we introduce two materials I and II with different properties occupying $\mathcal{B}_t^-$ and $\mathcal{B}_t^+$, respectively. The variables in both domains are characterized by the superscript I or II, e.g. the electric field in domain I is denoted by $\boldsymbol{E}^I$ and in domain II by $\boldsymbol{E}^{II}$. To derive

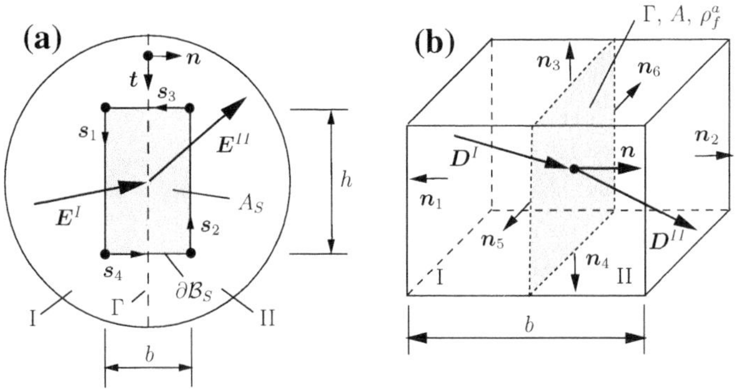

Fig. 9 Electrical interface conditions

the interface condition, we consider a closed line $\partial\mathcal{B}_S$ which surrounds the surface A_S, and the interface Γ that characterizes the separation plane of the two materials. Details are explained in Fig. 9a. Evaluating Faraday's law of induction for stationary magnetic fields leads to

$$\oint_{\partial\mathcal{B}_S} \boldsymbol{E}\cdot\mathrm{d}\boldsymbol{s} = 0 \quad \text{and} \quad \operatorname{curl}\boldsymbol{E} = \boldsymbol{0}\,, \tag{84}$$

cf. (61). Using the path of integration, as depicted in Fig. 9a, simplifies the calculation as follows

$$(\boldsymbol{E}^I\cdot\boldsymbol{s}_1 + \boldsymbol{E}^{II}\cdot\boldsymbol{s}_2)h + (\boldsymbol{E}^{II}\cdot\boldsymbol{s}_3 + \boldsymbol{E}^I\cdot\boldsymbol{s}_3 + \boldsymbol{E}^I\cdot\boldsymbol{s}_4 + \boldsymbol{E}^{II}\cdot\boldsymbol{s}_4)\frac{b}{2} = 0\,. \tag{85}$$

A limiting value $b \to 0$ implies with $\boldsymbol{t} = \boldsymbol{s}_1 = -\boldsymbol{s}_2$, $E_t^I = \boldsymbol{t}\cdot\boldsymbol{E}^I$ and $E_t^{II} = \boldsymbol{t}\cdot\boldsymbol{E}^{II}$, thus it follows

$$-\boldsymbol{t}\cdot(\boldsymbol{E}^{II} - \boldsymbol{E}^I) = 0 \quad\to\quad \boldsymbol{t}\cdot[\![\boldsymbol{E}]\!] = 0 \quad\to\quad E_t^I = E_t^{II}\,. \tag{86}$$

Alternatively, we could write

$$\boldsymbol{n}\times[\![\boldsymbol{E}]\!] = \boldsymbol{0} \quad \text{across} \quad \Gamma\,. \tag{87}$$

Résumé: The tangential component of the electric field has to be continuous across Γ.

Let us now analyze the behavior of the electric displacement across an intersection plane Γ with a surface charge density ρ_f^a, see Fig. 9b. The considered cube characterizes the body $\mathcal{B}$ with the surface $\partial\mathcal{B}$. The surface is defined by six subareas A_i for $i = 1, ...6$ with the associated outward normal vectors $\boldsymbol{n}_i$ for $i = 1, ...6$. The intersection plane is characterized by the unit normal vector $\boldsymbol{n}$. Evaluating Gauß's law (65) and performing the limiting value $b \to 0$ yields

$$(\boldsymbol{D}^I \cdot \boldsymbol{n}_1)\, A_1 + (\boldsymbol{D}^{II} \cdot \boldsymbol{n}_2)\, A_2 = \rho_f^a\, A\ , \tag{88}$$

which can be simplified, using $A_1 = A_2 = A$ and $\boldsymbol{n}_2 = -\boldsymbol{n}_1 = \boldsymbol{n}$, to

$$\boldsymbol{n} \cdot [\![\boldsymbol{D}]\!] = \rho_f^a\ . \tag{89}$$

In the case of a vanishing surface charge density, i.e. $\rho_f^a = 0$, the expression turns into

$$\boldsymbol{n} \cdot [\![\boldsymbol{D}]\!] = 0 \quad \text{across} \quad \Gamma\ , \quad \text{i.e.} \quad D_n^I = D_n^{II}\ , \tag{90}$$

with $D_n^I = \boldsymbol{n} \cdot \boldsymbol{D}^I$ and $D_n^{II} = \boldsymbol{n} \cdot \boldsymbol{D}^{II}$.

Résumé: In the case of vanishing surface charges, the normal component of the electric displacement has to be continuous across Γ.

Jump conditions for magnetical quantities. In analogy to the above derivations, we obtain: The normal components of the magnetic field $\boldsymbol{B}$ and the tangential component of the $\boldsymbol{H}$-field are continuous across a material interface:

$$[\![\boldsymbol{B} \cdot \boldsymbol{n}]\!] = 0 \quad \text{and} \quad [\![\boldsymbol{H} \cdot \boldsymbol{t}]\!] = 0\ . \tag{91}$$

The remaining components are not constrained, i.e.

$$[\![\boldsymbol{H} \cdot \boldsymbol{n}]\!] \neq 0 \quad \text{and} \quad [\![\boldsymbol{B} \cdot \boldsymbol{t}]\!] \neq 0\ . \tag{92}$$

2.8 Poynting's Theorem

Poynting's theorem can be interpreted as the work-energy theorem of electrodynamics. It states:

The work (W_{em}) done on the charges by the electromagnetic force is equal to the decrease in energy stored in the field reduced by the energy flowed out through the surface.

Let us multiply the differential form of Ampère's law with $\boldsymbol{E}$ and Faraday's law of induction with $-\boldsymbol{H}$. Next, both resulting equations are added, then we obtain

$$\boldsymbol{E} \cdot \mathrm{curl}\boldsymbol{H} - \boldsymbol{H} \cdot \mathrm{curl}\boldsymbol{E} = \boldsymbol{E} \cdot \boldsymbol{J} + \boldsymbol{E} \cdot \dot{\boldsymbol{D}} + \boldsymbol{H} \cdot \dot{\boldsymbol{B}}\ . \tag{93}$$

Substituting the identity

$$\boldsymbol{E} \cdot \mathrm{curl}\boldsymbol{H} - \boldsymbol{H} \cdot \mathrm{curl}\boldsymbol{E} = -\mathrm{div}(\boldsymbol{E} \times \boldsymbol{H}) \tag{94}$$

yields,

$$\boldsymbol{S}_P := \boldsymbol{E} \times \boldsymbol{H} \,, \tag{95}$$

which is the Poynting vector, an energy flux density. Then, the relation becomes

$$-\,\mathrm{div}\boldsymbol{S}_P = \boldsymbol{E} \cdot \boldsymbol{J} + \boldsymbol{E} \cdot \dot{\boldsymbol{D}} + \boldsymbol{H} \cdot \dot{\boldsymbol{B}} \,. \tag{96}$$

Performing an integration over a body $\mathcal{B}$ with the surface $\partial\mathcal{B}$ and applying Gauß's theorem for the divergence term leads to

$$-\int_{\partial\mathcal{B}} \boldsymbol{S}_P \cdot \boldsymbol{n}\, da = \underbrace{\int_{\mathcal{B}} \boldsymbol{E} \cdot \boldsymbol{J}\, dv}_{\mathcal{P}^{th}} + \underbrace{\int_{\mathcal{B}} \boldsymbol{E} \cdot \dot{\boldsymbol{D}}\, dv}_{\mathcal{P}^{el}} + \underbrace{\int_{\mathcal{B}} \boldsymbol{H} \cdot \dot{\boldsymbol{B}}\, dv}_{\mathcal{P}^{ma}} \,. \tag{97}$$

The volume integrals are associated to individual powers $\mathcal{P}^\alpha = \int_{\mathcal{B}} p^\alpha\, dv$ with $\alpha = \{th, el, ma\}$, for the thermal (th), electrical (el) and magnetical (ma) contributions. They are in terms of the power densities

$$p^\alpha = \frac{\partial \mathcal{P}^\alpha}{\partial V} = \frac{\partial}{\partial V}\frac{\partial \mathcal{W}^\alpha}{\partial t} = \frac{\partial}{\partial t}\frac{\partial \mathcal{W}^\alpha}{\partial V} = \frac{\partial w^\alpha}{\partial t} \,, \tag{98}$$

where energies and the volume specific energies are denoted by $\mathcal{W}^\alpha$ and w^α, respectively. The surface integral is performed over the Poynting vector. Obviously, the Poynting vector represents the flux of electromagnetic energy across the boundary of the considered body.

For the following analysis, we exploit the relations

$$\boldsymbol{E} \cdot \frac{\partial \boldsymbol{E}}{\partial t} = \frac{1}{2}\frac{\partial(\boldsymbol{E}\cdot\boldsymbol{E})}{\partial t} = \frac{1}{2}\frac{\partial\|\boldsymbol{E}\|^2}{\partial t} \,, \tag{99}$$

and

$$\boldsymbol{B} \cdot \frac{\partial \boldsymbol{B}}{\partial t} = \frac{1}{2}\frac{\partial(\boldsymbol{B}\cdot\boldsymbol{B})}{\partial t} = \frac{1}{2}\frac{\partial\|\boldsymbol{B}\|^2}{\partial t} \,. \tag{100}$$

Furthermore, we use the linear constitutive equations $\boldsymbol{D} = \epsilon\boldsymbol{E}$ and $\boldsymbol{B} = \mu\boldsymbol{H}$ and identify from

$$\boldsymbol{E} \cdot \frac{\partial \boldsymbol{D}}{\partial t} = \epsilon\boldsymbol{E} \cdot \frac{\partial \boldsymbol{E}}{\partial t} = \frac{\epsilon}{2}\frac{\partial\|\boldsymbol{E}\|^2}{\partial t} = \frac{\partial}{\partial t}\frac{1}{2}\boldsymbol{D}\cdot\boldsymbol{E} = \frac{\partial w^{el}}{\partial t} = p^{el} \,, \tag{101}$$

the electrical volume specific energy density. In a similar way, the magnetical volume specific energy density can be derived from

$$\boldsymbol{H} \cdot \frac{\partial \boldsymbol{B}}{\partial t} = \frac{1}{\mu}\boldsymbol{B} \cdot \frac{\partial \boldsymbol{B}}{\partial t} = \frac{1}{2\,\mu}\frac{\partial\|\boldsymbol{B}\|^2}{\partial t} = \frac{\partial}{\partial t}\left(\frac{1}{2}\boldsymbol{B}\cdot\boldsymbol{H}\right) = \frac{\partial w^{ma}}{\partial t} = p^{ma} \,. \tag{102}$$

For these linear and isotropic relations, the energy densities of the electric and the magnetic fields are given by

$$w^{el} = \frac{1}{2}\epsilon\|\boldsymbol{E}\|^2 \quad \text{and} \quad w^{ma} = \frac{1}{\mu_0}\|\boldsymbol{B}\|^2 \,, \tag{103}$$

respectively. Based on this derivations, Poynting's theorem can be represented as[10]

$$-\frac{\partial}{\partial t}\int_{\mathcal{B}} \left(w^{el} + w^{ma}\right) \mathrm{d}t = \oint_{\partial\mathcal{B}} \boldsymbol{S}_P \cdot \mathrm{d}\boldsymbol{a} + \mathcal{P}^{th} \,. \tag{104}$$

Thus, the decrease of electromagnetic energy $(\mathcal{W}^{el} + \mathcal{W}^{ma})$ is equal to the sum of the flux of electromagnetic energy across the surface and the thermal power of the electrical energy. The local form of (104) is

$$-\left(p^{el} + p^{ma}\right) = \mathrm{div}\boldsymbol{S}_P + p^{th} \,. \tag{105}$$

Specifications. Applying Ohm's law (56) to the thermal part in Poynting's theorem yields

$$\int_{\mathcal{B}} \boldsymbol{E} \cdot \boldsymbol{J} \,\mathrm{d}v = \int_{\mathcal{B}} \frac{1}{\sigma_c}||\boldsymbol{J}||^2 \,\mathrm{d}v = \int_{\mathcal{B}} p^{th} \,\mathrm{d}v \,. \tag{106}$$

If in addition we take into account forms of non-electric nature, denoted as impressed forces $\boldsymbol{E}_i$, the Ohm's law (56) has to be replaced by

$$\boldsymbol{E} + \boldsymbol{E}_i = \frac{1}{\sigma_c}\boldsymbol{J} \,. \tag{107}$$

With this modification, we obtain

$$\int_{\mathcal{B}} \boldsymbol{E} \cdot \boldsymbol{J}\mathrm{d}v = \int_{\mathcal{B}} \left(\frac{1}{\sigma_c}||\boldsymbol{J}||^2 - \boldsymbol{E}_i \cdot \boldsymbol{J}\right) \mathrm{d}v = \int_{\mathcal{B}} \left(p^{th} - p_i\right) \mathrm{d}v \,. \tag{108}$$

The modified local form of Poynting's theorem is then

$$-\left(p^{el} + p^{ma}\right) = \mathrm{div}\boldsymbol{S}_P + p^{th} - p_i \,. \tag{109}$$

Energy loss during polarization. To compute the change of electrical energy density, we have to integrate the power density p^{el}, see (101),

[10]For further details, see e.g. Griffiths, page 347 ff., Hofmann 1986.

$$\Delta w^{el} = \int_{t_1}^{t_2} p^{el}\, \mathrm{d}t = \int_{t_1}^{t_2} \boldsymbol{E} \cdot \frac{\partial \boldsymbol{D}}{\partial t}\, \mathrm{d}t = \int_{D_1}^{D_2} \boldsymbol{E} \cdot \mathrm{d}\boldsymbol{D} \,. \tag{110}$$

Applying now the constitutive model for the (differential) electric displacement $\mathrm{d}\boldsymbol{D}$ in matter, see (35), we obtain the expression

$$\Delta w^{el} = \underbrace{\int_{E_1}^{E_2} \epsilon_0 \boldsymbol{E} \cdot \mathrm{d}\boldsymbol{E}}_{=:\Delta w^{el}_E} + \underbrace{\int_{P_1}^{P_2} \boldsymbol{E} \cdot \mathrm{d}\boldsymbol{P}}_{=:\Delta w^{el}_P} \,. \tag{111}$$

Here, Δw^{el}_E and Δw^{el}_P represent the changes of the energy densities associated to the electric field $\boldsymbol{E}$ and the polarization $\boldsymbol{P}$, respectively.

Let us assume a unique relation between $\boldsymbol{D} = \boldsymbol{D}(\boldsymbol{E})$, then we obtain

$$\int_{D_1}^{D_2} \boldsymbol{E} \cdot \mathrm{d}\boldsymbol{D} = -\int_{D_2}^{D_1} \boldsymbol{E} \cdot \mathrm{d}\boldsymbol{D} \,. \tag{112}$$

We consider a reversible process. In this case, the energy increase during polarization $D_1 \rightarrow D_2$ is identical to the energy decrease during de-polarization $D_2 \rightarrow D_1$, see Fig. 10a.

However, a variety of materials exhibits a significant hysteresis during polarization and de-polarization. This is associated to the non-uniqueness of the function $\boldsymbol{D}(\boldsymbol{E})$, in this case we observe

$$\int_{D_1}^{D_2} \boldsymbol{E} \cdot \mathrm{d}\boldsymbol{D} \neq -\int_{D_2}^{D_1} \boldsymbol{E} \cdot \mathrm{d}\boldsymbol{D} \,. \tag{113}$$

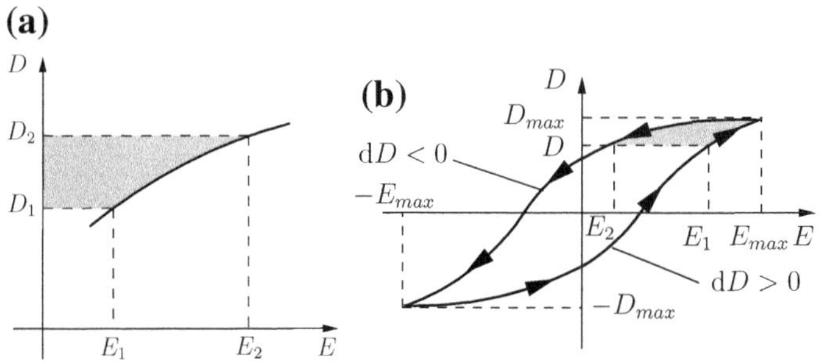

Fig. 10 **a** Unique $\boldsymbol{D}(\boldsymbol{E})$ relation between the electric displacement $\boldsymbol{D}$ and the electric field $\boldsymbol{E}$, **b** $D - E$-Hysteresis curve, non-unique $\boldsymbol{D}(\boldsymbol{E})$ relation

During de-polarization the decrease of energy is lower than the increase of energy during polarization. Using the notation in Fig. 10b, we compute the change of the energy density during the process

$$\boldsymbol{D} \rightarrow \boldsymbol{D}_{max} \rightarrow \boldsymbol{D} \tag{114}$$

by evaluating the integral

$$\Delta w^{el} = \int_{D}^{D^{max}} \boldsymbol{E} \cdot \mathrm{d}\boldsymbol{D} + \int_{D^{max}}^{D} \boldsymbol{E} \cdot \mathrm{d}\boldsymbol{D} > 0 \, . \tag{115}$$

This correlates to the shaded area in Fig. 10b. If we consider now a complete hysteresis cycle, i.e.

$$-\boldsymbol{E}_{max} \rightarrow +\boldsymbol{E}_{max} \rightarrow -\boldsymbol{E}_{max} \, , \tag{116}$$

the electrical energy density associated to the area inside the envelope of the $D-E$-hysteresis curve in Fig. 10b is dissipated into heat. The hysteresis loss $\Delta w_H^{el} \geq 0$ is given by the closed-loop integral

$$\Delta w_H^{el} = \oint \boldsymbol{E} \cdot \mathrm{d}\boldsymbol{D} = \oint \epsilon_0 \boldsymbol{E} \cdot \mathrm{d}\boldsymbol{E} + \oint \boldsymbol{E} \cdot \mathrm{d}\boldsymbol{P} \, . \tag{117}$$

The closed-loop integral of the reversible part of the electrical energy density

$$\Delta w_E^{el} = \oint \epsilon_0 \boldsymbol{E} \cdot \mathrm{d}\boldsymbol{E} = \frac{\epsilon_0}{2} \boldsymbol{E} \cdot \boldsymbol{E} \Big|_{-\boldsymbol{E}_{max}}^{-\boldsymbol{E}_{max}} = 0 \tag{118}$$

vanishes. Finally, the hysteresis loss appears as

$$\Delta w_H^{el} = \oint \boldsymbol{E} \cdot \mathrm{d}\boldsymbol{P} \, . \tag{119}$$

Analogous equations can be derived for magnetic materials. The hysteresis loss $\Delta w_H^{ma} \geq 0$ for a complete hysteresis cycle, i.e.

$$-\boldsymbol{H}_{max} \rightarrow +\boldsymbol{H}_{max} \rightarrow -\boldsymbol{H}_{max} \, , \tag{120}$$

is given by the closed-loop integral

$$\Delta w_H^{ma} = \oint \boldsymbol{H} \cdot \mathrm{d}\boldsymbol{B} \, . \tag{121}$$

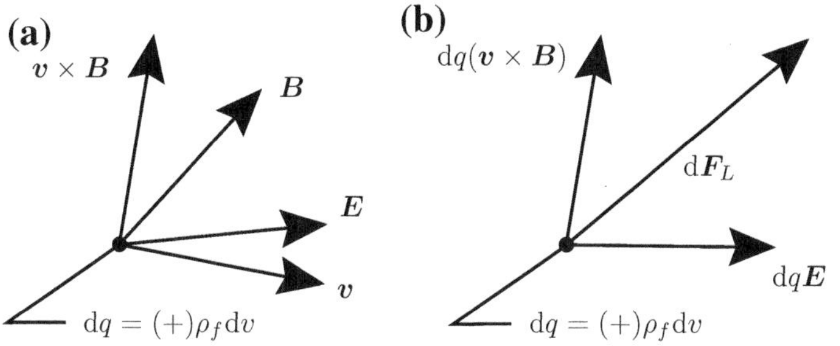

Fig. 11 **a** Electric and magnetic fields and **b** forces acting on a charged 'particle' dq

When substituting (54) into the integral expression, the integral changes to

$$\Delta w_H^{ma} = \oint (\frac{1}{\mu_0}\boldsymbol{B} - \boldsymbol{M}) \cdot d\boldsymbol{B} = -\oint \boldsymbol{M} \cdot d\boldsymbol{B} \,. \tag{122}$$

Additionally reformulating (54) leads to the differential constitutive equation

$$d\boldsymbol{B} = \mu_0 \, d\boldsymbol{H} + \mu_0 \, d\boldsymbol{M} \,. \tag{123}$$

Then, after inserting (123) into (121), the equation transforms into the alternative expression

$$\Delta w_H^{ma} = \oint \mu_0 \, \boldsymbol{H} \cdot d\boldsymbol{H} + \oint \mu_0 \, \boldsymbol{H} \cdot d\boldsymbol{M} = \oint \mu_0 \, \boldsymbol{H} \cdot d\boldsymbol{M} \,. \tag{124}$$

For further general considerations, we refer to Hofmann 1986, Lines and Glass (1977) and Fatuzzo and Merz (1967). A comprehensive treatment of hysteresis in magnetism as well as its mathematical aspects can be found in Bertotti (1998). An extensive overview of the "science of hysteresis" is given in the three-volume treatise of the same title in Bertotti and Mayergoyz (2006a, b, c).

2.9 *Maxwell Stress Tensor*

The so called Maxwell stress tensor, denoted by $\boldsymbol{\sigma}^M$, represents the influence of electromagnetic forces on the mechanical momentum. Several expressions for the Maxwell stress tensor have been proposed at the beginning of the last century. Well-known formulas go back to Minkowski (1908),[11]

[11] Hermann Minkowski, 1864–1909, German mathematician. He showed (1907) that the special theory of relativity (1905), proposed by his former student A. Einstein, could be interpreted geometrically in four-dimensional space time, the *Minkowski spacetime*.

$$\boldsymbol{\sigma}^M_{Minkowski} = \boldsymbol{E} \otimes \boldsymbol{D} + \boldsymbol{H} \otimes \boldsymbol{B} - \frac{1}{2}(\boldsymbol{E} \cdot \boldsymbol{D} + \boldsymbol{H} \cdot \boldsymbol{B})\mathbf{1} ,$$

Einstein and Laub (1908),

$$\boldsymbol{\sigma}^M_{Einstein\,\&\,Laub} = \boldsymbol{E} \otimes \boldsymbol{D} + \boldsymbol{H} \otimes \boldsymbol{B} - \frac{1}{2}(\epsilon_0 ||\boldsymbol{E}||^2 + \frac{1}{\mu_0}||\boldsymbol{B}||^2)\mathbf{1} ,$$

and Abraham (1909, 1910),

$$\boldsymbol{\sigma}^M_{Abraham} = \text{sym}[\boldsymbol{\sigma}^M_{Einstein\,\&\,Laub}] .$$

Let $\mathrm{d}\boldsymbol{F}_L$ denote the infinitesimal Lorentz force acting on an 'infinitesimal' volume $\mathrm{d}v$ with the charge density $mboxdq = \rho_f\, \mathrm{d}v$, see Fig. 11, i.e.

$$\mathrm{d}\boldsymbol{F}_L = \mathrm{d}q\ (\boldsymbol{E} + \boldsymbol{v} \times \boldsymbol{B}) . \tag{125}$$

The associated force density per unit volume is

$$\boldsymbol{f}_L = \rho_f(\boldsymbol{E} + \boldsymbol{v} \times \boldsymbol{B}) . \tag{126}$$

Substituting the motion of the charge density with the associated current density $\boldsymbol{J} = \rho_f \boldsymbol{v}$, yields

$$\boldsymbol{f}_L = \rho_f \boldsymbol{E} + \boldsymbol{J} \times \boldsymbol{B} \quad \rightarrow \quad \boldsymbol{F}_L = \int (\rho_f \boldsymbol{E} + \boldsymbol{J} \times \boldsymbol{B})\mathrm{d}v . \tag{127}$$

In $(127)_1$, we replace ρ_f by $\text{Div}\boldsymbol{D}$ (Gauß's law) and $\boldsymbol{J}$ by $\text{curl}\boldsymbol{H} - \dot{\boldsymbol{D}}$ (Ampère's law):

$$\boldsymbol{f}_L = (\text{div}\boldsymbol{D})\boldsymbol{E} + \text{curl}\boldsymbol{H} \times \boldsymbol{B} - \frac{\partial \boldsymbol{D}}{\partial t} \times \boldsymbol{B} . \tag{128}$$

Using the formula

$$\frac{\partial}{\partial t}(\boldsymbol{D} \times \boldsymbol{B}) = \frac{\partial \boldsymbol{D}}{\partial t} \times \boldsymbol{B} + \boldsymbol{D} \times \frac{\partial \boldsymbol{B}}{\partial t} = \frac{\partial \boldsymbol{D}}{\partial t} \times \boldsymbol{B} - \boldsymbol{D} \times \text{curl}\boldsymbol{E} ,$$

we obtain the expression

$$\boldsymbol{f}_L = \{(\text{div}\boldsymbol{D})\boldsymbol{E} - \boldsymbol{D} \times \text{curl}\boldsymbol{E}\} + \{\text{curl}\boldsymbol{H} \times \boldsymbol{B}\} - \frac{\partial}{\partial t}(\boldsymbol{D} \times \boldsymbol{B}) .$$

Substituting the identity $\text{curl}\boldsymbol{H} \times \boldsymbol{B} = -\boldsymbol{B} \times \text{curl}\boldsymbol{H}$ and adding $(\text{div}\boldsymbol{B})\boldsymbol{H} = \mathbf{0}$ to the expression in the second bracket, allows for the manipulation

$$\boldsymbol{f}_L = \{(\text{div}\boldsymbol{D})\boldsymbol{E} - \boldsymbol{D} \times \text{curl}\boldsymbol{E}\} + \{(\text{div}\boldsymbol{B})\boldsymbol{H} - \boldsymbol{B} \times \text{curl}\boldsymbol{H}\} - \frac{\partial}{\partial t}(\boldsymbol{D} \times \boldsymbol{B}) .$$

In order to get rid of the curl-vector operators, e.g. curl$\boldsymbol{E} = \nabla \times \boldsymbol{E}$, we apply the rule

$$\boldsymbol{D} \times (\nabla \times \boldsymbol{E}) = (\text{grad}\boldsymbol{E})^T \boldsymbol{D} - (\text{grad}\boldsymbol{E})\boldsymbol{D} \ .$$

Obviously, we get the expression

$$\begin{aligned} \boldsymbol{f}_L = & \underbrace{(\text{div}\boldsymbol{D})\boldsymbol{E} + (\text{grad}\boldsymbol{E})\boldsymbol{D}}_{\text{div}(\boldsymbol{E} \otimes \boldsymbol{D})} - (\text{grad}\boldsymbol{E})^T \boldsymbol{D} \\ & + \underbrace{(\text{div}\boldsymbol{B})\boldsymbol{H} + (\text{grad}\boldsymbol{H})\boldsymbol{B}}_{\text{div}(\boldsymbol{H} \otimes \boldsymbol{B})} \\ & - (\text{grad}\boldsymbol{H})^T \boldsymbol{B} - \frac{\partial}{\partial t}(\boldsymbol{D} \times \boldsymbol{B}) \ . \end{aligned}$$

After applying the linear and isotropic relations $\boldsymbol{D} = \epsilon_0 \boldsymbol{E}$ and $\boldsymbol{H} = \boldsymbol{B}/\mu_0$, we obtain the expression

$$\begin{aligned} \boldsymbol{f}_L = & \underbrace{\epsilon_0 \text{div}[\boldsymbol{E} \otimes \boldsymbol{E} - \frac{1}{2}||\boldsymbol{E}||^2 \mathbf{1}]}_{=:\ \text{div}\boldsymbol{\sigma}^M_{el}} \\ & + \underbrace{\frac{1}{\mu_0} \text{div}[\boldsymbol{B} \otimes \boldsymbol{B} - \frac{1}{2}||\boldsymbol{B}||^2 \mathbf{1}]}_{=:\ \text{div}\boldsymbol{\sigma}^M_{mag}} \\ & - \epsilon_0 \frac{\partial}{\partial t}(\boldsymbol{E} \times \boldsymbol{B}) \ , \end{aligned}$$

with the electric $\boldsymbol{\sigma}^M_{el}$ and the magnetic $\boldsymbol{\sigma}^M_{mag}$ part of the Maxwell stress tensor $\boldsymbol{\sigma}^M = \boldsymbol{\sigma}^M_{el} + \boldsymbol{\sigma}^M_{mag}$. Thus, the Maxwell stress tensor appears under the above mentioned assumptions as

$$\boldsymbol{\sigma}^M = \epsilon_0 \left[\boldsymbol{E} \otimes \boldsymbol{E} - \frac{1}{2}||\boldsymbol{E}||^2 \mathbf{1} \right] + \frac{1}{\mu_0} \left[\boldsymbol{B} \otimes \boldsymbol{B} - \frac{1}{2}||\boldsymbol{B}||^2 \mathbf{1} \right] \ . \tag{129}$$

Let us consider the Poynting vector

$$\boldsymbol{S}_P = \boldsymbol{E} \times \boldsymbol{H} = \frac{1}{\mu_0} \boldsymbol{E} \times \boldsymbol{B} \ , \tag{130}$$

see (95), then we are able to reformulate the formula for $\boldsymbol{f}_L$ as follows

$$\boldsymbol{f}_L = \text{div}\boldsymbol{\sigma}^M - \epsilon_0 \mu_0 \frac{\partial \boldsymbol{S}_P}{\partial t} \ . \tag{131}$$

The total force on the charges in the volume of the body $\mathcal{B}$ is therefore

$$\boldsymbol{F}_L = \oint_{\partial\mathcal{B}} \boldsymbol{\sigma}^M \cdot \mathrm{d}\boldsymbol{a} - \epsilon_0\mu_0 \int_{\mathcal{B}} \frac{\partial \boldsymbol{S}_P}{\partial t} \mathrm{d}v \,. \tag{132}$$

3 Thermodynamics

Let the body of interest $\mathcal{B} \subset \mathbb{R}^3$ be parameterized with the coordinate $\boldsymbol{x}$. The basic kinematic variable is the linear strain tensor

$$\varepsilon = \frac{1}{2}(\mathrm{grad}^T \boldsymbol{u} + \mathrm{grad}\boldsymbol{u}) \,, \tag{133}$$

where $\boldsymbol{u}$ is the displacement vector. The fundamental balance equation is the balance of linear momentum

$$\mathrm{div}\boldsymbol{\sigma} + \boldsymbol{f} - \rho\ddot{\boldsymbol{x}} = \boldsymbol{0} \,, \tag{134}$$

with the symmetric Cauchy stress tensor $\boldsymbol{\sigma}$, the mechanical body forces $\boldsymbol{f}$, the density ρ, and the acceleration $\ddot{\boldsymbol{x}}$.

3.1 *First Law of Thermodynamics, Balance of Energy*

The change of total energy $\mathcal{E}$, consisting of two parts, the change of internal energy $\mathcal{U}$ and the change of kinetic energy $\mathcal{K}$. It is equal to the external mechanical work $\mathcal{W}^e_{mech}$, further work contributions $\sum_\alpha \mathcal{W}_\alpha$, and the thermal energy $\mathcal{Q}$:

$$\dot{\mathcal{E}} = \dot{\mathcal{U}} + \dot{\mathcal{K}} = \mathcal{W}^e_{mech} + \sum_\alpha \mathcal{W}_\alpha + \mathcal{Q} \,. \tag{135}$$

The individual energies are defined by

$$\mathcal{E} = \int_{\mathcal{B}} \rho\, \bar{e}(\boldsymbol{x}, t)\, dv \,, \quad \mathcal{U} = \int_{\mathcal{B}} \rho\, \bar{U}(\boldsymbol{x}, t)\, dv \,, \tag{136}$$

based on *mass-specific* total energy $\bar{e}$ and internal energy $\bar{U}$ densities. The kinetic energy is defined by

$$\mathcal{K} = \frac{1}{2} \int_{\mathcal{B}} \rho\, \dot{\boldsymbol{x}} \cdot \dot{\boldsymbol{x}}\, dv \,. \tag{137}$$

Its rate can, exploiting the *balance of mass* $\dot{\rho} + \rho\,\text{div}\dot{\boldsymbol{x}} = 0$ and the *balance of linear momentum* $\text{div}\boldsymbol{\sigma} + \boldsymbol{f} - \rho\,\ddot{\boldsymbol{x}} = \boldsymbol{0}$, be expressed as

$$\dot{\mathcal{K}} = \int_{\mathcal{B}} \rho\,\ddot{\boldsymbol{x}} \cdot \dot{\boldsymbol{x}}\,dv = \underbrace{\int_{\partial\mathcal{B}} \boldsymbol{t} \cdot \dot{\boldsymbol{x}}\,da + \int_{\mathcal{B}} \boldsymbol{f} \cdot \dot{\boldsymbol{x}}\,dv}_{\mathcal{W}^e_{mech}} - \underbrace{\int_{\mathcal{B}} \boldsymbol{\sigma} : \text{grad}\dot{\boldsymbol{x}}\,dv}_{\mathcal{W}^i_{mech}}\,, \tag{138}$$

where $\mathcal{W}^e_{mech}$ and $\mathcal{W}^i_{mech}$ denote the external and internal mechanical work contributions. The thermal energy $\mathcal{Q}$, the increase of thermal work, is given by

$$\mathcal{Q} = \int_{\mathcal{B}} \rho\,\bar{r}\,dv - \int_{\partial\mathcal{B}} \boldsymbol{q} \cdot \boldsymbol{n}\,da = \int_{\mathcal{B}} \rho\,\bar{r}\,dv - \int_{\mathcal{B}} \text{div}\boldsymbol{q}\,dv\,, \tag{139}$$

in terms of a *mass-specific* heat production source $\bar{r}$ and the heat flux $\boldsymbol{q}$. Summarizing, we obtain the change of internal energy $\dot{\mathcal{U}} = \dot{\mathcal{E}} - \dot{\mathcal{K}}$ as

$$\dot{\mathcal{U}} = \mathcal{W}^a_{mech} + \sum_\alpha \mathcal{W}_\alpha + \mathcal{Q} - (\mathcal{W}^a_{mech} - \mathcal{W}^i_{mech}) = \mathcal{W}^i_{mech} + \sum_\alpha \mathcal{W}_\alpha + \mathcal{Q}\,. \tag{140}$$

For technical reasons, we introduce

$$\sum_\alpha \mathcal{W}_\alpha = \sum_\alpha \int_{\mathcal{B}} \boldsymbol{f}_\alpha \cdot \dot{\boldsymbol{s}}_\alpha\,dv \tag{141}$$

and obtain, identifying $\boldsymbol{\sigma} : \text{grad}\dot{\boldsymbol{x}} = \boldsymbol{\sigma} : \dot{\boldsymbol{\varepsilon}}$, the expression

$$\int_{\mathcal{B}} \rho\,\dot{\bar{U}}\,dv = \int_{\mathcal{B}} \boldsymbol{\sigma} : \dot{\boldsymbol{\varepsilon}}\,dv + \sum_\alpha \int_{\mathcal{B}} \boldsymbol{f}_\alpha \cdot \dot{\boldsymbol{s}}_\alpha\,dv + \int_{\mathcal{B}} \rho\,\bar{r}\,dv - \int_{\mathcal{B}} \text{div}\boldsymbol{q}\,dv\,. \tag{142}$$

Finally, the local form of the first law of thermodynamics appears as

$$\rho\,\dot{\bar{U}} = \boldsymbol{\sigma} : \dot{\boldsymbol{\varepsilon}} + \sum_\alpha \boldsymbol{f}_\alpha \cdot \dot{\boldsymbol{s}}_\alpha + \rho\,\bar{r} - \text{div}\boldsymbol{q}\,. \tag{143}$$

3.2 Second Law of Thermodynamics, Entropy Inequality

Clausius[12] laid the foundation for the second law of thermodynamics with his statement:

Heat can never pass from a colder to a warmer body without some other change, connected therewith, occurring at the same time.

[12]Rudolf Clausius, german scientist, 1822–1888.

The second law of thermodynamic states that the change of entropy is equal to the sum of entropy supply due to (i) heat production, (ii) heat flux across the surface, and (iii) entropy production within the body:

$$\int_{\mathcal{B}} \rho\, \dot{\bar{\eta}}(\boldsymbol{x}, t)\, dv = \int_{\mathcal{B}} \frac{1}{\vartheta}\, \rho\, \bar{r}\, dv - \int_{\partial\mathcal{B}} \frac{1}{\vartheta}\, \boldsymbol{q} \cdot \boldsymbol{n}\, da + \int_{\mathcal{B}} \rho\, s\, dv\, . \tag{144}$$

Here, we have introduced the *mass-specific* entropy density $\bar{\eta}$, the absolute temperature ϑ, and the internal entropy production s. For all physical processes $s \geq 0$ holds. Therefore, we can express the second law of thermodynamics as an inequality:

$$\int_{\mathcal{B}} \rho\, \dot{\bar{\eta}}(\boldsymbol{x}, t)\, dv \geq \int_{\mathcal{B}} \frac{1}{\vartheta}\, \rho\, \bar{r}\, dv - \int_{\partial\mathcal{B}} \frac{1}{\vartheta}\, \boldsymbol{q} \cdot \boldsymbol{n}\, da\, . \tag{145}$$

Applying Gauß's theorem to the surface integral, the local form

$$\dot{\bar{\eta}} \geq \frac{\bar{r}}{\vartheta} - \frac{1}{\rho} \mathrm{div} \frac{\boldsymbol{q}}{\vartheta}\, , \tag{146}$$

which is known as the *Clausius-Duhem-Inequality* emerge. Exploiting

$$\mathrm{div} \frac{\boldsymbol{q}}{\vartheta} = \frac{\mathrm{div} \boldsymbol{q}}{\vartheta} - \boldsymbol{q} \cdot \frac{\mathrm{grad} \vartheta}{\vartheta^2} \tag{147}$$

leads to the modified expression of the local form

$$\rho\, \vartheta\, \dot{\bar{\eta}} - \rho\, \bar{r} + \mathrm{div} \boldsymbol{q} - \boldsymbol{q} \cdot \frac{\mathrm{grad} \vartheta}{\vartheta} \geq 0\, . \tag{148}$$

Solving the local form of the balance of energy (143) with respect to $\rho\, \bar{r}$ and substituting this result in (148) states

$$-\rho\, \dot{\bar{U}} + \rho\, \vartheta\, \dot{\bar{\eta}} + \boldsymbol{\sigma} : \dot{\boldsymbol{\varepsilon}} + \sum_{\alpha} \boldsymbol{f}_{\alpha} \cdot \dot{\boldsymbol{s}}_{\alpha} - \boldsymbol{q} \cdot \frac{\mathrm{grad} \vartheta}{\vartheta} \geq 0\, . \tag{149}$$

3.3 Thermodynamic Potentials

Let us assume $\dot{\rho} = 0$ and define a *volume-specific* (i) internal energy, (ii) density of heat production, and (iii) entropy density

$$U = \rho\, \bar{U}\, , \quad r = \rho\, \bar{r}\, , \quad \eta = \rho\, \bar{\eta}\, , \tag{150}$$

respectively. Then, we restrict ourselves to isothermal processes, i.e. assuming a constant temperature ϑ in space, which implies $\mathrm{grad}\vartheta = \mathbf{0}$,

$$-\dot{U} + \vartheta\,\dot{\eta} + \boldsymbol{\sigma} : \dot{\varepsilon} + \sum_{\alpha} \boldsymbol{f}_{\alpha} \cdot \dot{\boldsymbol{s}}_{\alpha} \geq 0\,. \tag{151}$$

In order to take into account electrical contributions, we have to replace in Eq. (141) $\sum_{\alpha} \mathcal{W}_{\alpha}$ by $\mathcal{W}_{el}$:

$$\mathcal{W}_{el} = \int_{\mathcal{B}} \boldsymbol{E} \cdot \dot{\boldsymbol{D}}\, dv \quad \text{i.e. we set} \quad \boldsymbol{f}_{el} = \boldsymbol{E} \quad \text{and} \quad \dot{\boldsymbol{s}}_{el} = \dot{\boldsymbol{D}}\,. \tag{152}$$

Thus, the second law of thermodynamics takes the form

$$-\dot{U} + \vartheta\,\dot{\eta} + \boldsymbol{\sigma} : \dot{\varepsilon} + \boldsymbol{E} \cdot \dot{\boldsymbol{D}} \geq 0\,. \tag{153}$$

To obtain an expression based on a total differential of the thermodynamic potential, we set ε, $\boldsymbol{D}$ and η as independent process variables[13]:

$$U = \hat{U}(\varepsilon, \boldsymbol{D}, \eta) \quad \text{with} \quad \dot{U} = \frac{\partial U}{\partial \varepsilon} : \dot{\varepsilon} + \frac{\partial U}{\partial \boldsymbol{D}} \cdot \dot{\boldsymbol{D}} + \frac{\partial U}{\partial \eta}\dot{\eta}\,. \tag{154}$$

Substituting (154) in (153) yields

$$(\boldsymbol{\sigma} - \frac{\partial U}{\partial \varepsilon}) : \dot{\varepsilon} + (\boldsymbol{E} - \frac{\partial U}{\partial \boldsymbol{D}}) \cdot \dot{\boldsymbol{D}} + (\vartheta - \frac{\partial U}{\partial \eta})\dot{\eta} \geq 0\,. \tag{155}$$

Applying the standard argument of rational continuum mechanics, that (155) has to hold for all possible processes, leads to the following relations for the stresses, electric field, and temperature:

$$\boldsymbol{\sigma} = \frac{\partial U}{\partial \varepsilon}\,, \quad \boldsymbol{E} = \frac{\partial U}{\partial \boldsymbol{D}} \quad \text{and} \quad \vartheta = \frac{\partial U}{\partial \eta}\,. \tag{156}$$

It can be reasonable to use other thermodynamic potentials as the internal energy $U(\varepsilon, \boldsymbol{D}, \eta)$, though they depend on different process variables. The mathematical tool for the construction of new potentials is the Legendre transformation.

If we are interested in a potential Ψ, with the independent variables $\{\varepsilon, \boldsymbol{D}, \vartheta\}$, then we have to "substitute" the entropy η in the expression for the internal energy U with the temperature ϑ in order to obtain the free energy expression $\hat{\Psi}(\varepsilon, \boldsymbol{D}, \vartheta)$. Exploiting the Legendre transformation leads to the free energy

$$\Psi = \hat{\Psi}(\varepsilon, \boldsymbol{D}, \vartheta) = \hat{U}(\varepsilon, \boldsymbol{D}, \eta) - \vartheta\,\eta\,. \tag{157}$$

The time derivative of the free energy becomes

$$\dot{\Psi} = \dot{U} - \dot{\vartheta}\,\eta - \vartheta\,\dot{\eta} \quad \Rightarrow \quad -\dot{U} = -\dot{\Psi} - \dot{\vartheta}\,\eta - \vartheta\,\dot{\eta}\,. \tag{158}$$

[13]The independent process variables and the internal energy are denoted as extensive quantities.

When substituting $-\dot{U}$ in (153), the formula reads

$$-\dot{\Psi} - \eta\,\dot{\vartheta} + \boldsymbol{\sigma} : \dot{\boldsymbol{\varepsilon}} + \boldsymbol{E} \cdot \dot{\boldsymbol{D}} \geq 0\,. \tag{159}$$

Therefore, with the derivative of the free energy

$$\dot{\Psi} = \frac{\partial \Psi}{\partial \boldsymbol{\varepsilon}} : \dot{\boldsymbol{\varepsilon}} + \frac{\partial \Psi}{\partial \boldsymbol{D}} \cdot \dot{\boldsymbol{D}} + \frac{\partial \Psi}{\partial \vartheta}\,\dot{\vartheta}\,, \tag{160}$$

we obtain the expressions

$$\eta = -\frac{\partial \Psi}{\partial \vartheta}\,, \quad \boldsymbol{\sigma} = \frac{\partial \Psi}{\partial \boldsymbol{\varepsilon}}\,, \quad \boldsymbol{E} = \frac{\partial \Psi}{\partial \boldsymbol{D}}\,. \tag{161}$$

Beside the free energy, we can construct further thermodynamic potentials, e.g. the enthalpy function or the Gibbs functions. Table 3 summarizes the thermodynamic potentials, their independent process variables, as well as the associated total differentials.

For example, considering a Gibbs potential $G(\boldsymbol{\sigma}, \boldsymbol{E}, \vartheta)$, for linear constitutive relations between $\{\boldsymbol{\varepsilon}, \boldsymbol{D}, \Delta\eta\}$ and $\{\boldsymbol{\sigma}, \boldsymbol{E}, \Delta\vartheta\}$, the matrix representation

$$\begin{bmatrix} \underline{\boldsymbol{\varepsilon}} \\ \underline{\boldsymbol{D}} \\ \Delta\underline{\eta} \end{bmatrix} = \begin{bmatrix} \underline{\mathbb{D}} & \underline{\mathbb{d}}^T & \underline{\boldsymbol{\alpha}} \\ \underline{\mathbb{d}} & \underline{\boldsymbol{\kappa}} & \underline{\mathbb{p}} \\ \underline{\boldsymbol{\alpha}}^T & \underline{\mathbb{p}}^T & \dfrac{c}{\vartheta} \end{bmatrix} \begin{bmatrix} \underline{\boldsymbol{\sigma}} \\ \underline{\boldsymbol{E}} \\ \Delta\underline{\vartheta} \end{bmatrix} \tag{162}$$

Table 3 Thermodynamic potentials

Name and definition of thermodynamic potential	Independent variables	Total differential of thermodynamic potentials
Internal Energy U	$\boldsymbol{\varepsilon}, \boldsymbol{D}, \eta$	$\mathrm{d}U = \boldsymbol{\sigma} : \mathrm{d}\boldsymbol{\varepsilon} + \boldsymbol{E} \cdot \mathrm{d}\boldsymbol{D} + \vartheta\,\mathrm{d}\eta$
Free Energy $\Psi = U - \vartheta\,\eta$	$\boldsymbol{\varepsilon}, \boldsymbol{D}, \vartheta$	$\mathrm{d}\Psi = \boldsymbol{\sigma} : \mathrm{d}\boldsymbol{\varepsilon} + \boldsymbol{E} \cdot \mathrm{d}\boldsymbol{D} - \eta\,\mathrm{d}\vartheta$
Enthalpy $H = U - \boldsymbol{\sigma} : \boldsymbol{\varepsilon} - \boldsymbol{E} \cdot \boldsymbol{D}$	$\boldsymbol{\sigma}, \boldsymbol{E}, \eta$	$\mathrm{d}H = -\boldsymbol{\varepsilon} : \mathrm{d}\boldsymbol{\sigma} - \boldsymbol{D} \cdot \mathrm{d}\boldsymbol{E} + \vartheta\,\mathrm{d}\eta$
Elastic enthalpy $H_1 = U - \boldsymbol{\sigma} : \boldsymbol{\varepsilon}$	$\boldsymbol{\sigma}, \boldsymbol{D}, \eta$	$\mathrm{d}H_1 = -\boldsymbol{\varepsilon} : \mathrm{d}\boldsymbol{\sigma} + \boldsymbol{E} \cdot \mathrm{d}\boldsymbol{D} + \vartheta\,\mathrm{d}\eta$
Electric enthalpy $H_2 = U - \boldsymbol{E} \cdot \boldsymbol{D}$	$\boldsymbol{\varepsilon}, \boldsymbol{E}, \eta$	$\mathrm{d}H_2 = \boldsymbol{\sigma} : \mathrm{d}\boldsymbol{\varepsilon} - \boldsymbol{D} \cdot \mathrm{d}\boldsymbol{E} - \vartheta\,\mathrm{d}\eta$
Gibbs potential $G = U - \boldsymbol{\sigma} : \boldsymbol{\varepsilon} - \boldsymbol{E} \cdot \boldsymbol{D} - \eta\vartheta$	$\boldsymbol{\sigma}, \boldsymbol{E}, \vartheta$	$\mathrm{d}G = -\boldsymbol{\varepsilon} : \mathrm{d}\boldsymbol{\sigma} - \boldsymbol{D} \cdot \mathrm{d}\boldsymbol{E} - \eta\,\mathrm{d}\vartheta$
Elastic Gibbs potential $G_1 = U - \boldsymbol{\sigma} : \boldsymbol{\varepsilon} - \eta\vartheta$	$\boldsymbol{\sigma}, \boldsymbol{D}, \vartheta$	$\mathrm{d}G_1 = -\boldsymbol{\varepsilon} : \mathrm{d}\boldsymbol{\sigma} + \boldsymbol{E} \cdot \mathrm{d}\boldsymbol{D} - \eta\,\mathrm{d}\vartheta$
Electric Gibbs potential $G_2 = U - \boldsymbol{E} \cdot \boldsymbol{D} - \eta\vartheta$	$\boldsymbol{\varepsilon}, \boldsymbol{E}, \vartheta$	$\mathrm{d}G_2 = \boldsymbol{\sigma} : \mathrm{d}\boldsymbol{\varepsilon} - \boldsymbol{D} \cdot \mathrm{d}\boldsymbol{E} - \eta\,\mathrm{d}\vartheta$

can be identified. Here, we have used matrix notations where $\underline{\mathbb{D}}$ represents the elastic compliances, *underlined* the piezoelectric moduli, $\underline{\alpha}$ the thermal expansion coefficients, *underline*κ the permittivities, $\underline{\mathbb{p}}$ the pyroelectric coefficients and c the heat capacity.

The constitutive equations depend on the chosen thermodynamic potential; however, regardless of which potential is used, the number of material parameters will be the same. The material properties, that describe the linear relations between the intensive and extensive state variables, can be represented by a matrix of 10 by 10 entries with 55 independent material parameters in the most general case of crystal symmetry. This matrix consists of 1 thermal, 6 dielectric, 21 elastic, 6 thermoelastic, 18 piezoelectric and 3 pyroelectric constants.

In order to take into account magnetic effects, we have to add the term $\mathcal{W}_{ma}$ in Eq. (142), with with

$$\mathcal{W}^{ma} = \int_{\mathcal{B}} \boldsymbol{H} \cdot \dot{\boldsymbol{B}} \, \mathrm{d}v, \quad \text{i.e. we set} \quad \boldsymbol{f}_{ma} = \boldsymbol{H} \quad \text{and} \quad \dot{s}_{ma} = \dot{\boldsymbol{B}} \,. \tag{163}$$

4 Rotation, Spatial Reflection, Time-Reversal

The principle of relativity requires, that physical laws (the ones of classical mechanics as well as the Maxwell equations) hold in any inertial reference frame. Inertial means, that the system is at rest or moving with a constant speed. In classical mechanics the principle of relativity was understood since the works of Galileo.[14] In contrast to the applications in classical mechanics, we have some additional effects in electrodynamics with respect to a moving frame compared to an inertial system at rest. As an example, a charge in motion produces a magnetic field, in contrast a charge at rest does not. The requirements in order to fulfill several transformation properties of physical quantities lead to limitations and restrictions for the design of phenomenological constitutive equations.

A basis for several applications are Einstein's two famous postulates:

- The principle of relativity: the laws of physics apply in all inertial reference systems.
- The universal speed of light: the universal speed of light in vacuum ($c \approx 3.0 \cdot 10^8$ m/s) is the same for all inertial observers, regardless of the motion of the source.

First of all, we recapitulate the Lorentz transformation, which relates space and time coordinates in different inertial systems. In the following, we will discuss aspects of rotations, spatial reflection and inversion, and time reversal.

Summary: Maxwell's equations are applicable in any inertial systems, that means, that they do not change their form with respect to the Lorentz transformation.

---

[14]Galileo Galilei, 1564–1642, Italian physicist, mathematician, astronomer, and philosopher.

4.1 Lorentz Invariance

Let us consider two inertial coordinate frames $\mathcal{IF}$ and $\widetilde{\mathcal{IF}}$ with the two associated observers $\mathcal{O}$ and $\widetilde{\mathcal{O}}$, respectively. The frame $\widetilde{\mathcal{F}}$ moves with speed v_x relative to $\mathcal{F}$. Without any further restrictions, we set for simplicity $v_y = v_z = 0$, see Fig. 12.

In the following considerations, we assume for simplicity that the coordinate systems coincide at time $t = \widetilde{t} = 0$ and that the two inertial frames $\mathcal{IF}$ and $\widetilde{\mathcal{IF}}$ are related by

$$x(t) = \gamma\left(\widetilde{x} + v_x \widetilde{t}\,\right), \quad y = \widetilde{y}, \quad z = \widetilde{z}\,. \tag{164}$$

Here, v_x is the constant relative speed of the origins of the inertial frames and γ the just unknown parameter. The inverse transformation has to be of the form

$$\widetilde{x}(\widetilde{t}) = \gamma\,(x - v_x t\,)\,, \tag{165}$$

because of the equivalence of both frames. Einstein's principle of relativity requires the invariance of

$$(c\;\mathrm{d}t\,)^2 - \|\mathrm{d}\boldsymbol{x}\|^2\,, \tag{166}$$

which induces that the speed of light is constant in any inertial system. From the invariance requirement

$$(c\;\mathrm{d}t)^2 - \mathrm{d}x^2\Big|_{\mathcal{IF}} = \left(c\;\mathrm{d}\widetilde{t}\,\right)^2 - \mathrm{d}\widetilde{x}^{\,2}\Big|_{\widetilde{\mathcal{IF}}}\,, \tag{167}$$

we conclude

$$\left.\frac{\mathrm{d}x}{\mathrm{d}t}\right|_{\mathcal{IF}} = c \quad \text{and} \quad \left.\frac{\mathrm{d}\widetilde{x}}{\mathrm{d}\widetilde{t}}\right|_{\widetilde{\mathcal{IF}}} = c\,. \tag{168}$$

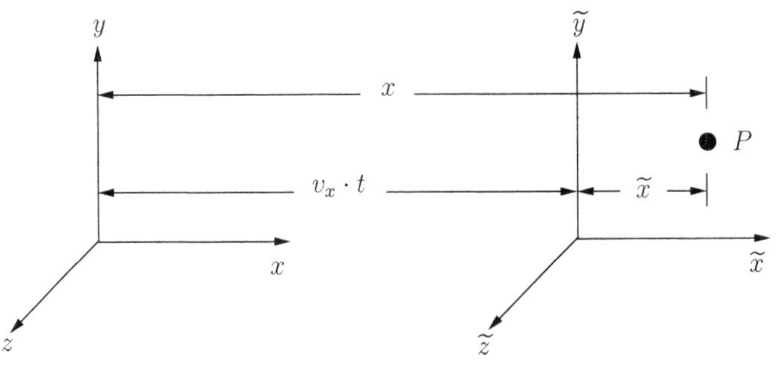

Fig. 12 Relative motion of the two inertial frames $\mathcal{IF}$ and $\widetilde{\mathcal{IF}}$ with the constant speed v_x

Let the light travel at time $t = \widetilde{t} = 0$ from the (common) origins of the coordinate frames. From $(164)_1$, we obtain the actual position of the light impulse in $\mathcal{IF}$

$$(x =)\; c\, t = \gamma\, (c + v_x)\, \widetilde{t}\,. \tag{169}$$

In analogy, we get from (165) the associated position with respect to $\widetilde{\mathcal{IF}}$

$$(\widetilde{x} =)\; c\, \widetilde{t} = \gamma\, (c - v_x)\, t \quad \rightarrow \quad \widetilde{t} = \gamma \left(1 - \frac{v_x}{c}\right) t\,. \tag{170}$$

This expression for $\widetilde{t}$ in (169) yields

$$c\, t = \gamma\, (c + v_x)\, \gamma \left(1 - \frac{v_x}{c}\right) t \quad \rightarrow \quad 1 = \gamma^2 \left(1 + \frac{v_x}{c}\right) \left(1 - \frac{v_x}{c}\right), \tag{171}$$

thus, we obtain

$$\gamma = \frac{1}{\sqrt{1 - v_x^2/c^2}}\,. \tag{172}$$

Summary: The *Lorentz transformation*[15] is given by

$$x(t) = \gamma(\widetilde{x} + v_x\, \widetilde{t}), \quad y = \widetilde{y}, \quad z = \widetilde{z}, \quad t = \gamma(\widetilde{t} + v_x\, \widetilde{x}/c^2)\,. \tag{173}$$

Example: A point P has a speed of $\dot{\widetilde{x}} = 0.6\, c$ relative to the moving frame $\widetilde{\mathcal{IF}}$. This frame moves with a constant speed $v_x = 0.6\, c$ relative to a fixed frame $\mathcal{IF}$. In order to compute the velocity of the point P with respect to the fixed frame, we compute the time derivative of $(173)_1$:

$$\dot{x} = \frac{\mathrm{d}x}{\mathrm{d}t} = \frac{\mathrm{d}}{\mathrm{d}\widetilde{t}} \left[\gamma(\widetilde{x} + v_x\, \widetilde{t}\,) \right] \frac{\mathrm{d}\widetilde{t}}{\mathrm{d}t} = \gamma \left[\frac{\mathrm{d}\widetilde{x}}{\mathrm{d}\widetilde{t}} + v_x \right] \frac{\mathrm{d}\widetilde{t}}{\mathrm{d}t} = \gamma \left[\dot{\widetilde{x}} + v_x \right] \frac{\mathrm{d}\widetilde{t}}{\mathrm{d}t}\,. \tag{174}$$

The time derivative of $(173)_4$ leads to

$$\frac{\mathrm{d}t}{\mathrm{d}\widetilde{t}} = \frac{\mathrm{d}}{\mathrm{d}\widetilde{t}} \left[\gamma(\widetilde{t} + v_x\, \widetilde{x}/c^2) \right] = \gamma \left(1 + \frac{v_x}{c^2} \frac{\mathrm{d}\widetilde{v}}{\mathrm{d}\widetilde{t}} \right) = \gamma \left(1 + \frac{v_x\, \dot{\widetilde{x}}}{c^2} \right), \tag{175}$$

from which we can gain the derivative

$$\frac{\mathrm{d}\widetilde{t}}{\mathrm{d}t} = \left[\gamma \left(1 + \frac{v_x\, \dot{\widetilde{x}}}{c^2} \right) \right]^{-1}. \tag{176}$$

While combining these results, we get

[15] H.A. Lorentz derived this transformation before A. Einstein did. However, Einstein was the first who interpreted this transformation as the natural relation between space and time, which is the foundation of relativity.

$$\dot{x} = \gamma\left[\dot{\tilde{x}} + v_x\right]\left[\gamma\left(1 + \frac{v_x\,\dot{\tilde{x}}}{c^2}\right)\right]^{-1} = \frac{\dot{\tilde{x}} + v_x}{1 + \dfrac{v_x\,\dot{\tilde{x}}}{c^2}} \; . \tag{177}$$

Inserting the numerical values, we recognize that the velocity, relative to the fixed frame, is

$$\dot{x} = \frac{0.6\,c + 0.6\,c}{1 + 0.36} = \frac{1.2}{1.36}c = 0.88\,c \; , \tag{178}$$

(and not $1.2\,c$, of course). The *Lorentz transformation* is valid for all velocities up to the speed of light.

4.2 Galilean Transformation

A Galilean transformation is used to relate the coordinates of two reference systems, where the reference frames differ by a constant relative motion. Let $\boldsymbol{Q}$ be a constant orthogonal tensor, i.e. $\dot{\boldsymbol{Q}} = \boldsymbol{0}$, and $\boldsymbol{c}(t)$ a displacement vector with $\ddot{\boldsymbol{c}}(t) = \boldsymbol{0}$. The Galilean transformation is given by

$$\widetilde{\boldsymbol{x}} = \boldsymbol{Q}\boldsymbol{x} + \boldsymbol{c}(t) \; . \tag{179}$$

A time-independent transformation appears for a constant $\boldsymbol{c}(t) = \boldsymbol{c}$, i.e. $\dot{\boldsymbol{c}}(t) = \boldsymbol{0}$. The transformation for a scalar-valued function $\alpha(\boldsymbol{x})$, a vector-valued function $\boldsymbol{a}(\boldsymbol{x})$ and a tensor-valued function of second order $\boldsymbol{A}(\boldsymbol{x})$ are defined by

$$\widetilde{\alpha}(\widetilde{\boldsymbol{x}}) = \alpha(\boldsymbol{x}) \; , \quad \widetilde{\boldsymbol{a}}(\widetilde{\boldsymbol{x}}) = \boldsymbol{Q}\boldsymbol{a}(\boldsymbol{x}) \; , \quad \widetilde{\boldsymbol{A}}(\widetilde{\boldsymbol{x}}) = \boldsymbol{Q}\boldsymbol{A}(\boldsymbol{x})\boldsymbol{Q}^T \; , \tag{180}$$

respectively. Attention must be made of the cross product of two vectors

$$\boldsymbol{a} = \boldsymbol{b} \times \boldsymbol{c} \; , \tag{181}$$

because it has some characteristics of a traceless antisymmetric second-order tensor. The cross-product transforms as

$$\widetilde{\boldsymbol{a}} = \det[\boldsymbol{Q}]\boldsymbol{Q}\,\boldsymbol{a} \; . \tag{182}$$

For a proper orthogonal transformation, i.e. $\det \boldsymbol{Q} = 1$, the cross product transforms like a vector.

A spatial reflection with respect to a plane, e.g. $x_1 - x_2$-plane, corresponds to a sign-change of the associated normal components

$$(x_1, x_2, x_3) \longmapsto (\widetilde{x}_1, \widetilde{x}_2, \widetilde{x}_3) = (x_1, x_2, -x_3) \; . \tag{183}$$

Space inversion corresponds to the reflection of all components

$$\boldsymbol{x} \longmapsto \widetilde{\boldsymbol{x}} = -\boldsymbol{x} \,. \tag{184}$$

Furthermore, the discrete transformations (183) and (184) are associated to

$$\boldsymbol{Q} = \mathrm{diag}[1, 1, -1] \text{ and } \boldsymbol{Q} = \mathrm{diag}[-1, -1, -1] \,, \text{ with } \det \boldsymbol{Q} = -1 \,, \tag{185}$$

respectively. Obviously, a vector (also called polar vector) $\boldsymbol{v}$ changes sign under spatial inversion, i.e.

$$\boldsymbol{v} \longmapsto \widetilde{\boldsymbol{v}} = -\boldsymbol{v} \,. \tag{186}$$

In contrast to that, a cross product (also called axial or pseudovector) $\boldsymbol{a}$ behaves as

$$\boldsymbol{a} \longmapsto \widetilde{\boldsymbol{a}} = \boldsymbol{a} \tag{187}$$

under spatial inversion. A similar definition is used to differentiate scalars and pseudoscalar. A pseudoscalar changes its sign under spatial inversion, but a scalar does not. Let $\boldsymbol{a}$, $\boldsymbol{b}$, $\boldsymbol{c}$ be polar vectors, then the triple scalar product

$$\boldsymbol{a} \cdot (\boldsymbol{b} \times \boldsymbol{c}) \tag{188}$$

is a pseudoscalar. Any scalar product between a pseudovector and a polar vector is a pseudoscalar.

A tensor (true tensor) of rank n transforms under spatial inversion with the factor $(-1)^n$, whereas a pseudotensor has a factor $(-1)^{n+1}$, summary is given in Table 4, see Jackson (2002).

4.3 $\mathcal{T}$-Symmetry

$\mathcal{T}$-Symmetry claims for consistent transformation of physical quantities under time reversal transformation

$$\mathcal{T} : t \mapsto \widetilde{t} = -t \,. \tag{189}$$

Let us consider the motion of a point from an initial state $\boldsymbol{x}_i$ to a final state $\boldsymbol{x}_f$. Then, we assume a reverse motion over the reversed path to the initial configuration. Obviously, we obtain

$$\widetilde{\boldsymbol{x}} = \boldsymbol{x} \,; \quad \dot{\boldsymbol{x}} = \frac{\mathrm{d}\boldsymbol{x}}{\mathrm{d}t} \,; \quad \dot{\widetilde{\boldsymbol{x}}} = \frac{\mathrm{d}\boldsymbol{x}}{\mathrm{d}\widetilde{t}} = \frac{\mathrm{d}\boldsymbol{x}}{\mathrm{d}(-t)} = -\dot{\boldsymbol{x}} \,. \tag{190}$$

Let $\boldsymbol{I}$ denote the linear momentum, m the mass of a particle, then Newton's law appears as

Table 4 Space inversion—transformation properties

Mechanical quantity:		Rank	Space inversion (name)
Coordinates	$\boldsymbol{x}$	1	Odd (vector)
Velocity	$\dot{\boldsymbol{x}}$	1	Odd (vector)
Momentum	$\boldsymbol{I}$	1	Odd (vector)
Angular momentum	$\boldsymbol{x} \times \boldsymbol{I}$	1	Even (pseudovector)
Force	$\boldsymbol{F}$	1	Odd (vector)
Torque	$\boldsymbol{x} \times \boldsymbol{F}$	1	Even (pseudovector)
Kinetic energy	$\mathcal{K}$	0	Even (scalar)
Potential energy	$\mathcal{U}$	0	Even (scalar)
Charge density	ρ	0	Even (scalar)
Current density	$\boldsymbol{J}$	1	Odd (vector)
Electric field	$\boldsymbol{E}$	1	Odd (vector)
Polarization	$\boldsymbol{P}$	1	Odd (vector)
Electric displacement	$\boldsymbol{D}$	1	Odd (vector)
Magnetic field	$\boldsymbol{B}$	1	Even (pseudovector)
Magnetization	$\boldsymbol{M}$	1	Even (pseudovector)
Auxiliary field	$\boldsymbol{H}$	1	Even (pseudovector)
Poynting vector	$\boldsymbol{S}_p = \boldsymbol{E} \times \boldsymbol{H}$	1	Odd (vector)
Maxwell stress tensor	$\boldsymbol{\sigma}^M$	2	Even (tensor)

$$\boldsymbol{F} = \frac{\mathrm{d}\boldsymbol{I}}{\mathrm{d}t} = \frac{\mathrm{d}m\dot{\boldsymbol{x}}}{\mathrm{d}t} = m\,\ddot{\boldsymbol{x}}\,. \tag{191}$$

Under time reversal, we obtain with $\widetilde{\boldsymbol{v}} = -\boldsymbol{v}$

$$\widetilde{\boldsymbol{F}} = \frac{\mathrm{d}\widetilde{\boldsymbol{I}}}{\mathrm{d}\widetilde{t}} = \frac{\mathrm{d}m\dot{\widetilde{\boldsymbol{x}}}}{\mathrm{d}\widetilde{t}} = m\frac{\mathrm{d}(-\dot{\boldsymbol{x}})}{\mathrm{d}(-t)} = m\,\ddot{\boldsymbol{x}}\,. \tag{192}$$

A comparison of (191) and (192) shows that the form of both equations is identical. Thus, Newton's law is invariant under time reversal, whereas the linear momentum transforms as $\widetilde{\boldsymbol{I}} = -\boldsymbol{I}$.

Under spatial inversion, the electric field $\boldsymbol{E}$ and the magnetic field $\boldsymbol{B}$ transforms as

$$\widetilde{\boldsymbol{E}}(\widetilde{\boldsymbol{x}}, t) = -\boldsymbol{E}(\boldsymbol{x}, t) \quad \text{and} \quad \widetilde{\boldsymbol{B}}(\widetilde{\boldsymbol{x}}, t) = +\boldsymbol{B}(\boldsymbol{x}, t)\,. \tag{193}$$

The time reversal yields

$$\widetilde{\boldsymbol{E}}(\boldsymbol{x}, \widetilde{t}) = +\boldsymbol{E}(\boldsymbol{x}, t) \quad \text{and} \quad \widetilde{\boldsymbol{B}}(\boldsymbol{x}, \widetilde{t}) = -\boldsymbol{B}(\boldsymbol{x}, t)\,. \tag{194}$$

Table 5 Time reversal—transformation properties

Mechanical quantity		Rank	Time reversal
Coordinates	$\boldsymbol{x}$	1	Even
Velocity	$\dot{\boldsymbol{x}}$	1	Odd
Momentum	$\boldsymbol{I}$	1	Odd
Angular momentum	$\boldsymbol{x} \times \boldsymbol{I}$	1	Odd
Force	$\boldsymbol{F}$	1	Even
Torque	$\boldsymbol{x} \times \boldsymbol{F}$	1	Even
Kinetic energy	$\mathcal{K}$	0	Even
Potential energy	$\mathcal{U}$	0	Even
Charge density	ρ	0	Even
Current density	$\boldsymbol{J}$	1	Odd
Electric field	$\boldsymbol{E}$	1	Even
Polarization	$\boldsymbol{P}$	1	Even
Electric displacement	$\boldsymbol{D}$	1	Even
Magnetic field	$\boldsymbol{B}$	1	Odd
Magnetization	$\boldsymbol{M}$	1	Odd
Auxiliary field	$\boldsymbol{H}$	1	Odd
Poynting vector	$\boldsymbol{S}_p = \boldsymbol{E} \times \boldsymbol{H}$	1	Odd
Maxwell stress tensor	$\boldsymbol{\sigma}^M$	2	Even

A summary of the transformation properties is given in Table 5, see Jackson (2002). The left term in Faraday's law of induction

$$\operatorname{curl} \boldsymbol{E} = -\dot{\boldsymbol{B}} \tag{195}$$

transforms under spatial inversion and rotations as a pseudovector and even under time reversal. In order to preserve form-invariance, the magnetic field has to be a pseudovector which is odd under time reversal.

Arguments of symmetry properties can be used for the construction of phenomenological constitutive equations. If we are interested in an isotropic, non-dissipative function of the polarization $\boldsymbol{P}$ depending on $\boldsymbol{E}$ and an external magnetic field $\boldsymbol{B}_0$, we have to consider that $\boldsymbol{P}$ is a polar vector and even under time reversal. Let us consider the first-order terms in $\boldsymbol{B}_0$

$$\boldsymbol{E} \times \boldsymbol{B}_0, \quad \frac{\partial \boldsymbol{B}}{\partial t} \times \boldsymbol{B}_0, \quad \frac{\partial^2 \boldsymbol{E}}{\partial t^2} \times \boldsymbol{B}_0\,, \ldots \tag{196}$$

and the second-order terms in $\boldsymbol{B}_0$

$$||\boldsymbol{B}_0||^2 \boldsymbol{E}, \quad (\boldsymbol{E} \cdot \boldsymbol{B}_0)\boldsymbol{B}_0, \quad ||\boldsymbol{B}_0||^2 \frac{\partial \boldsymbol{E}}{\partial t}\,, \ldots . \tag{197}$$

A general expression for $\boldsymbol{P}$ up to second-order is

$$\boldsymbol{P} = \alpha_0 \boldsymbol{E} + \alpha_1 \frac{\partial \boldsymbol{E}}{\partial t} \times \boldsymbol{B}_0 + \alpha_2 ||\boldsymbol{B}_0||^2 \boldsymbol{E} + \alpha_3 (\boldsymbol{E} \cdot \boldsymbol{B}) \boldsymbol{B}_0 + \ ... \tag{198}$$

with the real scalar coefficients $\alpha_i \big|_{i=0,1,2,...}$. In (196) only terms with odd time derivations fulfill the time reversal condition and in (197) only elements with zero and even time derivatives satisfy the transformation properties, which are required for the polarization. For a detailed discussion of time inversion symmetries of electromagnetic systems, we refer to Kiehn (1977), Fushchych (1969), Simonyi (2001) and Pauli (1979).

4.4 Crystal Classes and Magnetic Crystal Classes

Crystal classes. A crystalline material is characterized by a periodical arrangement of its unit cells, which form the appropriate crystal lattice. The unit cell is composed of positively and negatively charged ions, but the position and type of these ions depends on the considered material. The crystal structure is obtained by assigning a group of atoms to each node of the lattice, which can affect the symmetry. In contrast to a lattice, a crystal does not necessarily possess a center of symmetry. In classifying crystals according to the point symmetry of the lattice, we define seven crystal systems, in which each system is characterized by the geometrical form of the cell, and through repetition of this pattern all lattices are generated with the same symmetry. These seven crystal systems can be subdivided into fourteen Bravais lattices. These Bravais lattices arise from translational symmetry operations that do not change the point symmetry of the lattice. The set of point symmetry transformations for a crystal is called its point group, leading to a division of 32 point symmetry classes of crystals, denoted by $\mathcal{G}$. Here, 21 classes exhibit no center of symmetry. Curie showed experimentally that 20 of these 21 non-centrosymmetric classes are piezoelectric.

Magnetic crystal classes. The profound difference of the behavior of charges and currents with respect to time reversal yields a profound difference between electric and magnetic properties of crystals. A reversal of time changes the sign of the current and reverses the orientation of the atomic magnetic moments. In contrast to that, a polar vector is unaltered under time reversal. Therefore, we have to take into account, in addition to the spatial arrangements of atoms, the orientation of the atomic magnetic moments (spins). Conventional symmetry operations have been used to build the 32 crystal classes, that are the rotations and reflections. They have been used for the analysis of mechanical and electrical quantities. However, usual symmetry operations could match the geometrical structure with itself and reverse the orientation of the spins. A consequence of this observation is to apply a further transformation, which leads to a reversal of the spins. Such kinds of combined operations are called complementary symmetry operations, see Bhagavantam and Pantulu (1964) and the references herein. This new space-time symmetry operators correspond to proper and improper spatial rotations and reflections, combined with time reversal. The time reversal operation $\mathcal{T}$, causing a reversal of the spins, has to be combined with the usual symmetry operations $\boldsymbol{Q}$ of the 32 crystal classes $\mathcal{G}$.

An element of $\underline{\boldsymbol{Q}}_i$ of the magnetic group $\widetilde{\mathcal{M}}$ is given by the composition

$$\mathcal{T} \circ \boldsymbol{Q}_i = \underline{\boldsymbol{Q}}_i \,. \tag{199}$$

Obviously, we obtain

$$\boldsymbol{Q}_1 \cdot \boldsymbol{Q}_2 = \boldsymbol{Q}_3 \in \mathcal{G} \quad \text{since} \quad \{\boldsymbol{Q}_1, \boldsymbol{Q}_2\} \in \mathcal{G} \tag{200}$$

and

$$\left.\begin{array}{l} \underline{\boldsymbol{Q}}_1 \cdot \underline{\boldsymbol{Q}}_2 = \boldsymbol{Q}_3 \in \mathcal{G}, \\ \underline{\boldsymbol{Q}}_1 \cdot \boldsymbol{Q}_2 = \boldsymbol{Q}_1 \cdot \underline{\boldsymbol{Q}}_2 = \underline{\boldsymbol{Q}}_3 \in \widetilde{\mathcal{M}} \end{array}\right\} \text{since} \left\{\begin{array}{l} \{\boldsymbol{Q}_1, \boldsymbol{Q}_2\} \in \mathcal{G} \\ \{\underline{\boldsymbol{Q}}_1, \underline{\boldsymbol{Q}}_2\} \in \widetilde{\mathcal{M}} \,. \end{array}\right. \tag{201}$$

The combination of the 32 groups of $\mathcal{G}$ with the 58 additional groups of $\widetilde{\mathcal{M}}$ increases the number to possible point groups from 32 to 90. The magnetic crystal classes are defined by magnetic point groups $\mathcal{M}$ with

$$\mathcal{M} = \mathcal{G} \oplus \widetilde{\mathcal{M}} \,. \tag{202}$$

In this analysis, we have restricted ourselves to reversible processes. In order to describe dissipative effects, we have to introduce internal variables, which have to reflect the history-dependence of the material. For further details, we refer to Landau and Lifschitz (1985), Maugin (1988), Kiral and Eringen (1990) and Newham (2005).

5 Piezoelectricity, Piezomagnetism, Some Foundations

5.1 *Piezoelectricity*

The piezoelectric effect is based on the displacements of the charge carriers in the material, i.e. electric polarization occurs as a result of mechanical loads. Vice versa, mechanical deformations occur due to electrical loads. Only crystals without center of symmetry, e.g. BaT_iO_3, are piezoelectrics. Such an idealized crystal is illustrated in Fig. 13a, where an applied load leads to a relative displacement of the centroids of the positive and negative charge carriers; this is the microscopic origin of the piezoelectric effect. If a crystal has a center of symmetry, see Fig. 13b, then $-\mathbf{1}$ is

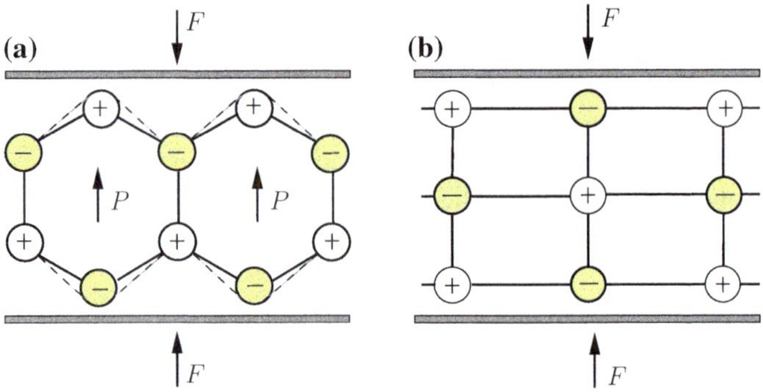

Fig. 13 **a** Crystal without a center of symmetry, **b** crystal with a center of symmetry

an element of the material symmetry group and the piezoelectric moduli transforms with

$$\mathbb{e}_{ijk} = (-1)^3 \mathbb{e}_{ijk} \quad \rightarrow \quad \mathbb{e}_{ijk} = 0 \,. \tag{203}$$

Obviously, the coupling effect vanishes for this type of crystal.

Constitutive equations of piezo-electric materials. A possible formulation of the fundamental piezoelectric constitutive relation in the polarization scheme in direct and index notation is given by

$$\left.\begin{aligned} \boldsymbol{\varepsilon} &= {}^E\mathbb{D} : \boldsymbol{\sigma} + \mathbb{d}^T \cdot \boldsymbol{E} \\ \boldsymbol{P} &= \quad \mathbb{d} : \boldsymbol{\sigma} + \quad \boldsymbol{\kappa} \cdot \boldsymbol{E} \end{aligned}\right\} \left\{\begin{aligned} \varepsilon_{ij} &= {}^E\mathbb{D}_{ijkl}\, \sigma_{kl} + \mathbb{d}_{kij}\, E_k \,, \\ P_i &= \quad \mathbb{d}_{ikl}\, \sigma_{kl} + \kappa_{ik}\, E_k \,. \end{aligned}\right. \tag{204}$$

Instead of using polarization as an extensive variable, it is more suitable to use the electric displacement $\boldsymbol{D}$ in a variety of applications. The relation between the electric displacements and the polarization has the form

$$\boldsymbol{D} = \epsilon_0 \boldsymbol{E} + \boldsymbol{P} = \epsilon_0(\boldsymbol{E} + \chi \boldsymbol{P}) = \epsilon_0 \epsilon_r\, \boldsymbol{E} = \boldsymbol{\epsilon}\, \boldsymbol{E} \,, \tag{205}$$

where the second-order dielectric modulus $\boldsymbol{\epsilon}$ is governed by the permittivity of a vacuum ϵ_0 and the susceptibility χ. An alternative constitutive setting of (204), in the electric displacement scheme with the independent variables $\boldsymbol{\varepsilon}$ and $\boldsymbol{E}$, is

$$\left.\begin{aligned} \boldsymbol{\sigma} &= {}^E\mathbb{C} : \boldsymbol{\varepsilon} - \mathbb{e}^T \cdot \boldsymbol{E} \\ \boldsymbol{D} &= \quad \mathbb{e} : \boldsymbol{\varepsilon} + \quad \boldsymbol{\epsilon} \cdot \boldsymbol{E} \end{aligned}\right\} \left\{\begin{aligned} \sigma_{ij} &= {}^E\mathbb{C}_{ijkl}\, \varepsilon_{kl} - \mathbb{e}_{kij}\, E_k \,, \\ D_i &= \quad \mathbb{e}_{ikl}\, \varepsilon_{kl} + \epsilon_{ik}\, E_k \,. \end{aligned}\right. \tag{206}$$

Alternative forms of equivalent constitutive equations can be obtained by simple algebraic manipulations of (206). Besides, fundamental relations between electrical and mechanical properties of a crystal are depicted in Fig. 14. Let ${}^E\mathbb{D} := {}^E\mathbb{C}^{-1}$ be the fourth-order compliance tensor at frozen electric field and analogously

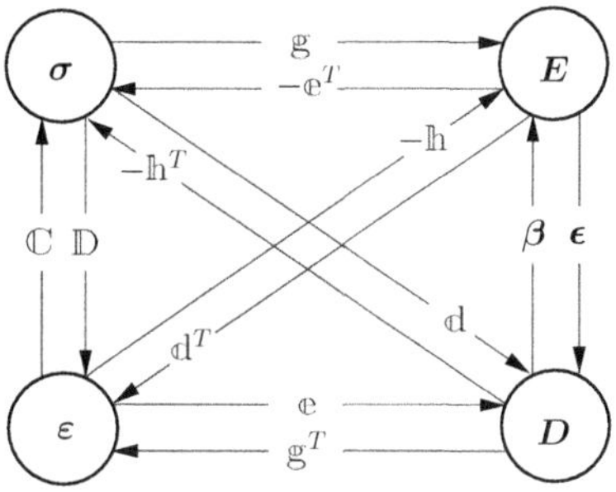

Fig. 14 Relations between electrical and mechanical properties

$$\mathbb{d} := \mathbb{e} : {}^E\mathbb{D} \quad \text{with} \quad \mathbb{d}_{ijk} := \mathbb{e}_{ilm}\, {}^E\mathbb{D}_{lmjk} \tag{207}$$

and

$$^{\sigma}\boldsymbol{\epsilon} := {}^{\varepsilon}\boldsymbol{\epsilon} + \mathbb{e} : {}^E\mathbb{D} : \mathbb{e}^T \quad \text{with} \quad ^{\sigma}\epsilon_{ij} := {}^{\varepsilon}\epsilon_{ij} + \mathbb{e}_{ilm}\, {}^E\mathbb{D}_{lmno}\mathbb{e}_{jon} \; . \tag{208}$$

Utilizing these definitions, we can derive the set of equations

$$\begin{aligned} \boldsymbol{\varepsilon} &= {}^E\mathbb{D} : \boldsymbol{\sigma} + \mathbb{d}^T \cdot \boldsymbol{E} \; , \\ \boldsymbol{D} &= \quad \mathbb{d} : \boldsymbol{\sigma} + \quad {}^{\sigma}\boldsymbol{\epsilon} \cdot \boldsymbol{E} \end{aligned} \tag{209}$$

with the independent variables $\boldsymbol{\sigma}$ and $\boldsymbol{E}$. Furthermore, we obtain from

$$^D\mathbb{C} := {}^E\mathbb{C} + \mathbb{e}^T \cdot {}^{\varepsilon}\boldsymbol{\epsilon}^{-1} \cdot \mathbb{e} \; , \quad \mathbb{h} := {}^{\varepsilon}\boldsymbol{\epsilon}^{-1} \cdot \mathbb{e} \quad \text{and} \quad {}^{\varepsilon}\boldsymbol{\beta} := {}^{\varepsilon}\boldsymbol{\epsilon}^{-1} \tag{210}$$

a set of linear equations in terms of the dependent variables $\boldsymbol{\varepsilon}$ and $\boldsymbol{D}$:

$$\begin{aligned} \boldsymbol{\sigma} &= {}^D\mathbb{C} : \boldsymbol{\varepsilon} - \mathbb{h}^T \cdot \boldsymbol{D} \; , \\ \boldsymbol{E} &= -\mathbb{h} : \boldsymbol{\varepsilon} + {}^{\varepsilon}\boldsymbol{\beta} \cdot \boldsymbol{D} \; . \end{aligned} \tag{211}$$

Finally, we introduce the abbreviations

$$^D\mathbb{D} := {}^D\mathbb{C}^{-1} \; , \quad \mathbb{g} := \mathbb{h} : {}^D\mathbb{C}^{-1} = {}^{\sigma}\boldsymbol{\epsilon}^{-1} \cdot \mathbb{d} \quad \text{and} \quad {}^{\sigma}\boldsymbol{\beta} := {}^{\sigma}\boldsymbol{\epsilon}^{-1} \; , \tag{212}$$

in order to derive response functions in terms of the agencies $\boldsymbol{\sigma}$ and $\boldsymbol{D}$:

$$\begin{aligned} \boldsymbol{\varepsilon} &= {}^D\mathbb{D} : \boldsymbol{\sigma} + \mathbb{g}^T \cdot \boldsymbol{D} \; , \\ \boldsymbol{E} &= -\mathbb{g} : \boldsymbol{\sigma} + {}^{\sigma}\boldsymbol{\beta} \cdot \boldsymbol{D} \; . \end{aligned} \tag{213}$$

In analogy to the definition of the Helmholtz free-energy per unit volume, we introduce the electric Gibbs function and the electric enthalpy per unit volume

$$\widetilde{G}_2(\boldsymbol{\varepsilon}, \boldsymbol{E}; \vartheta) = \rho_e \, G_2\big|_{\vartheta} \quad \text{and} \quad \widetilde{H}_2(\boldsymbol{\varepsilon}, \boldsymbol{E}; \eta) = \rho_e \, H_2\big|_{\eta} \; , \tag{214}$$

at fixed temperature or constant entropy, respectively. Using $(214)_2$, we obtain the constitutive expressions

$$\boldsymbol{\sigma} = \frac{\partial \widetilde{H}_2(\boldsymbol{\varepsilon}, \boldsymbol{E})}{\partial \boldsymbol{\varepsilon}} \quad \text{and} \quad \boldsymbol{D} = -\frac{\partial \widetilde{H}_2(\boldsymbol{\varepsilon}, \boldsymbol{E})}{\partial \boldsymbol{E}} , \tag{215}$$

for the stresses and electric displacements, respectively. For the partial derivatives of the stresses and electric displacements, we introduce the abbreviations

$$\mathbb{C} = \frac{\partial \boldsymbol{\sigma}(\boldsymbol{\varepsilon}, \boldsymbol{E})}{\partial \boldsymbol{\varepsilon}} , \quad \mathbb{e} = \frac{\partial \boldsymbol{D}(\boldsymbol{\varepsilon}, \boldsymbol{E})}{\partial \boldsymbol{\varepsilon}} \quad \text{and} \quad \boldsymbol{\epsilon} = \frac{\partial \boldsymbol{D}(\boldsymbol{\varepsilon}, \boldsymbol{E})}{\partial \boldsymbol{E}} , \tag{216}$$

where $\mathbb{C}$ denotes the fourth-order elasticity tensor, $\mathbb{e}$ the third-order tensor of piezoelectric moduli and $\boldsymbol{\epsilon}$ the second-order tensor of dielectric moduli. The potential structure induces

$$\left(\frac{\partial^2 \widetilde{H}_2}{\partial \boldsymbol{E}\, \partial \boldsymbol{\varepsilon}} \right)^T := \frac{\partial^2 \widetilde{H}_2}{\partial \boldsymbol{\varepsilon}\, \partial \boldsymbol{E}} . \tag{217}$$

Therefore, we can derive the relationship

$$\mathbb{e} := \frac{\partial \boldsymbol{D}}{\partial \boldsymbol{\varepsilon}} = -\frac{\partial^2 \widetilde{H}_2}{\partial \boldsymbol{\varepsilon}\, \partial \boldsymbol{E}} = -\left(\frac{\partial^2 \widetilde{H}_2}{\partial \boldsymbol{E}\, \partial \boldsymbol{\varepsilon}} \right)^T = -\left(\frac{\partial \boldsymbol{\sigma}}{\partial \boldsymbol{E}} \right)^T \tag{218}$$

and we conclude

$$\frac{\partial \boldsymbol{\sigma}}{\partial \boldsymbol{E}} = -\mathbb{e}^T . \tag{219}$$

This leads to the constitutive equations (206)

$$\left. \begin{aligned} \boldsymbol{\sigma} &= {}^E\mathbb{C} : \boldsymbol{\varepsilon} - \mathbb{e}^T \cdot \boldsymbol{E} \\ -\boldsymbol{D} &= -\mathbb{e} : \boldsymbol{\varepsilon} - \boldsymbol{\epsilon} \cdot \boldsymbol{E} \end{aligned} \right\} \left\{ \begin{aligned} \sigma_{ij} &= {}^E\mathbb{C}_{ijkl}\, \varepsilon_{kl} - \mathbb{e}_{kij}\, E_k , \\ -D_i &= -\mathbb{e}_{ikl}\, \varepsilon_{kl} - \epsilon_{ik}\, E_k . \end{aligned} \right. \tag{220}$$

In matrix notation the set of equations (220) appears as follows:

$$\begin{bmatrix} \sigma_{11} \\ \sigma_{22} \\ \sigma_{33} \\ \sigma_{12} \\ \sigma_{23} \\ \sigma_{13} \\ -D_1 \\ -D_2 \\ -D_3 \end{bmatrix} = \left[\begin{array}{cccccc|ccc} \mathbb{C}_{11} & \mathbb{C}_{12} & \mathbb{C}_{13} & \mathbb{C}_{14} & \mathbb{C}_{15} & \mathbb{C}_{16} & -\mathbb{e}_{11} & -\mathbb{e}_{12} & -\mathbb{e}_{13} \\ & \mathbb{C}_{22} & \mathbb{C}_{23} & \mathbb{C}_{24} & \mathbb{C}_{25} & \mathbb{C}_{26} & -\mathbb{e}_{21} & -\mathbb{e}_{22} & -\mathbb{e}_{23} \\ & & \mathbb{C}_{33} & \mathbb{C}_{34} & \mathbb{C}_{35} & \mathbb{C}_{36} & -\mathbb{e}_{31} & -\mathbb{e}_{32} & -\mathbb{e}_{33} \\ & & & \mathbb{C}_{44} & \mathbb{C}_{45} & \mathbb{C}_{46} & -\mathbb{e}_{41} & -\mathbb{e}_{42} & -\mathbb{e}_{43} \\ & & & & \mathbb{C}_{55} & \mathbb{C}_{56} & -\mathbb{e}_{51} & -\mathbb{e}_{52} & -\mathbb{e}_{53} \\ & & & & & \mathbb{C}_{66} & -\mathbb{e}_{61} & -\mathbb{e}_{62} & -\mathbb{e}_{63} \\ \hline & & & & & & -\epsilon_{11} & -\epsilon_{12} & -\epsilon_{13} \\ & sym. & & & & & & -\epsilon_{22} & -\epsilon_{23} \\ & & & & & & & & -\epsilon_{33} \end{array} \right] \begin{bmatrix} \varepsilon_{11} \\ \varepsilon_{22} \\ \varepsilon_{33} \\ 2\varepsilon_{12} \\ 2\varepsilon_{23} \\ 2\varepsilon_{13} \\ E_1 \\ E_2 \\ E_3 \end{bmatrix} \tag{221}$$

In order to find a more compact matrix notation, we use the following abbreviations for the constitutive equations

$$\begin{bmatrix} \underline{\boldsymbol{\sigma}} \\ -\underline{\boldsymbol{D}} \end{bmatrix} = \begin{bmatrix} \underline{\mathbb{C}} & -\underline{\mathbb{e}}^T \\ -\underline{\mathbb{e}} & -\underline{\epsilon} \end{bmatrix} \begin{bmatrix} \underline{\varepsilon} \\ \underline{\boldsymbol{E}} \end{bmatrix}. \tag{222}$$

As already mentioned, we have in general 21 elastic constants, 18 piezoelectric constants and 6 dielectric constants. The matrix notation (Voigt notation) for the elasticity tensor is

$$\underline{\mathbb{C}} = \begin{bmatrix} \mathbb{C}_{1111} & \mathbb{C}_{1122} & \mathbb{C}_{1133} & \mathbb{C}_{11(12)} & \mathbb{C}_{11(23)} & \mathbb{C}_{11(13)} \\ \mathbb{C}_{2211} & \mathbb{C}_{2222} & \mathbb{C}_{2233} & \mathbb{C}_{22(12)} & \mathbb{C}_{22(23)} & \mathbb{C}_{22(13)} \\ \mathbb{C}_{3311} & \mathbb{C}_{3322} & \mathbb{C}_{3333} & \mathbb{C}_{33(12)} & \mathbb{C}_{33(23)} & \mathbb{C}_{33(13)} \\ \mathbb{C}_{(12)11} & \mathbb{C}_{(12)22} & \mathbb{C}_{(12)33} & \mathbb{C}_{(12)(12)} & \mathbb{C}_{(12)(23)} & \mathbb{C}_{(12)(13)} \\ \mathbb{C}_{(23)11} & \mathbb{C}_{(23)22} & \mathbb{C}_{(23)33} & \mathbb{C}_{(23)(12)} & \mathbb{C}_{(23)(23)} & \mathbb{C}_{(23)(13)} \\ \mathbb{C}_{(13)11} & \mathbb{C}_{(13)22} & \mathbb{C}_{(13)33} & \mathbb{C}_{(13)(12)} & \mathbb{C}_{(13)(23)} & \mathbb{C}_{(13)(13)} \end{bmatrix}, \tag{223}$$

and the piezoelectric coupling terms are arranged as

$$\underline{\mathbb{e}} = \begin{bmatrix} \mathbb{e}_{111} & \mathbb{e}_{122} & \mathbb{e}_{133} & \mathbb{e}_{1(12)} & \mathbb{e}_{1(23)} & \mathbb{e}_{1(13)} \\ \mathbb{e}_{211} & \mathbb{e}_{222} & \mathbb{e}_{233} & \mathbb{e}_{2(12)} & \mathbb{e}_{2(23)} & \mathbb{e}_{2(13)} \\ \mathbb{e}_{311} & \mathbb{e}_{322} & \mathbb{e}_{333} & \mathbb{e}_{3(12)} & \mathbb{e}_{3(23)} & \mathbb{e}_{3(13)} \end{bmatrix}, \tag{224}$$

where we used the definitions $\mathbb{C}_{ij(kl)} = \frac{1}{2}(\mathbb{C}_{ijkl} + \mathbb{C}_{ijlk}), \mathbb{C}_{(ij)kl} = \frac{1}{2}(\mathbb{C}_{ijkl} + \mathbb{C}_{jikl})$, and $\mathbb{e}_{i(jk)} = \frac{1}{2}(\mathbb{e}_{ijk} + \mathbb{e}_{ikj})$. Furthermore, the coupling terms for the mechanical stresses and the electric field appear in matrix notation as

$$\underline{\mathbb{e}}^T = \begin{bmatrix} \mathbb{e}_{111} & \mathbb{e}_{112} & \mathbb{e}_{113} \\ \mathbb{e}_{221} & \mathbb{e}_{222} & \mathbb{e}_{223} \\ \mathbb{e}_{331} & \mathbb{e}_{332} & \mathbb{e}_{333} \\ \mathbb{e}_{(12)1} & \mathbb{e}_{(12)2} & \mathbb{e}_{(12)3} \\ \mathbb{e}_{(23)1} & \mathbb{e}_{(23)2} & \mathbb{e}_{(23)3} \\ \mathbb{e}_{(13)1} & \mathbb{e}_{(13)2} & \mathbb{e}_{(13)3} \end{bmatrix}, \tag{225}$$

with $\mathbb{e}_{(ij)k} = (\mathbb{e}_{ijk} + \mathbb{e}_{jik})/2$. Finally, the dielectric part, which relates the electric field with the electric displacements, is given as

$$\underline{\epsilon} = \begin{bmatrix} \epsilon_{11} & \epsilon_{12} & \epsilon_{13} \\ \epsilon_{21} & \epsilon_{22} & \epsilon_{23} \\ \epsilon_{31} & \epsilon_{32} & \epsilon_{33} \end{bmatrix}. \tag{226}$$

5.2 Piezomagnetism

In analogy to piezoelectricity, piezomagnetism describes a linear coupling between magnetization $\boldsymbol{M}$ and mechanical stresses $\boldsymbol{\sigma}$ (direct piezomagnetic effect). The converse effect relates the strains ε to the magnetic (auxiliary) field $\boldsymbol{H}$. The piezomagnetic coefficients of the associated third rank axial tensor $\mathbb{q}_{jkl}$ can be expressed in a 3×6 matrix:

$$\underline{\mathbb{q}} = \begin{bmatrix} \mathbb{q}_{111} & \mathbb{q}_{122} & \mathbb{q}_{133} & \mathbb{q}_{1(12)} & \mathbb{q}_{1(23)} & \mathbb{q}_{1(13)} \\ \mathbb{q}_{211} & \mathbb{q}_{222} & \mathbb{q}_{233} & \mathbb{q}_{2(12)} & \mathbb{q}_{2(23)} & \mathbb{q}_{2(13)} \\ \mathbb{q}_{311} & \mathbb{q}_{322} & \mathbb{q}_{333} & \mathbb{q}_{3(12)} & \mathbb{q}_{3(23)} & \mathbb{q}_{3(13)} \end{bmatrix} . \tag{227}$$

If we apply a symmetry operation involving mirror planes or inversion the properties transforms as

$$\widetilde{\mathbb{q}}_{jkl} = -Q_{ji}\, Q_{km}\, Q_{ln} \mathbb{q}_{imn} \, . \tag{228}$$

If we have pure rotations, then there is no change of sign for the tensor coefficients. Obviously, the classical crystallographic point groups do not show the effect; applying the time reversal operations yields

$$\mathbb{q} = -\mathbb{q} \quad \rightarrow \quad \mathbb{q} = \mathbf{0} \, . \tag{229}$$

There exist 90 magnetic point groups, 66 of them are piezomagnetic. For further details, we refer to Newham (2005).

5.3 Magnetoelectricity

The magnetoelectric effect is interesting for a variety of technical applications, because it involves a coupling between electric and magnetic quantities. A general linear constitutive relation between electric and magnetic quantities is

$$\begin{bmatrix} \underline{\boldsymbol{D}} \\ \underline{\boldsymbol{B}} \end{bmatrix} = \begin{bmatrix} \underline{\epsilon} & \underline{\alpha}^T \\ \underline{\alpha} & \underline{\mu} \end{bmatrix} \begin{bmatrix} \underline{\boldsymbol{E}} \\ \underline{\boldsymbol{H}} \end{bmatrix} . \tag{230}$$

Since $\underline{\boldsymbol{B}}$ is a pseudovector and $\underline{\boldsymbol{E}}$ is a polar vector, the electromagnetic effect has to be an axial second-order tensor. Therefore, the effect vanishes for

(i) all classical symmetry groups containing time reversal symmetry,
(ii) an inversion ($\boldsymbol{Q} = -\mathbf{1}$) operation.

The magnetoelectric effect only occurs in magnetic point groups. Furthermore, the effect is also permitted for space inversion accompanied by time inversion, i.e. $\mathcal{T} \circ (-\mathbf{1})$. 58 of the 90 magnetic point groups are magnetoelectric.

5.4 Anisotropic and Isotropic Tensor Functions

We now focus on the transverse isotropy group, which is characterized by one preferred direction $\boldsymbol{a}$.

The associated symmetry group can be classified into five subgroups

$$\left.\begin{array}{ll}\mathcal{G}_1 = \{\boldsymbol{Q} \in \mathcal{SO}(3), & \boldsymbol{Q}\boldsymbol{a} = \boldsymbol{a}\} \\ \mathcal{G}_2 = \{\boldsymbol{Q} \in \mathcal{O}(3), & \boldsymbol{Q}\boldsymbol{a} = \boldsymbol{a}\} \\ \mathcal{G}_3 = \{\boldsymbol{Q} \in \mathcal{O}(3), & \boldsymbol{Q} \in \mathcal{G}_1 \text{ or } -\boldsymbol{Q} \in \mathcal{G}_1\} \\ \mathcal{G}_4 = \{\boldsymbol{Q} \in \mathcal{SO}(3), & \boldsymbol{Q}\boldsymbol{a} = \boldsymbol{a} \text{ or } \boldsymbol{Q}\boldsymbol{a} = -\boldsymbol{a}\} \\ \mathcal{G}_5 = \{\boldsymbol{Q} \in \mathcal{O}(3), & \boldsymbol{Q}\boldsymbol{a} = \boldsymbol{a} \text{ or } \boldsymbol{Q}\boldsymbol{a} = -\boldsymbol{a}\}\end{array}\right\} , \tag{231}$$

see Liu (1982) and Spencer (1971). The smallest group $\mathcal{G}_1$ allows rotations only about the preferred direction; it is therefore denoted as rotational symmetry. The second class allows additional reflections with respect to planes that are characterized by normals lying in the isotropy plane. The remaining three classes include reflections with respect to the isotropy plane. Particularly $\mathcal{G}_5$ is the most suitable group for describing pure mechanical problems of transverse isotropy, e.g. materials with arranged uniaxial fibers. Hence, the material symmetry group for the mechanical behavior of this class can be represented by

$$\mathcal{G}_{ti}^{mech} := \mathcal{G}_5 \ . \tag{232}$$

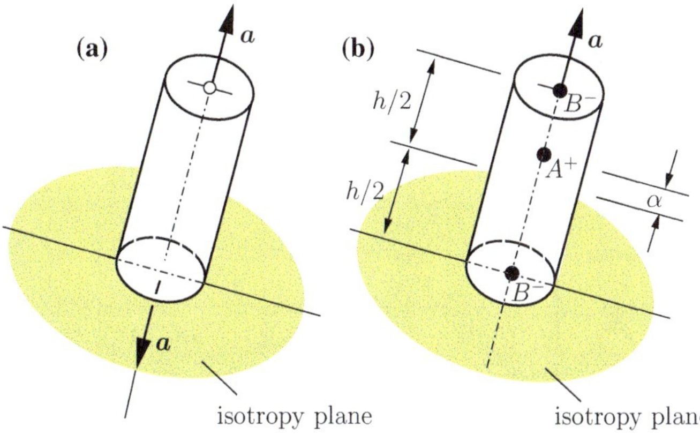

Fig. 15 Illustration of the principle of superposition of symmetries: **a** invariance of mechanical part with respect to $\pm\boldsymbol{a}$; **b** polarization of the electrical part due to the distance of the centers of gravity between the positively and negatively charged particles within a unit cell constitutes invariance with respect to $+\boldsymbol{a}$, respectively

This group can be expressed by introducing the second-order structural tensor $\boldsymbol{m}$, which reflects the invariance group of the material (but only for the mechanical response), see Fig. 15a. In piezoelectric solids, the unit cell of a transversely isotropic material is characterized by a spontaneous polarization, given by $\alpha\, q\, \boldsymbol{a}$. Here, q denotes the charge of the centered particle and α the distance between the centers of gravity of the positively and negatively charged particles within the unit cell, see Fig. 15b. Thus, it is obvious that the piezoelectric effect occurs only within systems that lack a center of symmetry. The suitable invariance group for the characterization of a polar vector is

$$\mathcal{G}_{ti}^{el} := \mathcal{G}_2 \,. \tag{233}$$

The purely electric part of the piezoelectric material is characterized by a potential function having the form $\widetilde{H}_{ti}^{el}(\boldsymbol{E}, \boldsymbol{a})$. The invariance group $\mathcal{G}_{ti}$ of the electro-mechanically coupled solid can be obtained by using the *Principle of the Superposition of Symmetries*, also known as the *Curie Symmetry Principle*. This principle states that the overall symmetry group of several objects is the highest common symmetry subgroup of these objects within consideration of the mutual orientations of their individual symmetry elements. This procedure reads as follows

$$\mathcal{G}_{ti}^{mech} \cap \mathcal{G}_{ti}^{el} = \mathcal{G}_2 =: \mathcal{G}_{ti} \,. \tag{234}$$

For a more exhaustive mathematical treatment of this topic, we refer to Boehler (1987a, b).

Remark: For the derivation of the polynomial basis, we have to distinguish between absolute and axial vectors. It is not possible to express the vector $\boldsymbol{E}$ by an associated (antisymmetric) second-order pseudotensor, because the electric field strength is a polar vector as is the electric displacement vector.

The further restrictions on the linear independent constants, induced by the associated symmetry group, are derived in a privileged frame where the x_3-axis is the preferred direction $\boldsymbol{a}$. Then, the coordinate-dependent representation of the mechanical moduli is

$$\underline{\mathbb{C}} = \begin{bmatrix} \mathbb{C}_{11} & \mathbb{C}_{12} & \mathbb{C}_{13} & 0 & 0 & 0 \\ \mathbb{C}_{12} & \mathbb{C}_{11} & \mathbb{C}_{13} & 0 & 0 & 0 \\ \mathbb{C}_{13} & \mathbb{C}_{13} & \mathbb{C}_{33} & 0 & 0 & 0 \\ 0 & 0 & 0 & \frac{1}{2}(\mathbb{C}_{11} - \mathbb{C}_{12}) & 0 & 0 \\ 0 & 0 & 0 & 0 & \mathbb{C}_{44} & 0 \\ 0 & 0 & 0 & 0 & 0 & \mathbb{C}_{44} \end{bmatrix} \tag{235}$$

with five independent constants. Furthermore, the matrix representation for the piezoelectric constants and dielectric constants is as follows

$$\underline{\mathbb{e}} = \begin{bmatrix} 0 & 0 & 0 & 0 & 0 & \mathbb{e}_{15} \\ 0 & 0 & 0 & 0 & \mathbb{e}_{15} & 0 \\ \mathbb{e}_{31} & \mathbb{e}_{31} & \mathbb{e}_{33} & 0 & 0 & 0 \end{bmatrix} \quad \text{and} \quad \underline{\epsilon} = \begin{bmatrix} \epsilon_{11} & 0 & 0 \\ 0 & \epsilon_{11} & 0 \\ 0 & 0 & \epsilon_{33} \end{bmatrix} , \tag{236}$$

respectively. In this case, we obtain three independent piezoelectric and two independent dielectric constants.

Coordinate-invariant formulation. The following discussion is mainly based on the representations given in Schröder and Gross (2004). For the coordinate invariant formulation of constitutive equations the representations of isotropic tensor functions are used. The governing equations must automatically represent the material symmetries of the body of interest. Let $\boldsymbol{a}$ be the preferred direction of the transversely isotropic material, then the material symmetry group is defined as

$$\mathcal{G}_{ti} = \{\pm\mathbf{1} \; ; \; \boldsymbol{Q}(\alpha, \boldsymbol{a}) | \; 0 \leq \alpha < 2\pi\} \; , \tag{237}$$

where $\boldsymbol{Q}(\alpha, \boldsymbol{a})$ represents all rotations about $\boldsymbol{a}$. The values of the enthalpy function must be scalar invariants under all transformations $\boldsymbol{Q} \in \mathcal{G}_{ti}$ of the set of tensorial variables involved in the scalar-valued function, i.e.

$$\widetilde{H}_2(\varepsilon, \boldsymbol{E}) = \widetilde{H}_2(\boldsymbol{Q}\varepsilon\boldsymbol{Q}^T, \boldsymbol{Q}\boldsymbol{E}) \quad \forall \quad \boldsymbol{Q} \in \mathcal{G}_{ti} \; . \tag{238}$$

Thus, $\mathcal{G}_{ti}$ represents the material symmetry group, which reflects the geometrical and physical symmetries of the anisotropic solid. The invariance conditions for the stresses and the electric displacements are given by

$$\left. \begin{aligned} \boldsymbol{Q}\boldsymbol{\sigma}(\varepsilon, \boldsymbol{E})\boldsymbol{Q}^T &= \boldsymbol{\sigma}(\boldsymbol{Q}\varepsilon\boldsymbol{Q}^T, \boldsymbol{Q}\boldsymbol{E}) \\ \boldsymbol{Q}\boldsymbol{D}(\varepsilon, \boldsymbol{E}) &= \boldsymbol{D}(\boldsymbol{Q}\varepsilon\boldsymbol{Q}^T, \boldsymbol{Q}\boldsymbol{E}) \end{aligned} \right\} \quad \forall \quad \boldsymbol{Q} \in \mathcal{G}_{ti} \; . \tag{239}$$

The main idea of the invariant theory is the extension of the $\mathcal{G}_{ti}$-invariant functions (238) and $(239)_{1,2}$ to functions that are invariant under a larger group of transformations, particularly under all elements of the orthogonal group $\mathcal{O}(3)$. For the invariant formulation of the constitutive expressions, we introduce additional argument tensors: the second-order structural tensor $\boldsymbol{m} = \boldsymbol{a} \otimes \boldsymbol{a}$ and the preferred direction $\boldsymbol{a}$. In the following, we neglect the explicit dependence of the potential function with respect to the second-order structural tensor. In this manner, we obtain, using the principle of isotropy of space, see e.g. Boehler (1987), the following representation for the enthalpy function:

$$\widetilde{H}_2(\varepsilon, \boldsymbol{E}, \boldsymbol{a}) = \widetilde{H}_2(\boldsymbol{Q}\varepsilon\boldsymbol{Q}^T, \boldsymbol{Q}\boldsymbol{E}, \boldsymbol{Q}\boldsymbol{a}) \quad \forall \quad \boldsymbol{Q} \in \mathcal{O}(3) \; . \tag{240}$$

For the second- and first-order tensor functions, we obtain

$$\left. \begin{aligned} \boldsymbol{Q}\boldsymbol{\sigma}(\boldsymbol{\varepsilon}, \boldsymbol{E}, \boldsymbol{a})\boldsymbol{Q}^T &= \boldsymbol{\sigma}(\boldsymbol{Q}\boldsymbol{\varepsilon}\boldsymbol{Q}^T, \boldsymbol{Q}\boldsymbol{E}, \boldsymbol{Q}\boldsymbol{a}) \\ \boldsymbol{Q}\boldsymbol{D}(\boldsymbol{\varepsilon}, \boldsymbol{E}, \boldsymbol{a}) &= \boldsymbol{D}(\boldsymbol{Q}\boldsymbol{\varepsilon}\boldsymbol{Q}^T, \boldsymbol{Q}\boldsymbol{E}, \boldsymbol{Q}\boldsymbol{a}) \end{aligned} \right\} \quad \forall \;\; \boldsymbol{Q} \in \mathcal{O}(3) \,. \tag{241}$$

The expressions (240) and (241) are the definitions of isotropic tensor functions, i.e. these relationships represent a set of isotropic functions with respect to the whole set of tensorial arguments $\{\boldsymbol{\varepsilon}, \boldsymbol{E}, \boldsymbol{a}\}$. Conversely, $(241)_{1,2}$ and (240) are anisotropic functions with respect to the arguments $\{\boldsymbol{\varepsilon}, \boldsymbol{E}\}$, i.e.

$$\left. \begin{aligned} \boldsymbol{Q}\boldsymbol{\sigma}(\boldsymbol{\varepsilon}, \boldsymbol{E}, \boldsymbol{a})\boldsymbol{Q}^T &= \boldsymbol{\sigma}(\boldsymbol{Q}\boldsymbol{\varepsilon}\boldsymbol{Q}^T, \boldsymbol{Q}\boldsymbol{E}, \boldsymbol{a}) \\ \boldsymbol{Q}\boldsymbol{D}(\boldsymbol{\varepsilon}, \boldsymbol{E}, \boldsymbol{a}) &= \boldsymbol{D}(\boldsymbol{Q}\boldsymbol{\varepsilon}\boldsymbol{Q}^T, \boldsymbol{Q}\boldsymbol{E}, \boldsymbol{a}) \end{aligned} \right\} \quad \forall \;\; \boldsymbol{Q} \in \mathcal{G}_{ti} \,. \tag{242}$$

Due to this observation, we conclude that the invariance group of the structural tensor, here the preferred direction $\boldsymbol{a}$, characterizes the type of anisotropy. For a detailed discussion concerning the minimal number of independent scalar variables that have to enter a general constitutive expression, we refer to Liu (1982) and the references therein.

6 Summary

In this contribution, we have discussed the fundamental aspects of the magneto-electro-mechanical coupling. Therefore, we recapitulated the basic aspects of electricity, magnetism and the Maxwell's equations, summerized the governing balance equations for electrostatics and magnetostatics and give an overview of some thermodynamical formulations. It is important to focus on transformation properties regarding physical laws and invariance conditions. Therefore, we discussed the properties of the physical quantities under coordinate-transformations including rotations, spatial reflections and time-reversal in order to get an overview of the mathematical modeling of coupled materials.

References

Abraham, M. (1909). Zur Elektrodynamik bewegter Körper. *Rendiconti del Circolo Matematico di Palermo*, *28*, 1–28.
Abraham, M. (1910). Sull'elettrodinamica di Minkowski. Rendiconti del Circolo Matematico di Palermo, *30*(0), 33–46. (see also: M. Abraham. (1923). Theorie der Elektrizität, Vol: II, Leibzig: Teubner).
Bertotti, G. (1998). *Hysteresis in magnetism for physicists, materials scientists, and engineers*. Elsevier-Academic Press.
Bertotti, G., & Mayergoyz, I. (2006a). *The science of hysteresis* (Vol. 1). Elsevier.
Bertotti, G., & Mayergoyz, I. (2006b). *The science of hysteresis* (Vol. 2). Elsevier.

Bertotti, G. & Mayergoyz, I. (2006c). *The science of hysteresis* (Vol. 3). Elsevier.
Bhagavantam, S., & Pantulu, P. V. (1964). Magnetic symmetry and physical properties of crystals. *Proceedings of the Indian Academy of Sciences-Section A*, *59*(1), 01–13.
Bobbio, S. (2000). *Electrodynamics of materials*. Elsevier Academic Press.
Boehler, J. P. (1987a). Introduction to the invariant formulation of anisotropic constitutive equations. In *Applications of tensor functions in solid mechanics* (pp. 13–30).
Boehler, J. P. (1987b) Representations for isotropic and anisotropic non-polynominal tensor functions. In *Applications of tensor functions in solid mechanics* (pp. 31–53).
Einstein, A., & Laub, J. (1908). über die elektromagnetischen Felder auf ruhende Körper ausgeübten ponderomotorischen Kräfte. *Annals of Physics (Leipzig)*, *26*, 541–550.
Eringen, A. C.. & Maugin, G. A. (1989). *Electrodynamics of continua i-foundations and solid media*. Springer Verlag.
Eringen, A. C., & Maugin, G. A. (1990). *Electrodynamics of continua II-fluids and complex media*. Springer Verlag.
Fabrizio, M., & Morro, A. (2003). *Electromagnetism of continuous media*. Oxford Science Publications.
Fatuzzo, E., & Merz, W. J. (1967). *Ferroelectricity*. John Wiley & Sons.
Fließbach, T. (1999). *Mechanik, Lehrbuch der theoretischen Physik 1*. Sprektrum Akademischer Verlag.
Fließbach, T. (2000). *Elektrodynamik, Lehrbuch der theoretischen Physik 2*. Spektrum Akademischer Verlag.
Fushchych, W. I. (1969). Equations of motion in odd-dimensional spaces and $T-$, $C-$invariance. *Preprint of Institute of Theoretical Physics, Kiev*, 69–17, 0 13ff.
Grehn, J., & Krause, J. (2007). *Metzler Physik*. Schroedel.
Griffiths, D. J. (2008). *Introduction to electrodynamics*. Pearson.
Hofmann, H. (1986). *Das elektormagnetische Feld-Theorie und grundlegende Anwendungen*. Springer.
Hutter K., Jöhnk K. (2004) Jump Conditions. In: *Continuum methods of physical modeling*. Berlin, Heidelberg: Springer.
Jackson, J. D. (2002). *Klassische Elektrodynamik*. John Wiley & Sons Ltd.
Kiehn, R. M. (1977). Parity and time inversion symmetries of electromagnetic systems. *Foundations of Physics*, *7*, 301–311.
Kiral, E., & Eringen, A. C. (1990). *Constitutive equations of nonlinear electromagnetic-elastic crystals*. Springer-Verlag.
Landau, L. D., & Lifschitz, E. M. (1985). *Lehrbuch der theoretischen Physik*. Akademie-Verlag.
Lines, M. E., & Glass, A. M. (1977). *Principles and applications of ferroelectrics and related materials*. Oxford University Press.
Liu, I.-S. (1982). On representations of anisotropic invariants. *International Journal of Engineering Science*, *20*, 1099–1109.
Maugin, G. A. (1988). *Continuum mechanics of electromagnetic solids*. Elsevier Science.
Maugin, G. A., Pouget, J., Drouot, R. & Collet, B. (1991). *Nonlinear electromechanical couplings*. John Wiley & Sons.
Maxwell, J. C. (1865). A dynamical theory of the electromagnetic field. *Philosophical Transactions of the Royal Society of London*.
Maxwell, J. C. (1873). *A treatise on electricity and magnetism* (Vol. 1 and 2). Dover.
Minkowski, H. (1908). Die Grundgleichungen für die elektromagnetischen Vorgänge in bewegten Körpern. *Göttinger Nachrichten* (pp. 53–111).
Nahin, P. J. (1992). Maxwell's grand unification. *IEEE Spectrum*, *29*(3), 0 45 ff.
Newham, R. E. (2005). *Properties of materials*. Oxford University Press.
Pauli, W. (1979). *Wissenschaftlicher Briefwechsel mit Bohr, Einstein, Heisenberg u.a. Band I: 1919-1929*. In A. Hermann, K.V. Meyenn, & V.F. Weisskopf.
Schröder, J., & Gross, D. (2004). Invariant formulation of the electromechanical enthalpy function of transversely isotropic piezoelectric materials. *Archive of Applied Mechanics*, *73*, 533–552.

Simonyi, K. (2001). *Kulturgeschichte der Physik: Von den Anfängen bis heute*. Harri Deutsch.
Spencer, A. J. M. (1971). Theory of invariants. Continuum. *Physics, 1*, 239–353.
Tipler, P. A. (1999). *Physics for scientists and engineers* (4th ed.). Freemann and Company: W.H.
Weile, D. S., Hopkins, D. A., Gazonas, G. A., & Powers, B. M. (2014). On the proper formulation of maxwellian electrodynamics for continuum mechanics. *Continuum Mechanics and Thermodynamics, 26*, 387–401.
Zohdi, T. (2012). *Electromagnetic properties of multiphase dielectrics-A primer on modeling theory and computation*. Springer.

Ferroelectric and Ferromagnetic Phase Field Modeling

Dorinamaria Carka and Christopher S. Lynch

Abstract This chapter provides an introduction to phase field modeling. It is directed at the level of a graduate student with some background in mechanics. It begins with a review of the electro-statics, magneto-statics, and mechano-statics that are needed when setting up phase field models. These yield the various conservation laws that are used. After this review, thermodynamics of materials is discussed. The first law of thermodynamics is used to equate work done on a material plus heat added to the material to the increase of internal energy. This is used with the second law, a statement that irreversible processes generate entropy. This leads to fundamental relations that must be followed in postulating forms for the internal energy. The ways that internal energy is stored in a material are determined through observation. Once the mechanisms have been identified, work conjugate internal variables are introduced to enable writing a specific form for the internal energy and its derivatives. Observations are invoked once again regarding material symmetry to reduce the number of constants that must be determined when modeling a specific material.

1 Introduction

Phase field models are based on the idea that there is some order parameter that describes the state of the material. In a ferroelectric material the order parameter is the polarization and in a ferromagnetic material the order parameter is the magnetization. Analysis of the behavior of the material when not in an equilibrium state gives the driving force for evolution of the order parameter. This driving force is then used with

D. Carka (✉)
Department of Mechanical Engineering, New York Institute of Technology, Old Westbury, USA
e-mail: dcarka@nyit.edu

C.S. Lynch
Department Chair, Mechanical and Aerospace Engineering, University of California, Los Angeles, CA, USA

J. Schröder and Doru C. Lupascu (eds.), *Ferroic Functional Materials*, CISM International Centre for Mechanical Sciences 581, https://doi.org/10.1007/978-3-319-68883-1_2

a kinetic relation that governs its rate of change. In the case of a ferroelectric material the kinetic relation is the time dependent Ginzburg Landau equation (TDGL) and in the case of a ferromagnetic material the kinetic relation is the Landau Lifshitz Gilbert equation (LLG). Detailed attention is given to the divergence of the second order tensors that are work conjugate to the polarization and magnetization gradients, as these are less familiar than the divergence of the second order stress tensor that is work conjugate to the strain tensor. The roles in balance laws are similar. This is followed by a mathematical treatment of the development of the equations used in phase field models and an example of their implementation and use.

2 Maxwell's Equations and Polarization

The discussion of electro-statics presented here is largely a reduced version of that presented by Panofsky and Phillips (2005). This discussion begins with a few comments about systems of units. Although the SI (rationalized MKS) system of units has been broadly adopted, much of the magnetics literature is in CGS units or in atomic units. As discussed in the chapter by J. Schöder, not only must the units be converted to SI from CGS or atomic units, Maxwell's equations must also be modified to correspond to the different systems of units. The equations discussed in this chapter will be those that correspond to rationalized SI units.

The differential form of Maxwell's equations in SI units is represented by four equations

$$\begin{aligned}
\nabla \cdot \boldsymbol{D} &= \rho_f && \text{(Gauss' law)}\\
\nabla \cdot \boldsymbol{B} &= 0 && \text{(Gauss' law for magnetism)}\\
\nabla \times \boldsymbol{E} &= -\frac{\partial \boldsymbol{B}}{\partial t} && \text{(Faraday's law of induction)}\\
\nabla \times \boldsymbol{H} &= \boldsymbol{J}_f + \frac{\partial \boldsymbol{D}}{\partial t} && \text{(Ampere's law of induction)}
\end{aligned} \tag{1}$$

where is $\boldsymbol{D}$ the electric displacement vector, ρ_f is the free charge density, $\boldsymbol{B}$ is the magnetic flux density, $\boldsymbol{E}$ is the electric field, $\boldsymbol{H}$ is the magnetic field, $\boldsymbol{J}_f$ is the free current density, and t is time. The subscript f on the volume charge density and the current density is used to indicate that these terms are associated only with the free charge. Separately considering the motion of charges associated with dipoles (bound charge for electric dipoles and bound current for magnetic dipoles) and motion of free charge carriers that can move over longer distances (free charge) provides a simpler description of a material at larger length scales. This simplification is at the expense of having to now consider electric displacement currents associated with oscillatory motion of the bound charge (important at higher frequencies) and the magnetic field associated with contributions from internal magnetic moments associated with the bound current loops. By bound we mean localized ionic charges or localized currents associated with electron spins and orbits that are tied to specific atoms or small groups of atoms in the material. The corresponding constitutive relations are

$$\boldsymbol{D} = \varepsilon_0 \boldsymbol{E} + \boldsymbol{P}$$
$$\boldsymbol{H} = \frac{\boldsymbol{B}}{\mu_0} - \boldsymbol{M} \tag{2}$$

where $\boldsymbol{P}$ is the polarization, and $\boldsymbol{M}$ is the magnetization, $\varepsilon_0 = 8.85 \times 10^{-12}(F/m)$ is the permittivity of free space, $\mu_0 = 4\pi \times 10^{-7}(N/A^2)$ is the magnetic permeability of free space.
Maxwell's equations are consistent with several observable forces.

• Coulomb's law gives the force between two charges.

$$\boldsymbol{F} = \frac{1}{4\pi\varepsilon_0}\frac{q_1 q_2 \boldsymbol{r}}{r^3} \tag{3}$$

where q_i are charges, $\boldsymbol{r}$ is the position vector connecting them.

• The Lorentz force (force on a charge) in a static electric field or moving through a magnetic field is given by

$$\boldsymbol{F} = q(\boldsymbol{E} + \boldsymbol{v} \times \boldsymbol{B}) \tag{4}$$

where $\boldsymbol{v}$ is the velocity of the charge.

• A current produces an associated magnetic flux density. A magnetic flux density increment generated by a current filament of magnitude I and length $d\boldsymbol{l}$ is given by the Biot-Savart law (in differential form)

$$d\boldsymbol{B} = \frac{\mu_0}{4\pi}\frac{I d\boldsymbol{l} \times \boldsymbol{r}}{r^3}\ . \tag{5}$$

2.1 *Electro-Statics*

In the electrostatic approximation the contributions of $\partial \boldsymbol{E}/\partial t$ and $\partial \boldsymbol{B}/\partial t$ in Maxwell's equations are neglected. In this case, the electric field can be represented by the negative gradient of the potential established by a distribution of point charges. It has units of V/m where 1V= 1 Nm/C is the work per unit charge to move the unit charge up a potential gradient. The electric field (N/C) is the force per unit charge on a charge held stationary in a potential gradient.

Certain clusters of point charges produce electric field distributions that are relatively easy to define mathematically. These include the field of two parallel sheets of charge which produces a uniform electric field within a parallel plate capacitor, the field of two equal magnitude but opposite sign charges held a fixed distance apart to create a dipole field, and the field of certain arrangements of charges that can be represented by quadrapoles, octapoles, and higher order poles.

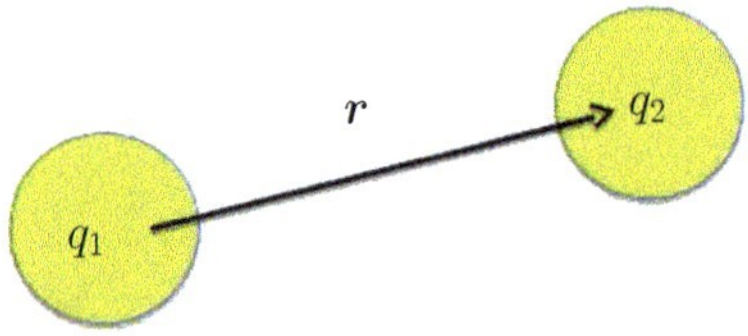

Fig. 1 Two charges separated by a distance, $\boldsymbol{r}$, experience an interaction force between them

The potential and electric field distributions in the vicinity of point charges and dipoles are of particular interest to the discussion of ferroelectric materials. The fields of higher order multi-poles fall off faster with distance than the fields of dipoles, enabling the representation of a continuum as a dipole moment density, the polarization. Although there is no point magnetic charge, from a distance the magnetic field associated with a current loop can be described by the same equations used to describe the electric dipole field. This leads to an often-used introduction of fictitious point magnetic charges to represent the magnetic field of current loops. This provides a convenient method of establishing a uniform external magnetic field when using computational methods.

Charge as the source of electric field. When two charges are placed a distance apart as shown in Fig. 1, they experience an interaction force. The force on charge 1 due to the presence of charge 2 is given by

$$\boldsymbol{F} = \frac{1}{4\pi\varepsilon_0}\frac{q_1 q_2 \boldsymbol{r}}{r^3}\,. \tag{6}$$

Like charges repel and unlike charges attract.

Electric field of a point charge. The electric field is the force per unit charge. The electric field at charge (1) due to the presence of charge (2) is given by

$$\boldsymbol{E}^{(1)} = \frac{\boldsymbol{F}}{q_2} = \frac{q_1 \boldsymbol{r}}{4\pi\varepsilon_0 r^3} = -\frac{q_1}{4\pi\varepsilon_0}\nabla\left(\frac{1}{r}\right)\,. \tag{7}$$

The electric field is defined as the negative gradient of the potential. This leads to the expression for the potential of a point charge,

$$\phi = \frac{q}{4\pi\varepsilon_0}\left(\frac{1}{r}\right)\,. \tag{8}$$

The gradient is most easily worked out in terms of Cartesian coordinates. As a reminder, write the distance "$\boldsymbol{r}$" as $(x_i x_i)^{1/2}$ (implied summation) and the operator dell $\nabla = \partial_k \hat{e}_k$ as (implied summation). Proficiency with this operation is important to understanding the dipole field.

Electric field of a dipole. Two equal charges q that are opposite in sign and held a fixed distance $\boldsymbol{d}$ apart define a dipole moment

$$\boldsymbol{p} = q\boldsymbol{d} \,. \tag{9}$$

The potential of a dipole is given by

$$\phi^{(2)} = \frac{1}{4\pi\varepsilon_0}\frac{\boldsymbol{p}\cdot\boldsymbol{r}}{r^3} \,. \tag{10}$$

We can derive the expression for the potential of a dipole by differentiating the potential of a point charge with respect to the position of the charge. The definition of the derivative is also the definition of the dipole field, i.e. $\lim_{\delta x \to 0} \frac{\phi(x+\delta x)-\phi(x)}{\Delta x}$ where ϕ is the potential field of the point charge. Further differentiation leads to quadrapoles, octopoles, etc. Their fields fall off faster than $1/r^2$ and thus can be (and are) neglected in continuum descriptions of volumes of material containing many atoms. The electric field in the vicinity of a single dipole is found by taking the negative gradient of the dipole field. This leads to

$$\boldsymbol{E}(\boldsymbol{p}) = -\frac{1}{4\pi\varepsilon_0}\nabla\left(\frac{\boldsymbol{p}\cdot\boldsymbol{r}}{r^3}\right) \,. \tag{11}$$

The field of a polarization distribution. Certain solids have naturally occurring dipoles. These are associated with a separation of positive and negative charge centers. The small length scale of the atomic dimensions associated with the dipoles makes finding the superposition of the fields of individual dipoles prohibitive in all but very small volumes, on the order of a few hundreds of lattice parameters. This leads to defining the polarization field as the average dipole moment per unit volume. The volume for defining the polarization needs to be large enough that the field fluctuations associated with individual dipoles cannot be detected and the contributions of the higher order poles can be neglected relative to the dipole field, yet small enough that gradients in the dipole fields can be neglected and the volume considered a "mathematical point". As the discussion progresses, this volume will be made even larger such that the gradient of the polarization field can be detected when domain walls are considered. The polarization is defined as the dipole moment density

$$\boldsymbol{P} = \lim_{V\to 0}\frac{\boldsymbol{p}}{V} \,. \tag{12}$$

Forces on a dipole in a uniform electric field. An electric field produces a force per unit charge on charges in the field. A dipole consists of a positive and a negative charge with a fixed separation. If a dipole is placed in a uniform electric field, the two charges will experience equal and opposite forces as shown in Fig. 2. The force on each charge is in the opposite direction and is given by

$$\begin{aligned} \boldsymbol{F} &= +q\boldsymbol{E} \\ \boldsymbol{F} &= -q\boldsymbol{E} \,. \end{aligned} \tag{13}$$

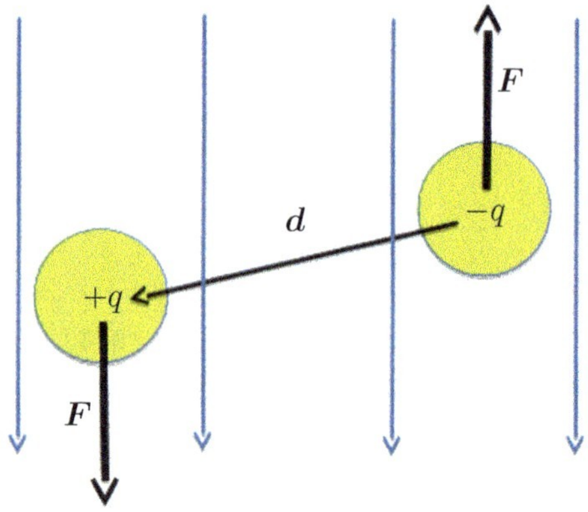

Fig. 2 A dipole placed in an electric field experiences equal magnitude and opposite sign forces on each of the charges. This results in a torque and stretching of the dipole

The resulting torque on a dipole of fixed magnitude is given by

$$\begin{aligned} \boldsymbol{\tau} &= \boldsymbol{d} \times \boldsymbol{F} \\ &= \boldsymbol{d} \times q\boldsymbol{E} \\ &= \boldsymbol{p} \times \boldsymbol{E} \,. \end{aligned} \tag{14}$$

If the charges are connected by a spring, the dipole will rotate to align with the field and will be stretched by the field. With certain asymmetric charge distributions bound to a crystal lattice this results in an electric field producing mechanical deformation, or converse piezoelectricity. Direct piezoelectricity is when mechanically deforming the lattice changes the charge separation of the dipoles causing a polarization change or, as in the case of piezoelectric polymers like PVDF, the dipole moments remain nearly constant and the volume is decreased; again leading to a polarization change.

The potential energy of a dipole of fixed charge separation in a uniform electric field can be found by imagining the dipole being initially aligned with the field. Perform a thought experiment where a torque is physically applied to rotate the dipole against the field. The torque will start small, go through a maximum when the dipole has been rotated 90°, and return to zero as the rotation reaches 180°. The work done is the integral of the torque over the angle of rotation. This is given by

$$dW = \boldsymbol{p} \times \boldsymbol{E} d\Theta = pE \sin\Theta d\Theta \,. \tag{15}$$

The work per unit volume is the work required to rotate the polarization in an electric field. This gives

$$dw = -d(PE\cos\Theta) = -d(\boldsymbol{P} \cdot \boldsymbol{E}) \,, \tag{16}$$

thus the potential energy of a polarized volume element in a uniform electric field is given by

$$\Psi = -\boldsymbol{P} \cdot \boldsymbol{E} \,. \tag{17}$$

The potential energy is at a minimum when $\boldsymbol{P}$ and $\boldsymbol{E}$ are aligned and at a maximum when $\boldsymbol{P}$ and $\boldsymbol{E}$ are at 180° to one another.

Dipole-dipole interaction energy. Two dipoles in proximity to one another interact. The electric field produced by one produces forces on the charges of the other. Panofsky and Phillips (2005) outline the calculation of the interaction energy. The result is obtained by placing one dipole, p_1 in the field of another dipole, p_2. The interaction energy is

$$U_{12} = \frac{1}{4\pi\varepsilon_0}\left[\frac{\boldsymbol{p}_1 \cdot \boldsymbol{p}_2}{r^3} + 3\frac{(\boldsymbol{p}_1 \cdot \boldsymbol{r})(\boldsymbol{p}_2 \cdot \boldsymbol{r})}{r^3}\right] . \tag{18}$$

When a solid is modeled as a continuum and the polarization of neighboring regions varies slightly, the neighboring polarization can be found from the gradient of the polarization field. The differences in neighboring polarizations interact just as the neighboring dipoles interact. The interaction energy per unit volume is expressed in terms of the polarization gradient squared in the phenomenological treatment discussed below.

Electrical boundary conditions. The electrical boundary conditions are found directly from Maxwell's equations. Integration of the expression

$$\nabla \cdot \boldsymbol{D} = \rho_f \tag{19}$$

over a volume that encloses an interface or a surface, applying the divergence theorem to convert to a surface integral, then shrinking the volume such that it closely follows a small portion of the surface leads to the statement that the jump in the normal component of electric displacement at a surface is equal to the surface charge density, i.e.

$$||\boldsymbol{D} \cdot \hat{n}|| = -\omega^s . \tag{20}$$

Similarly integrating using Stokes' theorem,

$$\nabla \times \boldsymbol{E} = 0 \tag{21}$$

indicates that the tangential component of electric field at an interface is continuous, i.e.

$$||\boldsymbol{E} \cdot \hat{t}|| = 0 . \tag{22}$$

Ferroelectric materials and multi-well behavior. Ferroelectric materials with the perovskite crystal structure have a spontaneous dipole moment per unit volume associated with the crystal structure when cooled below the Curie point. The dipole moment couples to the crystal structure in certain materials such as barium titanate and lead zirconate titanate such that it is at lower potential energy when it is aligned with certain crystallographic directions. When this spontaneous polarization occurs, a spontaneous strain simultaneously occurs. The ABO_3 Perovskite crystal structure

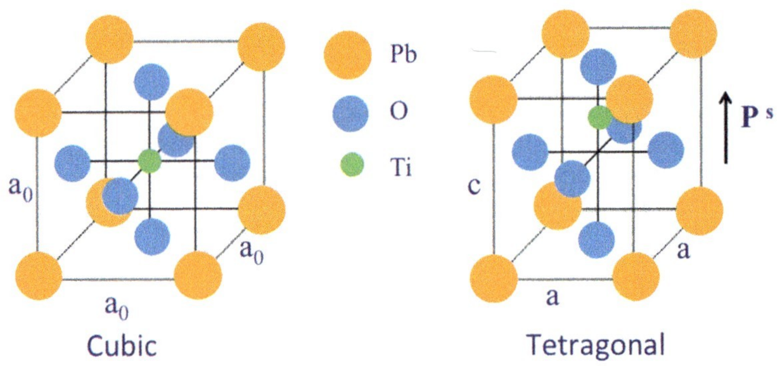

Fig. 3 A conceptual schematic showing the perovskite structure of lead titanate in the high temperature cubic structure (left) and the low temperature tetragonal structure (right). Note that the tetragonal structure has 6 variants with polarization along each of the six < 001 > directions

is shown in Fig. 3. The cubic referenced unit cell shown on the left is made up of Pb (+2) ions at the corners, O (−2) ions at each face center, and Zr (+4) or Ti (+4) at the center. When cooled below the Curie temperature, the structure distorts. The figure to the right conceptually shows the spontaneous strain and the spontaneous polarization. The polarization is the result of the A and B ions shifting relative to the O ions.

In a tetragonal crystal, the spontaneous polarization lies along one of the six < 001 > directions. The potential energy therefore has minima in these six directions. This is referred to as a crystal with a multi-well potential. An illustration of the energy wells is shown as a two dimensional plot of the potential energy as a function of the polarization in Fig. 4. Positive work must be done to rotate the polarization out of one well and negative work is done as it slides into another. Because the different energy wells have the same depth, no total work is done by changing the polarization from one well to another if the system is conservative. If the process is dissipative, positive work is done and heat is generated each time the polarization is forced to move from one well to another.

When the polarization is in a particular well and a small electric field is applied, it will change, but will not escape the well. This polarization change in response to an applied electric field contributes to the dielectric permittivity of the material. Because the polarization is the result of bound charge in the crystal structure and the bound charge is tied to the atomic structure in a piezoelectric, a change of polarization results in a change of strain. This is the converse piezoelectric effect. When a mechanical stress is applied to the lattice, the resulting strain deforms the lattice and changes the polarization. This is the direct piezoelectric effect.

There are many books describing piezoelectricity and ferroelectricity and the reader is referred to these for more information. The intent in this discussion is address how to identify the energy landscapes associated with particular crystals and to use them in the study of domain formation and domain wall motion.

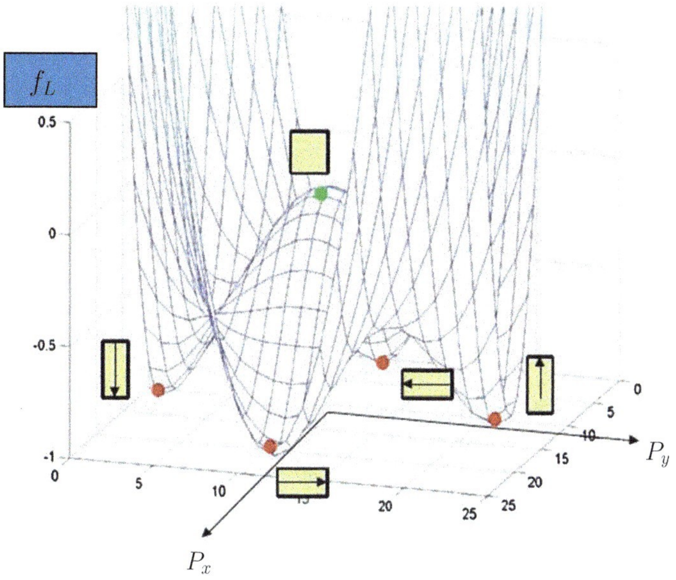

Fig. 4 Potential energy as a function of polarization in the $x - y$ plane shows four of the energy wells of the tetragonal structure

3 Magnetism

When modeling quasi-static magnetic material behavior, the focus is usually on understanding the magnetic dipole, the magnetization (magnetic moment per unit volume), the boundary conditions, and the multi-well potential. In many ways, modeling ferromagnetic material behavior is similar to modeling ferroelectric material behavior. The multi-well potentials appear remarkably similar. There is a significant difference in that the time rate of change of polarization is treated as a damped viscous behavior whereas the time rate of change of magnetization is treated as a damped time rate of change of the angular momentum associated with the electron spin and orbital motion. The latter is necessary to capture effects such as ferromagnetic resonance.

There are several fundamental differences between magnetization and polarization.

• Polarization is the result of the volume average of an approximation of a particular point charge distribution, the dipole moment (bound charge). The surface boundary conditions include the possibility of coating the surface with point charges to terminate the polarization.

The source of the magnetic dipole is a small current loop (bound current) associated with electron spin and orbital motion that gives rise to a magnetic field some distance away from the loop that looks just like the field of an electric dipole. The magnetization $\boldsymbol{M}$ is the magnetic moment per unit volume. The magnetic boundary conditions do not include the ability to spread magnetic charges on the surface. This means that some magnetic field will always exit a magnetized material and re-enter elsewhere. We can, however, introduce a fictitious magnetic charge on the surface to simulate the behavior of a magnetic material in a uniform external magnetic field.

• Another difference is in the way the magnetic moment couples to the lattice. The magnetic current loop that acts as the magnetization source comes from electron spin and electron orbital motion. Many ferromagnetic materials have unfilled d-orbitals that result in a net spin associated with each atom in the lattice. The spin couples to the orbital motion, and the orbital motion couples to the lattice. This results in "easy" directions and "hard" directions for the magnetization relative to the lattice. The electric dipole can be thought of as two charges connected by a spring. The dipole can rotate and it can stretch. The magnetization, being nothing more than the effect of electron spin and orbital motion, cannot change its magnitude much and is modeled as having a fixed magnitude. Only the orientation is allowed to change. This leads to representing magnetic moment in terms of the direction cosines of the magnetization vector (normalized magnetization).

3.1 *Magneto-Statics Review*

This section provides a brief overview of the equations that govern the behavior of magnetic current loops in an applied magnetic field for the quasi-static case.

Current loops as the source of magnetic field. If two current loops are in proximity to one another, the current in loop 1 will exert a force on loop 2. Ampere's law gives the magnitude of this force. In MKS units this is given by

$$\boldsymbol{F}_2 = \frac{\mu}{4\pi} J_1 J_2 \oint_1 \oint_2 \frac{d\boldsymbol{l}_2 \times (d\boldsymbol{l}_1 \times \boldsymbol{r}_{12})}{r_{12}^3} \,. \tag{23}$$

This can be written as the generalized Biot-Savart law,

$$\boldsymbol{F}_2 = J_2 \oint_2 d\boldsymbol{l}_2 \times \boldsymbol{B}_2 \tag{24}$$

where $\boldsymbol{B}_2$ is the magnetic flux density at the position the force is being evaluated and is given by,

$$\boldsymbol{B}_2 = \frac{\mu}{4\pi} J_1 \oint_1 \frac{(d\boldsymbol{l}_1 \times \boldsymbol{r}_{12})}{r_{12}^3} \quad (\text{Weber}/m^2) \,. \tag{25}$$

This can also be written as volume integrals over current densities. Under near quasi-static conditions, the magnetic flux density can be written as the negative gradient of a scalar potential with some restrictions on the line integral,

$$\boldsymbol{B} = -\mu_0 \nabla \phi_m \,. \tag{26}$$

More generally, $\boldsymbol{B}$ is derived from a vector potential and is given by

$$\boldsymbol{B} = \nabla \times \boldsymbol{A} \tag{27}$$

where the vector $\boldsymbol{A}$ is determined from

$$\boldsymbol{A} = \frac{\mu_0}{4\pi} \oint \frac{\boldsymbol{J}}{r} dv' = \frac{\mu_0}{4\pi} J \oint \frac{d\boldsymbol{l}}{r} \tag{28}$$

where r points from the current point to the observation point.

In developing expressions for magnetic field within a material, the problem can be formulated in terms of magnetization (the volume average of the atomic scale current loops) if the current within the material is separated into the current loops bound to the structure (electron spin and orbital current loops or bound current), and current loops associated with electrical conduction (free current). Several types of currents exist.

1. True of free currents associated with electron transport in response to an EMF.
2. Polarization currents $\frac{d\boldsymbol{P}}{dt}$.
3. Magnetization currents $\boldsymbol{j}_m$.
4. Convective currents associated with media in motion, $q\boldsymbol{v}$.

When we consider the design of multiferroic antennas or filters, all of these must be considered and care taken to justify neglecting terms for the particular problem.

The magnetic moment of a current loop. The magnetic moment as a function of position relative to a uniform current density within a small volume is defined as

$$\boldsymbol{m} = \frac{1}{2} \int (\boldsymbol{\xi} \times \boldsymbol{j}_m) dV \tag{29}$$

where the $\boldsymbol{\xi}$ are the position coordinates. If we consider the current densities as charge densities moving with a velocity $\boldsymbol{v}$, then the magnetic moment is given by

$$\boldsymbol{m} = \frac{1}{2} \int \rho_v (\boldsymbol{\xi} \times \boldsymbol{v}) dV . \tag{30}$$

Note the analogy with mechanical angular momentum,

$$\boldsymbol{s} = \frac{1}{2} \int \rho_m (\boldsymbol{\xi} \times \boldsymbol{v}) dV . \tag{31}$$

The gyromagnetic ratio is defined as the ratio of the magnetic moment to angular momentum,

$$\Gamma = \frac{\boldsymbol{m}}{\boldsymbol{s}} . \tag{32}$$

If the current is produced by an electron of mass m, and charge e, then the gyromagnetic ratio is given by

$$\Gamma = \frac{e}{2m} . \tag{33}$$

Note that any charged objects that are spinning about an axis will have a gyromagnetic ratio of

$$\Gamma = g\frac{e}{2m} . \tag{34}$$

The definition of magnetic moment becomes simpler when the current loop is so small that it is inaccessible to measurement, i.e. an electron circling an atom in its orbital. In this case the magnetic moment is given by the product of the current J, and the area of the loop S, with the direction determined by the right hand rule;

$$\boldsymbol{m} = JS . \tag{35}$$

The magnetization is a field quantity used in the continuum description of a ferromagnetic material. The magnetization is defined as the magnetic moment per unit volume,

$$\boldsymbol{M} = \frac{\boldsymbol{m}}{V} . \tag{36}$$

Working with the vector potential, it can be shown that the curl of the volume magnetization is equal to an equivalent current density Panofsky and Phillips (2005),

$$\boldsymbol{j}_m = \nabla \times \boldsymbol{M} . \tag{37}$$

Note that this equivalent current vanishes in regions where the magnetization is homogeneous and it represents the net current density produced in regions where the magnetization is inhomogeneous.

The descriptions of the three types of current lead to an important result. The total current density is given by

$$\boldsymbol{j}^{total} = \boldsymbol{j}^m + \boldsymbol{j}^P + \boldsymbol{j}^{true} = \nabla \times \boldsymbol{M} + \frac{\partial \boldsymbol{P}}{\partial t} + \boldsymbol{j}^{true} . \tag{38}$$

Charge continuity requires that

$$\nabla \cdot \boldsymbol{j}^{total} = -\frac{\partial \rho}{\partial t} = -\nabla \cdot \varepsilon_0 \frac{\partial \boldsymbol{E}}{\partial t} . \tag{39}$$

Combining these expressions gives

$$\nabla \cdot \left(\nabla \times \boldsymbol{M} + \frac{\partial \boldsymbol{D}}{\partial t} + \boldsymbol{j}^{true}\right) = 0 \tag{40}$$

where $\boldsymbol{D} = \varepsilon_0 \boldsymbol{E} + \boldsymbol{P}$. The solenoidal current $\boldsymbol{c}$, is given by

$$\nabla \times \boldsymbol{M} + \frac{\partial \boldsymbol{D}}{\partial t} + \boldsymbol{j}^{true} = \boldsymbol{c} = \boldsymbol{j}^{total} + \varepsilon_0 \frac{\partial \boldsymbol{E}}{\partial t} . \tag{41}$$

The true current is associated with the motion of free charges. The total current is associated with the sum of the effects of the free and bound charges. There is also an effective current associated with the time rate of change of electric field in free space that must be included. This term is present within the free space occupied by the material and outside of the material. The solenoidal current is used with the magnetic flux density expressions. It leads to

$$\begin{aligned} &\nabla \cdot \boldsymbol{B} = 0 \\ &\nabla \times \boldsymbol{B} = \mu_0 \boldsymbol{c} = \mu_0 \left(\nabla \times \boldsymbol{M} + \frac{\partial \boldsymbol{D}}{\partial t} + \boldsymbol{j}^{true} \right) . \end{aligned} \tag{42}$$

Rewriting the second equation with all of the magnetic terms on the LHS gives

$$\nabla \times (\boldsymbol{B} - \mu_0 \boldsymbol{M}) = \mu_0 \left(\frac{\partial \boldsymbol{D}}{\partial t} + \boldsymbol{j}^{true} \right) . \tag{43}$$

This leads to the definition of the magnetic field as

$$\boldsymbol{H} = \frac{1}{\mu_0} (\boldsymbol{B} - \mu_0 \boldsymbol{M}) \quad \text{(amp turns}/m) \tag{44}$$

often written as

$$\boldsymbol{B} = \mu_0 (\boldsymbol{H} + \boldsymbol{M}) . \tag{45}$$

Dipole character of a current loop. Panofsky and Phillips (2005) perform a multipole expansion of the vector magnetic potential. The result is that the first term in the expansion leads to the vector potential of a magnetic dipole, $\boldsymbol{m}$. This vector potential is given by

$$\boldsymbol{A} = -\frac{\mu_0}{4\pi} \frac{\boldsymbol{m} \times \boldsymbol{R}}{R^3} = \frac{\mu_0}{4\pi} \boldsymbol{m} \times \nabla \left(\frac{1}{R} \right) . \tag{46}$$

The field of a magnetization distribution. The magnetization is the magnetic moment per unit volume with the volume large enough that individual spins can be ignored, yet small enough that the magnetization can be defined at a point. This definition will be relaxed when larger volumes are used to capture the magnetization gradients within a domain wall.

Torque on a current loop in a magnetic field. The torque on a magnetic moment is given by

$$\boldsymbol{L} = \boldsymbol{m} \times \boldsymbol{B} . \tag{47}$$

For systems with angular momentum, the torque is equated to the time rate of change of angular momentum. This results in

$$\frac{ds}{dt} = \boldsymbol{m} \times \boldsymbol{B} = \Gamma(\boldsymbol{s} \times \boldsymbol{B}) . \tag{48}$$

If $\boldsymbol{s}$ makes an angle with $\boldsymbol{B}$, then the magnetic moment will precess about $\boldsymbol{B}$ with angular velocity

$$\omega = \Gamma \boldsymbol{B} . \tag{49}$$

Dipole-dipole magnetic interaction energy. Just as there is interaction energy between two electric dipoles, there is interaction energy between two magnetic moments. The equations and their derivations are similar to those for polarization and the reader is again referred to Panofsky and Phillips (2005).

Magnetic boundary conditions. The boundary conditions are analogous to those for polarization. Div $\boldsymbol{B} = 0$ leads to

$$||\boldsymbol{B} \cdot \hat{n}|| = 0 \tag{50}$$

and $\nabla \times \boldsymbol{H} = \boldsymbol{J}_f$ leads to

$$||\boldsymbol{H} \cdot \hat{t}|| = K \tag{51}$$

where K is the true surface current.

Ferromagnetic materials and multi-well behavior. As with ferroelectric materials, ferromagnetic crystalline materials can be described by a multi-well potential. Certain crystalline materials display a spin-orbital-lattice coupling wherein the lattice produces a torque on the magnetic moments of each atom. When the magnetic moments are aligned with the preferred lattice directions, the torque goes to zero. This defines an easy axis. In a cubic symmetry crystal there is typically an easy, an intermediate, and a hard axis. These can be in any of the $< 001 >$, $< 011 >$, or $< 111 >$ directions. The easy, intermediate, and hard directions are identified experimentally as described by Cullity and Graham (2011) and the results are fit to even order polynomials in magnetization. The equations for the energy wells in magnetization have a form identical to the energy wells in polarization, i.e. they are expressed in terms of even order powers of magnetization.

4 Mechano-Statics Review

The equations of mechano-statics describe mechanical equilibrium conditions (mechanical force balance) in terms of stress, strain, and displacement. Stress is a second order tensor that describes force per unit area within a solid. Strain is a second order tensor that represents the symmetric part of the displacement gradient.

In materials that undergo small deformation, geometric effects are neglected. This is the case for most of the ferroelectric and ferromagnetic materials of interest with the exception of ferroelectric polymers. Finite deformation formulations must be used when describing a force per unit area when the area changes appreciably during deformation, and in describing the deformation gradient when there is a significant difference between the un-deformed and deformed shape. The following discussion is focused on the small deformation case that is most often used for ferroelectric and ferromagnetic materials.

Stress and equilibrium. Stress is a second order tensor that describes internal forces per unit area within a body. The stress tensor is written as

$$\boldsymbol{\sigma} = \sigma_{ij}\hat{e}_i\hat{e}_j \tag{52}$$

where summation over the repeated indices is implied. The system is in dynamic equilibrium when the sum of forces on a volume element is equal to its mass times its acceleration. This leads to the equilibrium equation

$$\nabla \cdot \boldsymbol{\sigma} + \boldsymbol{f} = \rho_m \ddot{\boldsymbol{u}} \tag{53}$$

where $\boldsymbol{f}$ represents a vector force per unit volume, $\ddot{\boldsymbol{u}}$ the acceleration and ρ_m the mass density. The components of this vector equation are

$$\sigma_{ij,j} + f_i = \rho_m \ddot{u}_i \tag{54}$$

where ρ_m is the mass density. In the absence of acceleration the system is in quasi-static equilibrium. If the externally applied body forces are also zero, the stress can be derived from a potential that ensures the equilibrium equation will be satisfied by the second order tensor identity

$$\nabla \cdot (\nabla \times \boldsymbol{\phi} \times \nabla) = 0\,. \tag{55}$$

When $\boldsymbol{\phi} = \phi_{33}(x_1, x_2)\hat{e}_3\hat{e}_3$ we have Airy's stress function, and when the second order tensor has an off-diagonal element, we have a torsion stress function. Numerous special cases are discussed in books on the theory of elasticity.

Strain. Strain is the symmetric part of the displacement gradient. The displacement vector field is denoted

$$\boldsymbol{u}(\boldsymbol{x})\,. \tag{56}$$

The displacement gradient is given by

$$\nabla \boldsymbol{u}(\boldsymbol{x}) = u_{i,j}\hat{e}_i\hat{e}_j\,. \tag{57}$$

The second order displacement gradient tensor is separated into a symmetric part (strain) and a skew symmetric part (rigid body rotation),

$$\varepsilon_{ij}\hat{e}_i\hat{e}_j = \frac{1}{2}(u_{i,j} + u_{j,i})\hat{e}_i\hat{e}_j$$
$$\omega_{ij}\hat{e}_i\hat{e}_j = \frac{1}{2}(u_{i,j} - u_{j,i})\hat{e}_i\hat{e}_j \,. \tag{58}$$

In a material with linear constitutive behavior, the strain is proportional to the stress. This is expressed in component form as

$$\sigma_{ij} = C_{ijkl}\varepsilon_{kl} \quad \text{or} \quad \varepsilon_{kl} = S_{klij}\sigma_{ij} \,. \tag{59}$$

For historical reasons (German origin), $\boldsymbol{C}$ is called stiffness and $\boldsymbol{S}$ is called compliance.

In a common approach to 2-D elasticity problems, the equilibrium equation is satisfied by introducing a stress function. Stress is used to find strain using the constitutive law. The strain components are then integrated to find the displacement components. The six strain components are not independent. They must be derived from the gradient of a displacement field that has just three displacement components. A second order tensor identity, the compatibility condition, places a constraint on the strain components. This is given by

$$\nabla \times \varepsilon \times \nabla = 0 \,. \tag{60}$$

The dot product and cross product operations between the unit vectors are readily expanded using the relations

$$\hat{e}_i \cdot \hat{e}_j = \delta_{ij} \quad \text{and} \quad \hat{e}_i \times \hat{e}_j = e_{ijk}\hat{e}_k \tag{61}$$

where δ_{ij} represents the components of a unit second order tensor and e_{ijk} is the permutation symbol.

The mechanics that will be used to describe ferroelectric and ferromagnetic materials will require the definitions of stress and strain as well as the constitutive law, but will not make much use of the various available closed form solutions. The approach will be computational, where the partial differential equations and boundary conditions are identified and numerical methods are used to solve them.

Mechanical boundary conditions. The stress tensor just beneath a surface is related to the traction vector on that surface by the relation

$$\boldsymbol{\sigma} \cdot \hat{n} = \boldsymbol{t} \,, \tag{62}$$

with components

$$\sigma_{ij}\hat{n}_j = t_i \,. \tag{63}$$

Either the traction or the displacement can be prescribed on the surface.

5 Thermodynamics of Ferroelectric and Ferromagnetic Materials

The approach to describing ferroelectric and ferromagnetic material behavior will use the first and second laws of thermodynamics. The first law is a conservation of energy statement that includes thermal, mechanical, electrical, magnetic and internal energy. It is therefore a balance law wherein the energy added to a body, in the form of work performed on the body by external generalized set of forces and heat added and/or generated in the body, is stored within that body in the form of internal and kinetic energy. The second law says something about the conversion of mechanical, electrical, or magnetic energy to thermal energy. No energy is lost in this process, but some of the work is converted to heat in an irreversible process. Next, we focus on the phase-field modeling of ferroelectric and ferromagnetic and multiferroic materials.

5.1 *Ferroelectric Materials: External Mechanical, Electrical Work; and Heat Addition*

The description of material behavior within a thermodynamic framework begins with a description of all of the ways work can be done on the material through external surface and body forces. This work will be added to the description of the heat that can be added to the body and equated to the change of internal energy. The relations are formulated in a rate form. Figure 5 shows a body under general mechanical and electrical loading, including internal material interfaces.

Assessing the external mechanical work done on a material requires identifying the distributed forces on the surface of the body, the traction vector, and the associated displacements. It also requires the identification of any externally applied distributed body forces that act on displacements of the body elements. The work done by surface

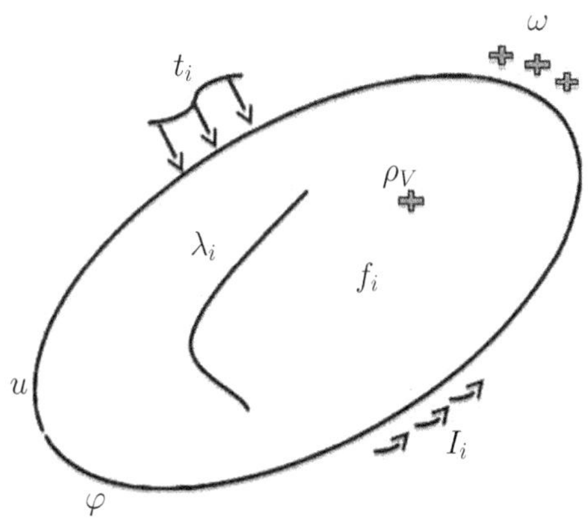

Fig. 5 A body subjected to mechanical, electrical, and magnetic boundary conditions

tractions (N/m^2) on a displacement increment (in this case the instantaneous power) is found by integrating the traction vector and displacement rate over the surface and the body forces over the volume as:

$$\dot{W}^M = \int_S t_i \dot{u}_i dS + \int_V f_i \dot{u}_i dV \; . \tag{64}$$

When surface and volume free charge densities are considered, the electrical work done must be evaluated by integrating over the surface such that:

$$\dot{W}^E = \int_T \varphi \dot{\omega}_s d\Gamma + \int_V \varphi \dot{\rho}_v dV \; . \tag{65}$$

Note here that the volume charge is a free charge distribution not equal to the bounded charge related to the polarization. Typically, in the treatment of ferroelectrics as insulators this volume charge is equal to zero unless an external charge distribution is considered.

Lastly, the thermal energy added to a body by heat transfer from the surrounding environment can be evaluated as the sum of the heat transferred to the volume (a material in a microwave oven is a good example) plus the heat transferred across the surface of the body by heat flux. The rate of heat addition is given by

$$\dot{\Theta} = \int_\Gamma \dot{r} dV - \int_{\partial\Gamma} \dot{q}_i n_i dS \tag{66}$$

where $\dot{r}$ is the rate at which heat is generated within the volume from an external source (to be distinguished from heat generated by an internal dissipative process), and $\dot{q}_i$ are components of the outward heat flux vector (heat leaving per unit area per unit time). The minus sign is to account for the heat entering the body. The expression for energy balance is thus

$$\dot{W}^M + \dot{W}^E + \dot{\Theta} = \dot{K} + \dot{U} \tag{67}$$

where each of the terms on the left hand side (LHS) of the equation represents work done on the body or heat added to the body, the LHS is a sum of all energy transferred to the body; and the terms on the right hand side (RHS) represent where this energy is going. The first term on the RHS is the rate of increase of kinetic energy of the body and the second term on the RHS is the increase of internal energy of the body. Note that the long wavelength assumption has been applied, which allows for the quasi-static decoupling of the electromagnetic fields. That is under quasi-static electrical behavior the effect of a changing electric field generating a magnetic field can be ignored and quasi-stationary magnetic fields have been assumed such that the radiation fields can be neglected.

Combining the expressions for mechanical, and electrical work gives

$$\begin{aligned}&\int_S t_i\dot{u}_i dS + \int_v f_i\dot{u}_i dV + \int_S \varphi\dot{\omega}_s dS + \int_V \varphi\dot{\rho}_\nu dV + \int_V \dot{r} dV\\&-\int_S \dot{q}_i n_i dS = \frac{d}{dt}\int_V \frac{1}{2}\rho\dot{u}_j\dot{u}_j dV + \int_V \rho\dot{e} dV\,.\end{aligned} \tag{68}$$

Using the definition of traction and surface charge the surface integrals on the LHS can be written as

$$\begin{aligned}&\int_S \sigma_{ij}n_j\dot{u}_i dS + \int_V f_i\dot{u}_i dV - \int_S \varphi\dot{D}_j n_j dS + \int_V \varphi\dot{\rho}_\nu dV + \int_V \dot{r} dV\\&-\int_S \dot{q}_i n_i dS = \frac{d}{dt}\int_V \frac{1}{2}\rho\dot{u}_j\dot{u}_j dV + \int_V \rho\dot{e} dV\end{aligned} \tag{69}$$

which, using the divergence theorem, becomes

$$\begin{aligned}&\int_V (\sigma_{ij}\dot{u}_i)_{,j} dV + \int_V f_i\dot{u}_i dV - \int_V (\varphi\dot{D}_j)_{,j} dV + \int_V \varphi\dot{\rho}_\nu dV\\&+\int_V \dot{r} dV - \int_V \dot{q}_{j,j} dV = \frac{d}{dt}\int_V \frac{1}{2}\rho\dot{u}_j\dot{u}_j dV + \int_V \rho\dot{e} dV\,.\end{aligned} \tag{70}$$

The partial derivatives are now expanded, it is noted that the stress does no work on the rigid body rotations (symmetric times anti-symmetric matrices give zero), and the terms are grouped.

$$\begin{aligned}&\int_V (\sigma_{ij,j}\dot{u}_i + \sigma_{ij}\dot{\varepsilon}_{ij}) dV + \int_V f_i\dot{u}_i dV - \int_V (\varphi_{,j}\dot{D}_j + \varphi\dot{D}_{j,j}) dV\\&+\int_V \varphi\dot{\rho}_\nu dV + \int_V \dot{r} dV - \int_V \dot{q}_{j,j} dV\\&=\frac{d}{dt}\int_V \frac{1}{2}\rho\dot{u}_j\dot{u}_j dV + \int_V \rho\dot{e} dV\,.\end{aligned} \tag{71}$$

Noting that $\sigma_{ij,j} + f_i = \frac{d}{dt}(\frac{1}{2}\rho_m\dot{u}_j\dot{u}_j)$ and $D_{j,j} - \rho_\nu = 0$ and $E_i = -\phi_{,i}$ the result is

$$\int_V \sigma_{ij}\dot{\varepsilon}_{ij} dV + \int_V E_j\dot{D}_j dV + \int_v \dot{r} dV - \int_V \dot{q}_{j,j} dV = \int_V \rho\dot{e} dV\,. \tag{72}$$

This is our desired expression for the rate of change of the internal energy. Note that up to this point along with the first law of thermodynamics we have used expressions of mechanical equilibrium and the Gauss equation and linear kinematics. Given a static domain structure, the solution of a boundary value problem within the context of linear piezoelectricity requires the introduction of the appropriate boundary conditions along with the constitutive equations required to connect the field quantities satisfying the fundamental balance laws (mechanical stress and electric field) to the kinematical fields describing the configuration of the body. The constitutive equation describing the material behavior can be derived through invoking a free energy that depends on the configurational quantities, the strain and the electric displacements in this case, through thermodynamic considerations.

However in this chapter we are not only interested in the distribution of the fields, but also in how these fields cause the domain structure to evolve. Thus at this point we introduce the notion of internal surfaces separating uniform domain structure in the material, which are related to the polarization (or magnetization for ferromagnetic materials) variable. The free energy of the material has to be able to represent the spatial distribution and evolution of these internal interfaces within the ferroelectric crystal and hence depend on the order parameter identifying the phases. The natural order parameter for the ferroelectric domain structure is the polarization vector. The relationship between the polarization vector, the electric displacement and the electric field is given by $D_i = P_i + \varepsilon_0 E_i$, where ε_0 is the permittivity of free space.

5.2 *Balance Laws for Internal Fields*

Just as we have balance laws for stress, electric field, and magnetic field; we can apply the concept of equilibrium on an internal surface in the material and have balance laws for the polarization and magnetization variables. These will first be written as an equilibrium equation, and then be modified to give a balance when there is a viscous force present that introduces damping proportional to polarization or magnetization rate. Polarization will be addressed first. The magnetization case is trickier when there is magnetization evolution present because of the angular momentum leading to precession.

Consider a volume element where the volume has been taken small enough that the electric dipole behavior of the element can be represented by the average polarization multiplied by the volume; but the neighboring elements can have small differences in polarization that give rise to interaction energy. This effect will be proportional to the polarization gradient that is present between neighboring elements. Lets consider the 2-D case of a 90 degree domain wall as an example. This is shown in Fig. 6. The polarization must rotate 90° across this domain wall and be in equilibrium at each

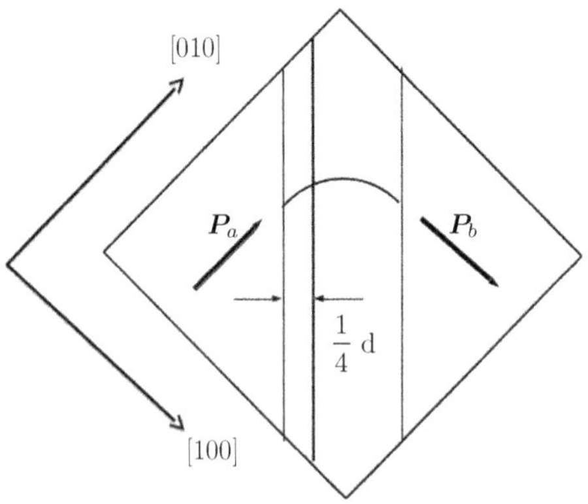

Fig. 6 A domain wall is a diffuse interface separating regions of uniform polarization (or magnetization). Within the wall there is a gradient in the order parameter

point within the wall. The first generalized forces to consider are those produced by the crystal structure on the polarization that tend to align the polarization with the crystal structure. This generalized force is the effective electric field. To the left and right of the wall, the polarization is in equilibrium with this force. Right at the center of the wall the polarization is also in equilibrium with this force, although it is in a metastable equilibrium state. At the $\frac{1}{4}$ and $\frac{3}{4}$ points through the wall, the polarization is not in equilibrium with the effective electric field. At the $\frac{1}{4}$ point the effective field is driving the polarization to rotate to the left to align with the polarization to the left, and at the $\frac{3}{4}$ point the effective field is driving the polarization to rotate to the right to align with the polarization to the right. There must be another source of effective field that can hold the polarization in an equilibrium position that is mis-aligned with the local effective field. This is the result of divergence of the polarization gradient. The effect is easily understood in terms of the interaction energy of two dipoles. Neighboring dipoles interact through their dipole fields. When they are held a fixed distance apart, they are in a low energy state when they are aligned head to tail. If one is rotated, the energy goes up. The energy is reduced if the polarization is rotated back into alignment. Consider a small volume element with average polarization $\boldsymbol{P}(\boldsymbol{x})$. The forces acting on this polarization are the effective field and the gradient relative to neighboring elements. The element size in this continuum approximation must be selected such that when the nearest neighbor elements are used to compute the gradient effects, a good approximation of the interaction energy is obtained. In the 2-D case being considered, the polarization is only changing in the x-direction. The force on the dipole is taken to be proportional to the polarization gradient in each direction. This force is given by the change of gradient in that direction, i.e. the gradient is

$$\nabla \boldsymbol{P}(\boldsymbol{x}) = P_{i,j} \hat{e}_i \hat{e}_j \tag{73}$$

where the gradient represents the rate of change of polarization with position. The rate of change of polarization in a particular direction is found by taking the dot product of the gradient with a unit normal in that direction. This gives

$$\nabla \boldsymbol{P}(\boldsymbol{x}) \cdot \hat{n} = P_{i,j} \hat{e}_i \hat{e}_j n_j \ . \tag{74}$$

The generalized force associated with the polarization gradient is given by

$$\boldsymbol{\xi} = \boldsymbol{G} \cdot \cdot \nabla \boldsymbol{P}(\boldsymbol{x}) \quad \text{or} \quad \xi_{ij} = G_{ijkl} P_{k,l} \tag{75}$$

and the force or the generalized traction vector in a particular direction is given by

$$\boldsymbol{\lambda} = \boldsymbol{\xi} \cdot \hat{n} = \boldsymbol{G} \cdot \cdot \nabla \boldsymbol{P}(\boldsymbol{x}) \cdot \hat{n} \ . \tag{76}$$

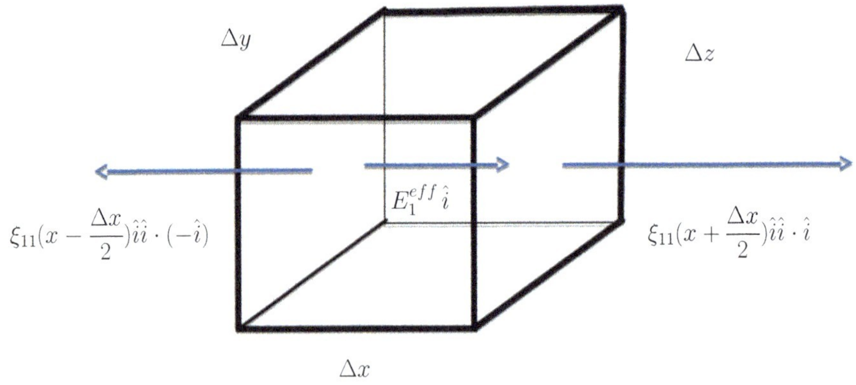

Fig. 7 A volume element used to describe the balance of generalized forces acting on the polarization (or magnetization) within

The fourth order ***G*** tensor will have cubic symmetry of the cubic phase and will be most easily presented in terms of a coordinate system aligned with the crystallographic directions. Note that in our example the coordinate system has been rotated 45° and thus the ***G***-matrix would have to be rotated.

Lets perform a balance of the generalized forces acting on a volume element in which the polarization is in equilibrium with the effective field and the polarization gradient. The volume element is shown in Fig. 7.

The generalized forces in this force balance are the fields that put torque on the dipole moment per unit volume within the element, the polarization. They therefore must have units that will contract with polarization and when multiplied by the volume or area will give work. The effective field has units of volts/meter. Contract with polarization (C/m^2) and multiply by the volume (m^3) and the result is $(N - m)$. The effective field associated with the gradient has units of volts $(N - m/C)$. This is multiplied by area (m^2) and polarization (C/m^2) to get work $(N - m)$.

The term ξ_{11} is related to the polarization gradient by

$$\xi_{11} = G_{1111} P_{1,1} \,. \tag{77}$$

Note that this is not the only non-zero component, but it is the only one being addressed in this 1-D force balance example. The force balance becomes

$$\xi_{11}\left(x + \frac{\Delta x}{2}\right)\Delta y \Delta z - \xi_{11}\left(x - \frac{\Delta x}{2}\right)\Delta y \Delta z + E_1^{\text{eff}} \Delta x \Delta y \Delta z = 0 \tag{78}$$

leading to

$$\frac{\xi_{11}\left(x + \frac{\Delta x}{2}\right) - \xi_{11}\left(x - \frac{\Delta x}{2}\right)}{\Delta x} + E_1^{\text{eff}} = 0 \tag{79}$$

or

$$\xi_{11,1} + E_1^{\text{eff}} = 0\,. \tag{80}$$

The same force balance can be performed with the possibility of polarization gradients in each direction. This leads to

$$\nabla \cdot \boldsymbol{\xi} + \boldsymbol{E}^{\text{eff}} = 0\,. \tag{81}$$

If there is an externally applied electric field, this will also put an effective body force on the polarization in the element. This can be added by superposition to obtain

$$\nabla \cdot \boldsymbol{\xi} + \boldsymbol{E}^{\text{eff}} + \boldsymbol{E}^{\text{ext}} = 0\,. \tag{82}$$

This set of generalized forces acts on the polarization. If the polarization is not in equilibrium with the generalized forces, there will be a driving force for it to change and a kinetic law needs to be added. An assumption that this polarization change is governed by a viscous type behavior is typically made such that the rate of change of polarization is proportional to the driving force leading to:

$$\nabla \cdot \boldsymbol{\xi} + \boldsymbol{E}^{\text{eff}} + \boldsymbol{E}^{\text{ext}} = \boldsymbol{\beta} \cdot \dot{\boldsymbol{P}}\,. \tag{83}$$

This result is the generalized form of the TDGL equation developed by Su and Landis using the micro-force balance approach of Gurtin. In their analysis (discussed next) the force balance for the polarization internal variable along with mechanical equilibrium and Gauss law is adopted a priori and the dissipative, viscous term follows as a restriction of the second law of thermodynamics for non-equilibrium processes along with the restriction of the constitutive equations. Although most points of the analysis are described next, the interested reader should refer to the original text for the full formal approach.

5.3 *Phase-Field Model of Ferroelectrics; Dissipative Evolution of Domains; Second Law of Thermodynamics*

The phase-field modeling approach has been used successfully to study several different features of ferroelectric domain switching behavior including the structure of domain walls Cao and Cross (1991), switching of polycrystals and single crystals Choudhury et al. (2005); Zhang and Bhattacharya (2005), the interactions of domain walls with charge defects Su and Landis (2007) and dislocations Kontsos and Landis (2009), domain switching Song et al. (2007) and nucleation and growth near crack tips Li and Landis (2011). Su and Landis (2007), working along the lines of Fried and Gurtin (1993, 1994); Gurtin (1996), developed a continuum mechanics, non-equilibrium, thermodynamic framework that distinguishes the fundamental balance laws which are universal from the material constitutive response.

In the phase field setting we are interested in how the electromechanical fields can cause the domain structure to evolve through the dissipative motion of internal material interfaces separating uniform domain structure. Given that position and evolution of the internal interface is now part of the solution and the free energy is permitted to depend on polarization, the internal micro-forces introduced that are work-conjugate to the order parameter have to satisfy their own balance equation. However in order to account for the internal dissipation in the material along with the ξ_{ij} tensor we introduce a micro-force vector π_i akin to $\boldsymbol{E}^{\text{eff}}$ which can be thought as the equilibrium internal force. The integral balance of this set of configurational forces,

$$\nabla \cdot \boldsymbol{\xi} + \boldsymbol{\pi} = 0\,. \tag{84}$$

The work associated with this set of internal micro-forces is then written as:

$$\int_S \lambda_i \dot{P}_i dS = \int_S \xi_{ji} n_j \dot{P}_i dS\,. \tag{85}$$

Here note that anticipating the dissipative nature of domain wall motion, the internal micro-force vector π_i enters the balance of the micro-forces but does not contribute to the external power. With this addition the first law of thermodynamics accounting for the thermal, electromechanical and domain wall energy conversion of the continuum is given as:

$$\begin{aligned}&\int_S \sigma_{ij} n_j \dot{u}_i dS + \int_V f_i \dot{u}_i dV + \int_S \xi_{ji} n_j \dot{P}_i dS - \int_S \varphi \dot{D}_j n_j dS \\ &+ \int_V \varphi \dot{\rho}_v dV + \int_V \dot{r} dV - \int_S \dot{q}_i n_i dS = \frac{d}{dt} \int_V \frac{1}{2} \rho \dot{u}_j \dot{u}_j dV + \int_V \rho \dot{e} dV.\end{aligned} \tag{86}$$

Using the traction and surface charge definition equations and applying the divergence theorem and eliminating terms that satisfy the balance laws we are left with the rate of change of internal energy as:

$$\dot{U} = \sigma_{ij} \dot{\varepsilon}_{ij} + E_i \dot{D}_i - \pi_i \dot{P}_i + \xi_{ij} \dot{P}_{i,j} - q_{i,i} + r\,. \tag{87}$$

The constraint of the second law of thermodynamics in the entropy production is given through the pointwise local Clausius-Duhem dissipation inequality as

$$\frac{\dot{q}_i T_{,i}}{T} - \dot{q}_{i,i} \leq T \dot{S} \tag{88}$$

where S is the total entropy and T the absolute temperature. Using the equation of rate of change of internal energy to eliminate the divergence of the heat flux vector we have

$$T \dot{S} \geq \dot{U} - \sigma_{ij} \dot{\varepsilon}_{ij} - E_i \dot{D}_i + \pi_i \dot{P}_i - \xi_{ij} \dot{P}_{i,j} - r + \frac{\dot{q}_i T_{,i}}{T}\,. \tag{89}$$

In terms of the free energy ψ the second law can be written as

$$\dot{\psi} \leq \sigma_{ij}\dot{\varepsilon}_{ij} + E_i\dot{D}_i - \pi_i\dot{P}_i + \xi_{ij}\dot{P}_{i,j} + r - \frac{\dot{q}_i T_{,i}}{T} - \dot{T}S \,. \tag{90}$$

To simplify calculation let's restrict the problem to spatially homogeneous, temperature independent microstructural evolution, below the Curie temperature. Therefore under isothermal conditions, the dissipation inequality reads

$$\dot{\psi} \leq \sigma_{ij}\dot{\varepsilon}_{ij} + E_i\dot{D}_i - \pi_i\dot{P}_i + \xi_{ij}\dot{P}_{i,j} \,. \tag{91}$$

Concluding the thermodynamics based formulation of the constitutive equations, we need to specify the independent state variables on which the thermodynamic functional and the thermodynamic conjugate forces, i.e. stresses, electric field microforce tensor and internal micro-force vector can depend on.

For isothermal behavior, below the Curie temperature the thermodynamic functional is the Helmholtz free energy and the independent configurational/state variables are components of strain, electric displacement, polarization vector, polarization vector gradients and time rate. The constitutive response is written as

$$\psi = \psi(\varepsilon_{ij}, D_i, P_i, P_{i,j}, \dot{P}_i) \,. \tag{92}$$

Then using the above functional form of the Helmholtz free energy we have:

$$\begin{aligned} &\frac{\partial\psi}{\partial\varepsilon_{ij}}\dot{\varepsilon}_{ij} + \frac{\partial\psi}{\partial D_i}\dot{D}_i + \frac{\partial\psi}{\partial P_i}\dot{P}_i + \frac{\partial\psi}{\partial P_{i,j}}\dot{P}_{i,j} + \frac{\partial\psi}{\partial\dot{P}_i}\ddot{P}_i \\ &\leq \sigma_{ij}\dot{\varepsilon}_{ij} + E_i\dot{D}_i - \pi_i\dot{P}_i + \xi_{ij}\dot{P}_{i,j} \end{aligned} \tag{93}$$

or equivalently:

$$\begin{aligned} &\left(\sigma_{ij} - \frac{\partial\psi}{\partial\varepsilon_{ij}}\right)\dot{\varepsilon}_{ij} + \left(E_i - \frac{\partial\psi}{\partial D_i}\right)\dot{D}_i \\ &-\left(\pi_i + \frac{\partial\psi}{\partial P_i}\right)\dot{P}_i + \left(\xi_{ij} - \frac{\partial\psi}{\partial P_{i,j}}\right)\dot{P}_{i,j} - \frac{\partial\psi}{\partial\dot{P}_i}\ddot{P}_i \geq 0 \,. \end{aligned} \tag{94}$$

The above inequality must hold for any permissible thermodynamic process for arbitrary levels of $\dot{\varepsilon}_{ij}$,$\dot{D}_i$,$\dot{P}_i$,$\dot{P}_{i,j}$, $\ddot{P}_i$ through the appropriate control of the external sources. A way not to violate the inequality for arbitrary combination of $\dot{\varepsilon}_{ij}$, $\dot{D}_i$,$\dot{P}_i$,$\dot{P}_{i,j}$,$\ddot{P}_i$ is to set the coefficients of $\dot{\varepsilon}_{ij}$, $\dot{D}_i$,$\dot{P}_i$,$\dot{P}_{i,j}$,$\ddot{P}_i$ equal to zero and ensure that the term

$$\left(\pi_i + \frac{\partial\psi}{\partial P_i}\right)\dot{P}_i \leq 0 \,. \tag{95}$$

Therefore we have

$$\frac{\partial\psi}{\partial\dot{P}_i} = 0 \rightarrow \psi = \psi(\varepsilon_{ij}, D_i, P_i, P_{i,j}) \tag{96}$$

and

$$\sigma_{ij} = \frac{\partial \psi}{\partial \varepsilon_{ij}}, \quad E_i = \frac{\partial \psi}{\partial D_i}, \quad \xi_{ij} = \frac{\partial \psi}{\partial P_{i,j}} . \tag{97}$$

Finally setting internal micro-force equal to

$$\pi_i = -\frac{\partial \psi}{\partial P_i} - \beta_{ij} \dot{P}_j \tag{98}$$

the inequality is satisfied for any level of $\dot{P}_i$ and for β_{ij} positive definite. For a cubic high temperature phase the tensor β_{ij} is usually taken to be $\beta_{ij} = \beta\delta_{ij}$ where $\beta > 0$. Substitution of the definitions of internal micro-force vector and tensor into the micro-force balance yields a generalized form of the Ginzburg-Landau equation governing the evolution of the material polarization in a ferroelectric material as:

$$\left(\frac{\partial \psi}{\partial P_{i,j}} \right)_{,j} - \frac{\partial \psi}{\partial P_i} + \gamma_i = \beta_{ij} \dot{P}_j . \tag{99}$$

The first term can be understood by considering three dipoles in a row. If the center one is vertical, the one to the left tilts left, and the one to the right tilts right by an equal but opposite angle, the dipoles on either side will put equal and opposite forces on the dipole in the middle. This corresponds to a polarization gradient with zero divergence. If the left dipole is vertical, the center dipole is vertical and the one on the right is tilted, there will be unequal forces on the center dipole such that the center dipole will want to tilt to make the polarization gradient divergence free. The second term in this expression is the effective electric field. This term includes all of the various torques and stretches the lattice puts on a particular volume with polarization. The third term represents external forces that drive changes of polarization, possibly an externally applied electric field. The term on the RHS is the kinetic (viscous) response of the polarization. It governs the rate at which the polarization will respond to the net effective electric field at a point in the solid.

5.4 Internal Energy

The next question to be addressed is how energy can be stored in ferroelectric and ferromagnetic materials. We have already seen that if we place an electric dipole in a uniform electric field, the dipole will experience a torque. If the dipole orientation is held in place by some kind of internal spring (atomic bonds of the lattice), then the spring will be stretched and the internal energy will go up. This suggests we need a way to describe these spring forces in terms of the lattice behavior. The same argument arises with respect to the storage of magnetic energy. The stored magnetic energy must be the result of winding up some kind of spring when the magnetic moment is torqued. Several means of storing energy in a lattice are identified and these are used as variables in our formulation.

Independent variables associated with internal energy and associated work-conjugate forces.

• Temperature (average stretch or torsion of the springs associated with vibration amplitude),
The temperature of a material is a function of lattice vibrations. It is the work conjugate variable to entropy and plays the role of a generalized force. The work conjugate pair is thus

$$(T, S) . \tag{100}$$

• Elastic strain (stretch of the springs by mechanical force),
The elastic strain is that fraction of the symmetric part of the displacement gradient that is induced by mechanical forces. It does not include thermal strain, electrostrictive strain, or magnetostrictive strain. The work conjugate pair for the internal energy is given by

$$(\sigma, \varepsilon^{el}) . \tag{101}$$

• Polarization (stretch or rotation of a dipole),
When the polarization is in a uniform electric field and is rotated against that electric field, the field will produce a restoring torque on that polarization. Within the material there are a number of sources of torque on the polarization. These include externally applied electric field, the combined fields produced by all other dipoles in the material (also expressed as combined fields of all other polarizations associated with small volume elements), and changes of the polarization gradient with position (the divergence of the polarization gradient applies a torque to the local polarization, an effect associated with a length scale for the particular set of internal generalized forces). The work conjugate pair associated with internal energy density for this term is the effective electric field and the polarization,

$$(\boldsymbol{E}^{\text{eff}}, \boldsymbol{P}) . \tag{102}$$

• Electric displacement (this term must be included to account for free space contribution),
Any time there is an electric field present, there will be a polarization of free space. This term is the component associated with the permittivity of free space. The work conjugate pair representing the polarization of free space is given by

$$(\boldsymbol{E}, \boldsymbol{P}^{FS}) \quad \text{or} \quad (\boldsymbol{E}, (\boldsymbol{D} - \boldsymbol{P})) . \tag{103}$$

• Polarization gradient (interaction energy of mis-aligned dipoles),
The divergence of the polarization gradient makes a contribution to the effective electric field. It can be easier to keep this as a separate term rather than to combine it with the effective electric field. The work conjugate variables associated with the

polarization gradient are a second order tensor with units of volts and the gradient of polarization,

$$(\boldsymbol{\xi}, \nabla \boldsymbol{P}) \,. \tag{104}$$

5.5 Series Expansions for the Energy Functions

The expression for internal energy indicates that the internal energy is a function of each of the independent variables. At this point the function is unknown, but a well behaved function can be approximated using a Taylor's series expansion. As mentioned, internal energy is also stored in polarization gradients. The electric displacement density is written as

$$D_i = \varepsilon_o E_i + P_i = P_i^{FS} + P_i \tag{105}$$

giving

$$U = U(\varepsilon_{ij}, P_i^{FS}, P_i, P_{i,j}) \quad \text{and} \quad \psi = \psi(\varepsilon_{ij}, P_i^{FS}, P_i, P_{i,j}) \,. \tag{106}$$

The series expansion will give the energy relative to a reference state. Many of the ferroelectric materials of interest have the perovskite structure. This structure is cubic in the parent phase and experiences spontaneous polarization when cooled below the Curie point. The reference energy state will be taken as a cubic state at a given temperature. This will be a metastable state when below the Curie point, but maintains all of the cubic symmetry. If the temperature is held constant, the Helmholtz energy density can be expanded about the cubic state to obtain

$$\begin{aligned}
&\Delta\psi(\varepsilon_{ij}, P_i, P_i^{FS}, P_{i,j}) = \\
&\left.\frac{\partial\psi}{\partial\varepsilon_{ij}}\right|_{P,T}\Delta\varepsilon_{ij} + \left.\frac{\partial\psi}{\partial P_k}\right|_{\varepsilon,T}\Delta P_k + \left.\frac{\partial\psi}{\partial P_k^{FS}}\right|_{\varepsilon,T}\Delta P_k^{FS} + \left.\frac{\partial\psi}{\partial P_{i,j}}\right|_{\varepsilon,T}\Delta P_{i,j} \\
&\frac{1}{2!}\left(\left.\frac{\partial^2\psi}{\partial\varepsilon_{ij}\partial\varepsilon_{kl}}\right|_{\varepsilon,T}\Delta\varepsilon_{ij}\Delta\varepsilon_{kl} + 2\frac{\partial^2\psi}{\partial\varepsilon_{ij}\partial P_k}\Delta\varepsilon_{ij}\Delta P_k\right. \\
&\left. + \left.\frac{\partial^2\psi}{\partial P_k\partial P_l}\right|_{\varepsilon,T}\Delta P_k\Delta P_l + \left.\frac{\partial^2\psi}{\partial\varepsilon_{ij}\partial P_k^{FS}}\right|_{\varepsilon,T}\Delta\varepsilon_{ij}\Delta P_k^{FS} + \ldots\right) \\
&+\frac{1}{3!}\left(\frac{\partial^3\psi}{\partial\varepsilon_{ij}\partial\varepsilon_{kl}\partial\varepsilon_{mn}}\Delta\varepsilon_{ij}\Delta\varepsilon_{kl}\Delta\varepsilon_{kl}\right. \\
&+3\frac{\partial^3\psi}{\partial\varepsilon_{ij}\partial\varepsilon_{kl}\partial P_k}\Delta\varepsilon_{ij}\Delta\varepsilon_{kl}\Delta P_k + 3\frac{\partial^3\psi}{\partial\varepsilon_{ij}\partial P_k\partial P_l}\Delta\varepsilon_{ij}\Delta P_k\Delta P_l \\
&\left.+\frac{\partial^3\psi}{\partial P_k\partial P_l\partial P_s}\Delta P_k\Delta P_l\Delta P_s + \ldots\right) + \ldots\,.
\end{aligned} \tag{107}$$

We can eliminate a number of terms by making some observations about the symmetry of the energy function.

1. There can be no linear terms. That is because the system is in a cubic state with symmetry in all directions about this state. A positive polarization change should give the same effect on energy as a negative polarization change. This eliminated the linear terms in the series.
2. The free space polarization is only a quadratic term that gives rise to the linear contribution of electric field to electric displacement. The time rate of change of this term gives the displacement current correction of free space. This term is not coupled to other terms.
3. The polarization gradient term is associated with dipole-dipole interactions. It will be taken as a quadratic term only. This term could be coupled to other terms, requiring higher order contributions. But for now it will be taken as quadratic.
4. Constant temperature will be assumed. In general, temperature dependence needs to be included. The temperature coupling terms are used to model Curie-Weiss behavior.
5. Making all measurements relative to the cubic configuration where the strain and polarization are zero enables dropping all of the deltas.

This simplified system will be considered as a means of presenting the approach. With these restrictions, the Helmholtz energy can be expressed as

$$
\begin{aligned}
&\Delta\psi(\varepsilon_{ij}, P_i, P_i^{FS}, P_{i,j}) \approx \\
&\frac{1}{2!}\left(\left.\frac{\partial^2\psi}{\partial\varepsilon_{ij}\partial\varepsilon_{kl}}\right|_{\varepsilon,T}\varepsilon_{ij}\varepsilon_{kl} + 2\frac{\partial^2\psi}{\partial\varepsilon_{ij}\partial P_k}\Delta\varepsilon_{ij}P_k + \left.\frac{\partial^2\psi}{\partial P_k\partial P_l}\right|_{\varepsilon,T}P_kP_l\right. \\
&\left.+\left.\frac{\partial^2\psi}{\partial P_j^{FS}\partial P_k^{FS}}\right|_{\varepsilon,T}P_j^{FS}P_k^{FS} + \left.\frac{\partial^2\psi}{\partial P_{k,r}\partial P_{l,s}}\right|_{\varepsilon,T}P_{k,r}P_{l,s}\right) \\
&+\frac{1}{3!}\left(\frac{\partial^3\psi}{\partial\varepsilon_{ij}\partial\varepsilon_{kl}\partial\varepsilon_{mn}}\varepsilon_{ij}\varepsilon_{kl}\varepsilon_{kl} + 3\frac{\partial^3\psi}{\partial\varepsilon_{ij}\partial\varepsilon_{kl}\partial P_k}\varepsilon_{ij}\varepsilon_{kl}P_k\right. \\
&\left.+3\frac{\partial^3\psi}{\partial\varepsilon_{ij}\partial P_k\partial P_l}\varepsilon_{ij}P_kP_l + \frac{\partial^3\psi}{\partial P_k\partial P_l\partial P_s}P_kP_lP_s + \ldots\right) + \ldots .
\end{aligned}
\tag{108}
$$

Note that all partial derivatives are taken with all other independent variables fixed. This was explicitly denoted in the second terms but from here forward will not be indicated.

The energy function is a perfect differential such that

$$
\begin{aligned}
d\psi &= \frac{\partial\psi}{\partial\varepsilon_{ij}}d\varepsilon_{ij} + \frac{\partial\psi}{\partial P_i}dP_i + \frac{\partial\psi}{\partial P_i^{FS}}dP_i^{FS} + \frac{\partial\psi}{\partial P_{i,j}}dP_{i,j} \\
&= \sigma_{ij}d\varepsilon_{ij} + E_idP_i + E_idP_i^{FS} + \xi_{ij}dP_{i,j}
\end{aligned}
\tag{109}
$$

where $P_i^{FS} = \varepsilon_0 E_i$ is notation used for convenience so as not to introduce both polarization and electric field as independent variables. This expression leads to the coupled constitutive behavior. Taking the partial derivatives of the energy function leads to

$$\begin{aligned}\sigma_{ij} = \frac{\partial \psi}{\partial \varepsilon_{ij}} \approx \frac{1}{2!}\left(2\left.\frac{\partial^2 \psi}{\partial \varepsilon_{ij}\partial \varepsilon_{kl}}\right|_{P,T} \varepsilon_{kl} + 2\frac{\partial^2 \psi}{\partial \varepsilon_{ij}\partial P_k} P_k\right) \\ +\frac{1}{3!}\left(3\frac{\partial^3 \psi}{\partial \varepsilon_{ij}\partial \varepsilon_{kl}\partial \varepsilon_{mn}}\varepsilon_{kl}\varepsilon_{kl} + 6\frac{\partial^3 \psi}{\partial \varepsilon_{ij}\partial \varepsilon_{kl}\partial P_k}\varepsilon_{kl}P_{kl}\right. \\ \left. +3\frac{\partial^3 \psi}{\partial \varepsilon_{ij}\partial P_k \partial P_l}P_k P_l\right) + \ldots\end{aligned} \tag{110}$$

$$\begin{aligned}E_k = \frac{\partial \psi}{\partial P_k} \approx \frac{1}{2!}\left(2\frac{\partial^2 \psi}{\partial \varepsilon_{ij}\partial P_k}\varepsilon_{ij} + 2\left.\frac{\partial^2 \psi}{\partial P_k \partial P_l}\right|_{\varepsilon,T} P_l\right) \\ +\frac{1}{3!}\left(3\frac{\partial^3 \psi}{\partial \varepsilon_{ij}\partial \varepsilon_{kl}\partial P_k}\varepsilon_{ij}\varepsilon_{kl} + 6\frac{\partial^3 \psi}{\partial \varepsilon_{ij}\partial P_k \partial P_l}\varepsilon_{ij}P_l\right. \\ \left. +3\frac{\partial^3 \psi}{\partial P_k \partial P_l \partial P_s}P_l P_s + \ldots\right) + \ldots\end{aligned} \tag{111}$$

$$\xi_{kr} = \frac{\partial \psi}{\partial P_{k,r}} \approx \left.\frac{\partial^2 \psi}{\partial P_{k,r}\partial P_{l,s}}\right|_{\varepsilon,T} P_{l,s} \,. \tag{112}$$

With the polarization fixed, a tensile strain should be associated with a tensile stress and a compressive strain with a compressive stress. The first term in the stress equation is the linear elastic stiffness. If the strain is held fixed and the polarization changed, coupling between the polarization and strain should give rise to stress. This is the second term in the stress equation. The term in strain squared provides a correction for a non-linear stiffness. This will be taken as zero. At this point the higher order terms in stress are dropped. This requires that the same terms be removed from the electric field equation. The result is

$$\begin{aligned}\sigma_{ij} &= C_{ijkl}^{P,T}\varepsilon_{kl} + \frac{1}{2}Q_{ijkl}^{\varepsilon,T}P_k P_l \\ E_k &= \kappa_{kl}^{\varepsilon,T}P_l + Q_{ijkl}^{\varepsilon,T}\varepsilon_{ij}P_l \\ \xi_{kr} &= G_{krls}P_{l,s}\end{aligned} \tag{113}$$

and the energy function is written as

$$\begin{aligned}\Delta\psi(\varepsilon_{ij}, P_i, P_i^{FS}, P_{i,j}) = \frac{1}{2}\Big(&C_{ijkl}^{P,T}\varepsilon_{ij}\varepsilon_{kl} + \kappa_{kl}^{\varepsilon,T} P_k P_l \\ &+\frac{1}{\varepsilon_0} P_j^{FS} P_j^{FS} + G_{krls}^{\varepsilon,T} P_{k,r} P_{l,s}\Big) \\ &+\frac{1}{2}(Q_{ijkl}^{\varepsilon,T}\varepsilon_{ij} P_k P_l) + \ldots\,.\end{aligned} \tag{114}$$

Higher order terms than 3rd order are still needed to be able to generate a multi-well energy function in polarization. This requires additional even order terms in $\boldsymbol{P}$ (4th, 6th, 8th, etc.). The easy way to visualize this is to consider a quadratic polynomial with a negative coefficient and a quartic polynomial with a positive coefficient. Add the two together and the resulting function will initially be negative, the slope will be zero at the bottom of the well, then the function will go positive. Lets add some even order terms in polarization. Remember that these are just the last terms in the 4th, 6th, etc. order terms in the Taylor's series expansion.

$$\begin{aligned}\Delta\psi(\varepsilon_{ij}, P_i, P_i^{FS}, P_{i,j}) = \frac{1}{2}\Big(&C_{ijkl}^{P,T}\varepsilon_{ij}\varepsilon_{kl} + \kappa_{kl}^{\varepsilon,T} P_k P_l \\ &+\frac{1}{\varepsilon_0} P_j^{FS} P_j^{FS} + G_{krls}^{\varepsilon,T} P_{k,r} P_{l,s}\Big) \\ &+\frac{1}{2}(Q_{ijkl}^{\varepsilon,T}\varepsilon_{ij} P_k P_l) \\ &+\frac{1}{4!}(\kappa_{klmn}^{\varepsilon,T} P_k P_l P_m P_n) \\ &+\frac{1}{6!}(\kappa_{klmnrs}^{\varepsilon,T} P_k P_l P_m P_n P_r P_s) + \ldots\,.\end{aligned} \tag{115}$$

This energy function can be further modified with additional terms to better represent ferroelectric crystals. For example, the fourth order elasticity tensor must have cubic symmetry with this energy function regardless of the polarization, but a real material has spontaneous polarization and will display an elastic symmetry when the polarization is fixed at the spontaneous polarization value that may have tetragonal, orthorhombic, or rhombohedral symmetry. In this case a higher order term should be kept that represents the elastic constant dependence on polarization, i.e.

$$\frac{\partial^3\psi}{\partial\varepsilon_{ij}\partial\varepsilon_{kl}\partial P_m}\varepsilon_{ij}\varepsilon_{kl} P_m \tag{116}$$

which leads to an elasticity tensor that is a function of polarization,

$$C_{ijkl}^{P,T} + \frac{\partial C_{ijkl}^{P,T}}{\partial P_m} P_m\,. \tag{117}$$

The same can be done for the quadratic electrostrictive stress tensor that has cubic symmetry in this formulation. Just add the higher order dependence on polarization to enable matching the lower symmetry piezoelectric constants associated with the polarization induced tetragonal, orthorhombic, or rhombohedral symmetry.

In conclusion, for a relatively accurate representation of material properties the general form of the free energy must contain a sufficient number of parameters to allow for independent fitting of the spontaneous polarization, spontaneous strain, dielectric permittivity, piezoelectric coefficients and the elastic properties near the zero stress and zero electric field free spontaneous polarization and strain states. To accomplish this task, Su and Landis (2007) introduced the following form for the free energy to fit the material parameters to the tetragonal single crystal properties of Barium Titanate.

$$
\begin{aligned}
\psi(\varepsilon_{ij}, P_i, P_{i,j}, D_i) = & \frac{1}{2}a_{ijkl}P_{i,j}P_{k,l} + \frac{1}{2}\bar{a}_{ij}P_iP_j \\
& + \frac{1}{2}\bar{\bar{a}}_{ijkl}P_iP_jP_kP_l \\
& + \frac{1}{2}\bar{\bar{\bar{a}}}_{ijklmn}P_iP_jP_kP_lP_mP_n \\
& + \frac{1}{8}\bar{\bar{\bar{\bar{a}}}}_{ijklmnrs}P_iP_jP_kP_lP_mP_nP_rP_s \\
& + b_{ijkl}\varepsilon_{ij}P_kP_l \\
& + \frac{1}{2}c_{ijkl}\varepsilon_{ij}\varepsilon_{kl} + f_{ijklmn}\varepsilon_{ij}\varepsilon_{kl}P_mP_n \\
& + g_{ijklmn}\varepsilon_{ij}P_kP_lP_mP_n \\
& - \frac{1}{2\varepsilon_0}(D_i - P_i)(D_i - P_i) \,.
\end{aligned}
\tag{118}
$$

The first term of the free energy penalizes large gradients of polarization and gives domain walls thickness and energy within the model. The four terms on the second and third lines are used to create the non-convex energy landscape of the free energy with minima located at the spontaneous polarization states. The four terms on the fourth line are then used to fit the material's spontaneous strain along with the dielectric, elastic and piezoelectric properties about the spontaneous state. The final term represents the energy stored within the free space occupied by the material. The eighth rank term on the third line was introduced in order to allow for adjustments of the dielectric properties and the energy barriers for 90 switching Zhang and Bhattacharya (2005). The sixth rank terms introduced on the fourth line allow us to fit the elastic, piezoelectric and dielectric properties of the low symmetry phase at the spontaneous state. Without these terms the elastic properties of the material arise only from the c_{ijkl} tensor, which must have the symmetry of the high temperature phase. With regard to the piezoelectric coefficients, b_{ijkl} is used to fit the spontaneous strain components associated with the stress and electric field free spontaneous polarization state and by introducing the f_{ijklmn} and g_{ijklmn} tensors the full tetragonal structure of the piezoelectric tensor can be matched.

5.6 Phase-Field Modeling of Ferromagnetics

Following the same approach as in the case of ferroelectric materials, ferromagnetic materials are analyzed in terms of micro-forces associated with the magnetization order parameter. Within the micro-magneto-mechanical approach taken here the free energy of the material under consideration depends on the magnetization and its gradients. The relationship between the magnetization components, the magnetic field and the magnetic induction is given as:

$$B_i = \mu_0(M_i + H_i) \tag{119}$$

where μ_0 is the permittivity of free space. Following Landis, Landis (2008) a micro-force system is introduced, work conjugate to the dependent variable introduced. Specifically, let ζ_{ij} be a micro-force tensor such that $\zeta_{ij}n_j\dot{M}_i$ is the power expended on a surface by neighboring configurations, an internal micro-force vector π_i such that $\pi_i\dot{M}_i$ is the power expended by the material internally and accounts for any dissipation as in the case of ferroelectrics. Then the angular momentum balance, stating that rate of change of angular momentum associated with changes in magnetization equals the torque associated with the moment of the magnetization and the micro-forces) reads as:

$$\frac{1}{\mu_0}\left(\int_S \varepsilon_{ijk}M_j\zeta_{lk}n_l dS + \int_V \varepsilon_{ijk}M_j\pi_k dV\right) = \frac{1}{\gamma_0}\int_V \dot{M}_i dV \tag{120}$$

where $\gamma_0 = 2.21 \times 10^5$ m/As is the gyromagnetic ratio of the electron spin. Applying the divergence theorem on the first term, leads to an equivalent local form of the angular momentum balance given as:

$$\varepsilon_{ijk}M_{j,l}\zeta_{lk} + \varepsilon_{ijk}M_j(\zeta_{lk,l} + \pi_k) = \frac{\mu_0}{\gamma_0}\dot{M}_i\,. \tag{121}$$

Taking the cross product of the third equation with the magnetization we have:

$$\begin{aligned}
&(\varepsilon_{jmn}M_{m,p}\zeta_{pn} + \varepsilon_{jmn}M_m(\zeta_{pn,p} + \pi_n))\varepsilon_{ijk}M_k = \frac{\mu_0}{\gamma_0}\varepsilon_{ijk}M_k\dot{M}_j \Rightarrow \\
&\varepsilon_{jki}M_k\varepsilon_{jmn}M_{m,p}\zeta_{pn} + \varepsilon_{jki}M_k\varepsilon_{jmn}M_n(\zeta_{pn,p} + \pi_n) \\
&\qquad = \frac{\mu_0}{\gamma_0}\varepsilon_{ijk}M_k\dot{M}_j \Rightarrow \\
&\varepsilon_{jki}M_k\varepsilon_{jmn}M_{m,p}\zeta_{pn} + (\delta_{km}\delta_{in} - \delta_{kn}\delta_{im})M_kM_m(\zeta_{pn,p} + \pi) \\
&\qquad = \frac{\mu_0}{\gamma_0}\varepsilon_{ijk}M_j\dot{M}_k \Rightarrow \\
&\varepsilon_{jki}M_k\varepsilon_{jmn}M_{m,p}\zeta_{pn} + M_kM_k(\zeta_{pi,p} + \pi_i) - M_kM_i(\zeta_{pk,p} + \pi_k) \\
&\qquad = \frac{\mu_0}{\gamma_0}\varepsilon_{ijk}M_j\dot{M}_k\,.
\end{aligned} \tag{122}$$

Note that the angular momentum balance does not provide any information for the changes along the magnetization axis. However it is assumed that the magnetization magnitude changes under balance of micro-force in the magnetization direction at every point in time and in shorter time than magnetization rotation. Mathematically then we have that:

$$M_k(\zeta_{lk,l} + \pi_k) = 0\,. \tag{123}$$

Therefore substituting the above result in the momentum balance and noting that $M^2 = M_k M_k$ we have:

$$(\zeta_{pi,p} + \pi_i) = \frac{\mu_0}{M^2\gamma_0}\varepsilon_{ijk} M_j \dot{M}_k - \frac{1}{M^2}\varepsilon_{jki}\varepsilon_{jmn} M_{m,p}\zeta_{pn} M_k\,. \tag{124}$$

Next consider the free energy of the material and the free space depending on strain, magnetic induction, the magnetization order parameter and time and spatial derivatives as:

$$\psi = \psi(\varepsilon_{ij}, B_i, M_i, M_{i,j}, \dot{M}_i)\,. \tag{125}$$

Also all field quantities are allowed to depend on the same set of independent variables. The energy balance is then written as

$$\begin{aligned}\int_V \dot{U} dV + \frac{d}{dt}\int_V \rho\dot{u}_i\dot{u}_i\, dV = &\int_V (b_i\dot{u}_i + J_i\dot{A}_i) dV \\ &+ \int_S (t_i\dot{u}_i + K_i\dot{A}_i + \zeta_{ij}n_j\dot{M}_i)\, dS \\ &+ \int_V r dV + \int_S -q_i n_i dS\end{aligned} \tag{126}$$

where U is the internal energy density, q_i are the components of the outward heat flux vector per unit area and r is the supply of heat per unit volume from external source. Also where J_i and K_i are the components of the volume and surface current density respectively and A_i is a magnetic vector potential arising from $B_{i,i} = 0 \rightarrow B_i = \varepsilon_{ijk} A_{k,j}$. Next, applying the divergence theorem on the surface integral term and substituting the balance equations the local equivalent form can be written as:

$$\begin{aligned}\dot{U} = &\,\sigma_{ij}\dot{\varepsilon}_{i,j} + \dot{B}_i H_i + \zeta_{ji}\dot{M}_{i,j} \\ &+ \left(\frac{1}{M^2}\frac{\mu_0}{\gamma_0}\varepsilon_{ijk} M_j \dot{M}_k - \frac{1}{M^2}\varepsilon_{ijk}\varepsilon_{jmn} M_{m,p}\zeta_{pn} M_k - \pi_i\right)\dot{M}_i \\ &+ r - q_{i,i}\,.\end{aligned} \tag{127}$$

The constraint of the second law of thermodynamics in the entropy production is given through the pointwise local Clausius-Duhem dissipation inequality as:

$$\frac{\dot{q}_i T_{,i}}{T} - \dot{q}_{i,i} \leq T\dot{S}\,. \tag{128}$$

Eliminating the divergence of the heat flux vector we have

$$\begin{aligned} T\dot{S} \geq & \frac{q_i T_{,i}}{T} + \dot{e} - \sigma_{ij}\dot{\varepsilon}_{ij} - \dot{B}_i H_i - \zeta_{ji}\dot{M}_{i,j} - r \\ & - \left(\frac{1}{M^2}\frac{\mu_0}{\gamma_0}\varepsilon_{ijk}M_j\dot{M}_k - \frac{1}{M^2}\varepsilon_{ijk}\varepsilon_{jmn}M_{m,p}\zeta_{pn}M_k - \pi_i\right)\dot{M}_i . \end{aligned} \tag{129}$$

In terms of the free energy $\psi = U - \theta s \rightarrow \dot{\psi} = \dot{U} - \dot{\theta} s - \theta\dot{s} \Rightarrow \dot{U} - \theta\dot{s} = \dot{\psi} + \dot{\theta} s$ the second law can be written as

$$\begin{aligned} & \dot{\psi} + \dot{T}S - \frac{\dot{q}_i T_{,i}}{T} - \sigma_{ij}\dot{\varepsilon}_{ij} - \dot{B}_i H_i - \zeta_{ji}\dot{M}_{i,j} - r \\ & - \left(\frac{1}{M^2}\frac{\mu_0}{\gamma_0}\varepsilon_{ijk}M_j\dot{M}_k - \frac{1}{M^2}\varepsilon_{ijk}\varepsilon_{jmn}M_{m,p}\zeta_{pn}M_k - \pi_i\right)\dot{M}_i \leq 0 . \end{aligned} \tag{130}$$

Finally, under isothermal conditions and using the definition of the free energy and for an arbitrary volume element the pointwise form of the above the inequality can be written as:

$$\begin{aligned} & \left(\frac{\partial\psi}{\partial\varepsilon_{ij}} - \sigma_{ij}\right)\dot{\varepsilon}_{ij} + \left(\frac{\partial\psi}{\partial B_i} - H_i\right)\dot{B}_i + \left(\frac{\partial\psi}{\partial M_{i,j}} - \zeta_{ji}\right)\dot{M}_{i,j} \\ & + \frac{\partial\psi}{\partial\dot{M}_i}\ddot{M}_i + \left(\frac{\partial\psi}{\partial M_i} + \frac{1}{M^2}\varepsilon_{ijk}\varepsilon_{jmn}M_{m,p}\zeta_{pn}M_k + \pi_i\right)\dot{M}_i \leq 0 . \end{aligned} \tag{131}$$

Linearity in $\dot{\varepsilon}_{ij}$, $\dot{B}_i$, $\dot{M}_i$, $\dot{M}_{i,j}$, $\ddot{M}_i$ implies that:

$$\frac{\partial\psi}{\partial\dot{M}_i} = 0 \rightarrow \psi \neq \psi(\dot{M}_i), \quad \frac{\partial\psi}{\partial\varepsilon_{ij}} = \sigma_{ij}, \quad \frac{\partial\psi}{\partial B_i} = H_i, \quad \frac{\partial\psi}{\partial M_{i,j}} = \zeta_{ji} \tag{132}$$

and

$$\begin{aligned} & \left(\frac{\partial\psi}{\partial M_i} + \frac{1}{M^2}\varepsilon_{ijk}\varepsilon_{jmn}M_{m,p}\zeta_{pn}M_k + \pi_i\right)\dot{M}_i \leq 0 \\ & \rightarrow \pi_i = -\left(\frac{\partial\psi}{\partial M_i} + \frac{1}{M^2}\varepsilon_{ijk}\varepsilon_{jmn}M_{m,p}\zeta_{pn}M_k\right) - \beta\dot{M}_i . \end{aligned} \tag{133}$$

Substituting the above result for the form of the internal micro-force vector in the balance of angular momentum we have:

$$\begin{aligned} & (\zeta_{pi,p} + \pi_i) = \frac{\mu_0}{M^2\gamma_0}\varepsilon_{ijk}M_j\dot{M}_k - \frac{1}{M^2}\varepsilon_{jki}\varepsilon_{jmn}M_{m,p}\zeta_{pn}M_k \\ & \Rightarrow \frac{1}{\mu_0}\left(\zeta_{pi,p} - \frac{\partial\psi}{\partial M_i}\right) = \beta\dot{M}_i + \frac{1}{M^2\gamma_0}\varepsilon_{ijk}M_j\dot{M}_k . \end{aligned} \tag{134}$$

The resulting equation is a generalized form of the Landau-Lifshitz-Gilbert equation describing the micromagnetic dynamics. Following Landis (2008), recognizing that

$$H_i^{\text{eff}} = \frac{1}{\mu_0}\left(\zeta_{pi,p} - \frac{\partial \psi}{\partial M_i}\right) \tag{135}$$

and taking the cross product of the equation with the magnetization vector we have:

$$\boldsymbol{M} \times (\boldsymbol{H}^{\text{eff}} - \beta\dot{\boldsymbol{M}}) = \frac{1}{\gamma_0 M^2}\boldsymbol{M} \times (\boldsymbol{M} \times \dot{\boldsymbol{M}})\,. \tag{136}$$

As the magnitude of the magnetization vector approaches the saturation value of magnetization, $M \to M_s$ we have

$$\boldsymbol{M} \times (\boldsymbol{H}^{\text{eff}} - \beta\dot{\boldsymbol{M}}) = \frac{1}{\gamma_0}\dot{\boldsymbol{M}} \tag{137}$$

where the viscous parameter β is related to the Gilbert damping parameter as

$$\beta = \frac{\alpha}{\gamma_0 M_s}\,. \tag{138}$$

An apparent advantage of using the generalized Landau-Lifshitz-Gilbert equation that results from the micro-force analysis is that the restriction of the value of the magnetization magnitude to be close to the saturation magnetization value can be applied directly without the use of special techniques to ensure the constraint in a numerical setting. Landis introduced an energy penalty term for deviations of the magnetization magnitude from the saturation value in the Helmholtz free energy of the ferromagnetic material of the form:

$$\psi_{\text{constraint}} = \frac{\mu_0(1+\chi)}{2\chi}(M - M_s)^2\,. \tag{139}$$

As the magnetic susceptibility $\chi \to 0$ the constraint $M = M_s$ is enforced. A general form of the free energy for ferromagnetic materials can be written as

$$\begin{aligned}\psi(\varepsilon_{ij}, M_i, M_{i,j}, B_j) = {} & \frac{1}{2}A_{ijkl}M_{i,j}M_{k,l} + K_{ijkl}M_iM_jM_kM_l \\ & + \bar{K}_{ijklmn}M_iM_jM_kM_lM_mM_n \\ & + \frac{1}{2}c_{ijkl}(\varepsilon_{ij} - \varepsilon_{ij}^0)(\varepsilon_{kl} - \varepsilon_{kl}^0) \\ & + \frac{\mu_0(1+\chi)}{2\chi}(M - M_s)^2 \\ & - \frac{1}{2\mu_0}B_iB_i - M_iB_i\,.\end{aligned} \tag{140}$$

The first term in the free energy expansion is the exchange energy between magnetic spins that penalizes large gradients of magnetization and gives the magnetic domain walls the thickness and the corresponding energy in the theory. The second and third terms are used to fit the magnetocrystalline anisotropy associated with the symmetry required for different materials that creates the energy wells for easy directions of magnetization. The next terms describe the energy associated with the elastic and magnetostrictive strain and the energy penalty that accounts for deviations of the magnetization magnitude from the saturation magnetization respectively. The last two terms are magnetostatic energy associated with the free energy.

5.7 Finite Element Implementation

To summarize, the governing equations for ferroelectric materials are

$$\begin{aligned} D_{i,i} &= 0 \quad && \text{in } V \\ D_i n_i &= -\omega && \text{on } S \\ E_i &= -\phi_{,i} \end{aligned} \tag{141}$$

for the electrical fields,

$$\begin{aligned} \sigma_{ij,j} &= \rho \ddot{u}_i \quad && \text{in } V \\ \sigma_{ij} n_j &= t_i && \text{on } S \\ \varepsilon_{ij} &= \frac{1}{2}(u_{i,j} + u_{j,i}) \end{aligned} \tag{142}$$

for the mechanical fields, and for the internal forces associated with the polarization

$$\begin{aligned} \xi_{ji,j} - \frac{\partial \psi}{\partial P_i} &= \beta \dot{P}_i \ \text{ in } \ V \\ \xi_{ij} n_j &= \lambda_i \quad \text{on } \ S \,. \end{aligned} \tag{143}$$

The relationship between the electric displacement, electric field and polarization vectors $D_i = \varepsilon_0 E_i + P_i$. Let δP_i, $\delta\phi$, δu_i be test functions. Then multiplying the equations with the test functions and integrating over the volume we have:

$$\int_V \left(\xi_{ij,j} - \frac{\partial \psi}{\partial P_i} - \beta \dot{P}_i \right) \delta P_i dV + \int_V D_{i,i} \delta\phi dV + \int_V \sigma_{ij,j} \delta u_i dV = 0\,. \tag{144}$$

Applying the divergence theorem and taking into account kinematics and boundary charge and traction definition we have:

$$\begin{aligned} &\int_V \left(\beta \dot{P}_i \delta P_i + \frac{\partial \psi}{\partial P_i} \delta P_i + \xi_{ji} \delta P_{i,j} \right) dV \\ &+ \int_V \sigma_{ij} \delta\varepsilon_{ij} dV - \int_V D_i \delta E_i dV = \int_S t_i \delta u_i - \omega \delta\phi + \xi_{ij} n_j \delta P_i dS\,. \end{aligned} \tag{145}$$

The above equation implies that the components of mechanical displacement, electric polarization and the electric potential are used as nodal degrees of freedom. Defining the array of degrees of freedom as $\boldsymbol{d}$ each of the field quantities is interpolated from the nodal quantities with the same set of shape functions such that

$$\{\ u_i \ \ \phi \ \ P_i\ \}^T = [N]\{d\}\,. \tag{146}$$

We note that even though polarization gradient appears in the free energy, the shape function matrix $\boldsymbol{N}$ must meet only the requirements for standard C^0 continuous elements. This is due to the fact that both electric field and polarization can be taken as independent variables. Therefore, the polarization components take the same status as mechanical displacement and electric potential and the polarization gradient takes the same status as strain and electric field. If, for example, the electric field were the order parameter, then higher order elements would be required in the formulation. Hence, the displacements, electric potential and polarization components are approximated by continuous functions throughout the mesh, but strains, electric fields, and polarization gradients will have jumps in certain components along element boundaries. The stress, electric displacement and micro-forces are computed as

$$\sigma_{ij} = \frac{\partial h}{\partial \varepsilon_{ij}}, \quad D_i = -\frac{\partial h}{\partial E_i}, \quad \xi_{ji} = \frac{\partial h}{\partial P_{i,j}} \tag{147}$$

where $h(\varepsilon_{ij}, P_i, P_{i,j}, E_i) = \psi(\varepsilon_{ij}, P_i, P_{i,j}, D_i) - E_i D_i$ is the enthalpy of the material is used in order to accommodate the electric potential as a degree of freedom. The discretized formulas for the polarization rates and the basic solution fields during a given time step are as follows:

$$\dot{P}_i = \frac{P_i^{t+\Delta t} - P_i^t}{\Delta t} \quad \text{and} \quad d_i = d_i^{t+\Delta t}\,. \tag{148}$$

Here, the superscript indicates the time step at which the field is evaluated which result in the first order accurate backward Euler scheme that allows for enhanced numerical stability with larger time increments. Then given a known set of nodal degrees of freedom at time t, a set of nonlinear algebraic equations results for the nodal degrees of freedom at $t + \Delta t$ that can be written in the form

$$\boldsymbol{B}(\boldsymbol{d}^{\,t+\Delta t}) = \boldsymbol{F}\,. \tag{149}$$

These equations can be solved incrementally with the Newton-Raphson method.

$$\left.\frac{\partial \boldsymbol{B}}{\partial \boldsymbol{d}}\right|_{\boldsymbol{d}_i^{t+\Delta t}} \Delta \boldsymbol{d}_i = \boldsymbol{F} - \boldsymbol{B}(\boldsymbol{d}_i^{t+\Delta t}) \tag{150}$$

where i is the current step counter in the Newton-Raphson sequence and $\Delta \boldsymbol{d}_i$ is the increment computed for $\boldsymbol{d}_i^{t+\Delta t}$ such that $\boldsymbol{d}_i^{t+\Delta t} = \boldsymbol{d}_{i-1}^{t+\Delta t} + \Delta \boldsymbol{d}_i$. The Newton-

Raphson procedure is carried out until a suitable level of convergence is obtained yielding a solution for the displacement, electric potential, and polarization fields. Note that the viscous parameter β should be equal to zero for equilibrium domain structure solution. However, in practice we usually need to allow the domain structure to evolve along non-equilibrium paths so this parameter is used as a free numerical parameter to drive the solution and is gradually reduced and set to zero at equilibrium. Similarly, the governing equations for ferromagnetic materials are

$$\begin{aligned} B_{i,i} &= 0 \quad \text{in } V \\ \varepsilon_{ijk} H_{j,k} &= J_i \quad \text{in } V \\ \varepsilon_{ijk} H_j n_k &= K_i \quad \text{in } S \end{aligned} \tag{151}$$

for the magnetic fields, where J_i and K_i are the volume and surface current density respectively and ε_{ijk} stands for the permutation symbol.

$$\begin{aligned} \sigma_{ij,j} &= \rho \ddot{u}_i \quad \text{in } V \\ \sigma_{ij} n_j &= t_i \quad \text{on } S \\ \varepsilon_{ij} &= \frac{1}{2}(u_{i,j} + u_{j,i}) \end{aligned} \tag{152}$$

for the mechanical fields, and for the internal forces associated with the magnetization

$$\frac{1}{\mu_0}\left(\zeta_{pi,p} - \frac{\partial \psi}{\partial M_i}\right) = \frac{\alpha}{\gamma_0 M_S}\dot{M}_i + \frac{1}{M_S^2 \gamma_0}\varepsilon_{ijk} M_j \dot{M}_k \quad \text{in} \quad V\,. \tag{153}$$

The relationship between the magnetic induction, magnetic field and magnetization vector is $B_i = \mu_0(H_i + M_i)$. Neglecting inertial terms and considering zero volume and surface current densities, the magnetic field can be written is terms of a scalar magnetic potential as $H_i = -\phi^M_{,i}$ and a variational form is then written as

$$\begin{aligned} &\int_V \sigma_{ij}\delta\varepsilon_{ij} dV - \int_V B_i \delta H_i dV \\ &+ \int_V \left(\frac{\alpha}{\gamma_0 M_S}\dot{M}_i + \frac{\partial \psi}{\partial M_i} + \frac{1}{\gamma M_S^2}\varepsilon_{ijk}\dot{M}_j M_k\right)\delta M_i dV \\ &+ \int_V \zeta_{ji}\delta M_{i,j} dV = \int_S t_i \delta u_i + \zeta_{ij} n_j \delta M_i dS\,. \end{aligned} \tag{154}$$

The above equation implies that the components of mechanical displacement, magnetization and the magnetic potential are used as nodal degrees of freedom and the nodal quantities with the same set of shape functions such that

$$\{\ u_i \ \ \phi^M \ \ M_i \ \}^T = [N]\{d\} \tag{155}$$

where the shape function matrix $\boldsymbol{N}$ must meet the requirements for standard C^0 continuous elements. In this case, the magnetization components take the same status as mechanical displacement and magnetic potential and the magnetization gradient takes the same status as strain and magnetic field. The stress, magnetic induction and micro-forces are computed as

$$\sigma_{ij} = \frac{\partial h}{\partial \varepsilon_{ij}}, \quad B_i = -\frac{\partial h}{\partial H_i}, \quad \zeta_{ji} = \frac{\partial h}{\partial M_{i,j}} \tag{156}$$

where $h(\varepsilon_{ij}, M_i, M_{i,j}, H_i) = \psi(\varepsilon_{ij}, M_i, M_{i,j}, B_i) - B_i H_i$ is the enthalpy of the material is used in order to accommodate the magnetic potential as a degree of freedom. The resulting nonlinear problem can be solved using the Newton-Raphson method as in the case of ferroelectrics. For an equilibrium domain structure the parameter α can either be used as a free parameter to drive the numerical solution or taken to corresponding values of the Gilbert damping constant for a realistic representation of the micromagnetic dynamics.

5.8 *Example of Strain-Mediated Multiferroic Phase-Field Modeling*

The above formulation of the phase field method for ferroelectric and ferromagnetic materials can be extended to the case of strain mediated multiferroic composites. This is demonstrated through an example case study as follows. Consider a ferroelectric and a ferromagnetic static domain structure as shown in the Fig. 8 where the ferroelectric and the ferromagnetic materials share an interface at the common boundary.

Then given a static domain structure the free energy of the system can be written as

$$\psi^{FE/M} = \int_{V^{FE}} \psi^{FE}(\varepsilon_{ij}, D_i, P_i, P_{i,j}) + \int_{V^{FM}} \psi^{FM}(\varepsilon_{ij}, B_i, M_i, M_{i,j}) dV \tag{157}$$

where V^{FE} and V^{FM} corresponds to the volume of the ferroelectric and ferromagnetic part respectively. In the strain-mediated approach the total strain is written as the sum of the elastic and electrostrictive/magnetostrictive strain as

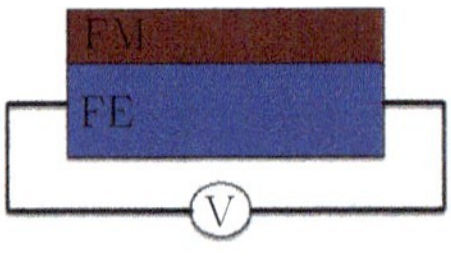

Fig. 8 Schematic of strain-mediated ferroelectric (FE) and ferromagnetic (FM) heterostructures

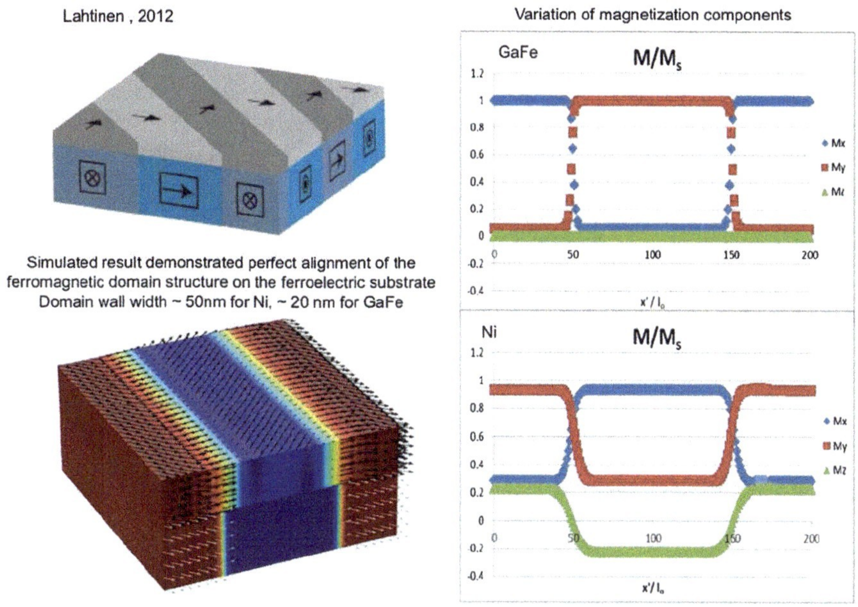

Fig. 9 Imprinting of ferromagnetic domains (GaFe and Ni) on FE domains ($BaTiO_3$) by strain coupling between the ferroic orders (Carka and Lynch 2013)

$$\varepsilon_{ij} = \varepsilon_{ij}^{el} + \varepsilon_{ij}^{0} = \varepsilon_{ij}^{el} + AQ_{ijkl}P_k P_l + (1 - A)\lambda_{ijkl}M_k M_l \tag{158}$$

where the parameter A takes the value 1 in the ferroelectric volume and 0 in the ferromagnetic. A variational form for the coupled problem is then written as:

$$\begin{aligned} &\int_{V^{FE}} \left(\xi_{ji,j} - \frac{\partial\psi}{\partial P_i} - \beta\dot{P}_i\right)\delta P_i dV + \int_{V^{FE}} D_{i,i}\delta\phi dV \\ &+ \int_{V^{FE}\&V^{FM}} \sigma_{ij,j}\delta u_i dV + \int_{V^{FM}} B_{i,i}\delta\phi_i^M dV \\ &+ \int_{V^{FM}} \left(\zeta_{ji,j} - \frac{\partial\psi}{\partial M_i} - \frac{\alpha\mu_0}{\gamma_0 M_S}\dot{M}_i - \frac{\mu_0}{\gamma_0 M_S^2}\varepsilon_{ijk}\dot{M}_j M_k\right)\delta M_i = 0. \end{aligned} \tag{159}$$

The finite element formulation described above is used to simulate the experimental results of Lahtinen et al. (2012) who demonstrated full imprinting of ferroelastic $BaTiO_3$ domains into CoFe during thin film growth (Fig. 9).

We first apply appropriate periodic boundary conditions based on the spontaneous strains and polarization distribution to allow for the formation of an equilibrium 90° domain pattern using the phase field formulation described above. The energy function and the parameters correspond to barium titanate and are taken from Su and Landis. Once the equilibrium periodic domain pattern of the ferroelectric substrate is obtained we consider a random thin ferromagnetic layer and allow for the evolution of the ferromagnetic domain structure using the multiferroic formulation. Evolution of the domain pattern towards minimizing the energy of the system results in full imprinting of the ferroelectric domain structure explained by strain transfer at the heterostructure interface. The ferroelastic strains of the ferroelectric substrate interact with the magnetoelastic anisotropy of the ferromagnetic material via the inverse magnetostriction effect. Two ferromagnetic materials of different magnetic anisotropies are considered. The resulting domain imprinting is demonstrated in the Figure for Galfanol (GaFe) and Nickel (Ni).

References

Cao, W., & Cross, L. E. (1991). Theory of tetragonal twin structures in ferroelectric perovskites with a first-order phase transition. *Physical Review B*, *44*, 5–12.

Carka, D., & Lynch, C. S. (2013). Phase-field modelling of ferroelectric/ferromagnetic bilayer system: The effect of domain structure on the magnetization reorientation. ASME Conference on Smart Materials and Adaptive Structures (SMASIS), September 2013.

Choudhury, S., Li, Y. L., Krill, C. E., & Chen, L. Q. (2005). Phase-field simulation of polarization switching and domain evolution in ferroelectric polycrystals. *Acta Materialia*, *53*, 5313–5321.

Cullity, B.D. & Graham, C.D. (2011) *Introduction to magnetic materials*. Wiley: New York.

Fried, E., & Gurtin, M. E. (1993). Continuum theory of thermally induced phase transitions based on an order parameter. *Physica D*, *68*, 326–343.

Fried, E., & Gurtin, M. E. (1994). Dynamic solid-solid transition with phase characterized by an order parameter. *Physica D*, *72*, 287–308.

Gurtin, M. E. (1996). Generalized Ginzburg-Landau and Cahn-Hilliard equations based on a mircoforce balance. *Physica D*, *92*, 178–192.

Kontsos, A., & Landis, C. M. (2009). Computational modeling of domain wall interactions with dislocations in ferroelectric single crystals. *International Journal of Solid and Structures*, *46*, 1491–1498.

Lahtinen, T., Franke, K., & van Dijken, S. (2012). Electric-field control of magnetic domain wall motion and local magnetization reversal. *Scientific Reports*, *2*, 258.

Landis, C. M. (2008). A continuum thermodynamics formulation for micromagnetomechanics with applications to ferromagnetic shape memory alloys. *Journal of the Mechanics and Physics of Solids*, *56*, 3059–3076.

Li, W., & Landis, C. M. (2011). Nucleation and growth of domains near crack tips in single crystal ferroelectrics. *Engineering Fracture Mechanics*, *78*, 1505–1513.

Panofsky, W., Phillips, M. (2005). *Classical electricity and magnetism*. Springer–Verlag.

Song, Y. C., Soh, A. K., & Ni, Y. (2007). Phase field simulation of crack tio domain switching in ferroelectrics. *Journal of Physics D: Applied Physics*, *40*, 1175–1182.

Su, Y., & Landis, C. M. (2007). Continuum thermodynamics of ferroelectric domain evolution: Theory, finite element implementation and alpplication to domain wall pinning. *Journal of the Mechanics and Physics of Solids*, *55*, 280–305.

Zhang, W., & Bhattacharya, K. (2005). A computational model of ferroelectric domains. Part I. *Acta Materialia*, *53*, 185–198.

Semiconductor Effects in Ferroelectrics

Doru C. Lupascu, Irina Anusca, Morad Etier, Yanling Gao, Gerhard Lackner, Ahmadshah Nazrabi, Mehmet Sanlialp, Harshkumar Trivedi, Naveed Ul-Haq and Jörg Schröder

Abstract In this textbook ferroelectrics have so far been dealt with as insulators. External electric fields can and will induce polarization in any insulating material. This is dielectricity. On top of this, pyroelectrics exhibit a temperature dependent spontaneous electric polarization, namely a crystallographic phase transition which is polar. It disappears above the Curie-point. Below the Curie point, external electric fields can rotate or alter the direction of this spontaneous polarization. If this becomes a remanent state, the material is ferroelectric and exhibits electric hysteresis. Another aspect in these materials is the fact that electrical insulation is a stretchable term. While metals are well defined and offer conductivity down to very low temperatures, already semi-metals will turn partly insulating at low temperature. Semiconductors are typically insulating in a certain low temperature range (energetically $\lesssim 1/10kT$) above which thermal excitation of charge carriers into the conduction band will induce a finite conductivity. The energetic band gap determines this barrier and the exponential tail of the Fermi-Dirac distribution determines the number of charge carriers in the conduction band as well as the missing electrons (termed holes) in the valence band. Typical ferroelectrics exhibit band gaps that turn the material insulating at room temperature. This is the case for most oxides. Ferroelectric sulfides typically display much lower band gaps and turn conducting at or even below room temperature already. Another aspect enters when one considers that external electroding is always necessary to drive a ferroelectric capacitor. In the context of a semiconductor picture we deal with a classical Schottky barrier. Grain boundaries play another particular role in polycrystalline materials. This may even lead to positive temperature coefficient resistor (PTCR) characteristics. In this lecture we will draw the connection between a ferroelectric, its semiconductor character, point defects, and their overall interactions. Particularly the inner and outer boundaries of crystallites

D.C. Lupascu (✉) · I. Anusca · M. Etier · Y. Gao · G. Lackner · A. Nazrabi · M. Sanlialp · H. Trivedi · N. Ul-Haq
Institute for Materials Science and Center for Nanointegration Duisburg-Essen (CENIDE), University of Duisburg-Essen, Essen, Germany
e-mail: doru.lupascu@uni-due.de

J. Schröder
Institute of Mechanics, University of Duisburg-Essen, Duisburg, Germany

J. Schröder and Doru C. Lupascu (eds.), *Ferroic Functional Materials*,
CISM International Centre for Mechanical Sciences 581,
https://doi.org/10.1007/978-3-319-68883-1_3

become subject to band bending, 2D-conducting planes, space charge regions, and diverse other effects. Also optical effects as well as fatigue depend on the semiconductor and defect induced properties. We intend to give the newcomer access to this complex field which has seen a peak in understanding in the late 70ies of the 20th century experiencing a certain revival recently due to a number of exciting findings associated with domain walls. Furthermore, magnetoelectric composites have recently been found to display peculiar electrical effects related to their semiconductor character rather than the magnetic part of their properties.

1 Introduction

This lecture is divided into four major sections. Section 2 revisits the fundamental description of ferroelectrics in a thermodynamic energetic approach. The reader who is well familiar with the thermodynamic description of a ferroelectric may skip this section. Only Sect. 2.4 should be read by everyone, as it is an effect which is very relevant but not so often discussed particularly in the mechanics society. Section 3 very briefly introduces the basic understanding of any semiconductor. Section 4 then combines both approaches to yield a joint energetic description of the semiconducting ferroelectric. Section 5 opens the perspective to the many different contexts where semiconductor properties are encountered in ferroelectrics. The manuscript will end in outlining challenges to young students in the field.

This article intends to draw the attention to the multiple facets that influence the behavior of functional ceramics. It can merely outline the different aspects that one may encounter in real world materials. Particularly, the many sources of heterogeneity in realistic ceramics generate a challenge in material description. Thin film devices are typically more homogenous, but hetero-interfaces to substrates and electrodes play a much larger role than in a ceramic.

At this moment we want to draw the reader's attention to a number of excellent textbooks that deal with ceramics and their electronic properties: Stoneham (1975), Kingery et al. (1976), Moulson and Herbert (1989), Hench and West (1990), Chiang et al. (1997). On the side of textbooks for ferroelectrics a long list is available with different foci: Jona and Shirane (1993), Burfoot (1967), Jaffe et al. (1971), Sonin and Strukow (1974), Mitsui et al. (1976), Lines and Glass (1977), Yuhuan (1991), Uchino (1997), Strukov and Levanyuk (1998), Scott (2000), Wadhawan (2001) and Hong (2004). Furthermore, two monographs concerning semiconductor effects in ferroelectrics have seen the literature arena: Fridkin (1980), Fridkin (1979). Another standard reference is the collection of major papers on piezoelectricity by Rosen et al. (1992), which is a constant source of improved understanding. For those who have found pleasure in the approach by Lines and Glass (1977) an in depth study of the book by Smolenskii et al. (1984) will lead the reader into a deep understanding on how to design thermodynamic potentials in the different crystal classes, a microscopic description of ferroelectricity, as well as order-disorder phenomena. Special attention on domain structures and switching dynamics is found in the excellent textbook by Tagantsev et al. (2010). Tensorial properties of ferroelectrics and many other phenomena in solids are surveyed in the beautiful collection by Newnham (2005). For many other oxides electronic transport is highly relevant. Tsuda (1991) provided a comprehensive overview on the topic. If you find access to it in your library, the book by Bunget and Popescu (1984) is a smooth read for approaching the many facets of dielectrics. Piezoelectricity and its use as ultrasonic and resonating device is well explained in the book by Ikeda (1990). Also a good read for the overall understanding of functional materials in electronic applications is the excellent textbook by Waser (2012) and his many co-authors therein. The physics of semiconductors is well described in several textbooks. The Blakemore (1962) has become my personal favorite.

2 Thermodynamics of a Ferroelectric

The description of material properties must obey two major principles: It must follow the laws of thermodynamics and it must match the symmetry requirements imposed by the crystal structure. Thus, talking about mostly crystalline systems in the context of ferroelectrics—we here exclude ferroelectric liquid crystals as well as ferroelectric polymers for simplicity—tensorial properties must be mapped onto a mathematical formalism and certain ways must be found to describe energy. Both have long been established, but nevertheless, they are so vital to material understanding that they still dominate daily discussions on material behavior in research as well as application. This introduction tries to outline the major lines of thought. Nonetheless, it cannot replace full textbooks on the issue.

For tensor properties there is the eternal monograph by Nye (1985) which is irreplaceable in understanding tensorial properties and their relation to thermodynamics in solids. For a thorough introduction to thermodynamics, we refer the reader to Callen (1985), Anderson (2012), Stowe (2014).

2.1 Material Properties, Tensors, and Summation Rules

Some properties of materials such as temperature or density are independent of direction. Mathematically these non-directional properties are scalars. They can also be called zero rank tensors, because they are represented by a single point or a single number. Other physical quantities are directional such as electric field, mechanical force, or magnetic field. These quantities are vectors. In three dimensions vectors are represented by the vector components which are projections of the vector onto the three axes of a coordinate system of choice. Taking the electric field vector as an example, Nye (1985):

$$\boldsymbol{E} = [E_1, E_2, E_3], \tag{1}$$

is representative of a force onto a charge. This vector is a first rank tensor and can be specified by three numbers. Bold face letters will represent vectors in the remainder of this text, while indexed components of a vector will remain italic in order to display their scalar character, each as a single entry in a vector. An insulating material will respond to the electric field by building up a polarization $\boldsymbol{P}$ also represented by a first rank tensor

$$\boldsymbol{P} = [P_1, P_2, P_3]. \tag{2}$$

the dielectric displacement unites material polarization and the polarization of free space $\epsilon_0 \boldsymbol{E}$. As material response to the external fields is not necessarily collinear, this directional variation must be taken into account leading to a second rank tensor representation of dielectric constant $\boldsymbol{D} = \epsilon_0 \boldsymbol{E} + \boldsymbol{P}$. It contains nine numbers explicitly given by:

$$\begin{aligned} D_1 &= \epsilon_{11} E_1 + \epsilon_{12} E_2 + \epsilon_{13} E_3 \\ D_2 &= \epsilon_{21} E_1 + \epsilon_{22} E_2 + \epsilon_{23} E_3 \\ D_3 &= \epsilon_{31} E_1 + \epsilon_{32} E_2 + \epsilon_{33} E_3 \end{aligned} \tag{3}$$

each entry $\epsilon_{11}, \epsilon_{12}, \epsilon_{13}, \epsilon_{21}, \epsilon_{22}, \epsilon_{23}, \epsilon_{31}, \epsilon_{32}, \epsilon_{33}$ reflects a mapping of a field component onto an electric displacement component, each given in coordinates of the same coordinate system. This is true, if the polarization itself is linearly dependent on electric field: $P_i = \chi_{ij} E_j$. Thus, nine components of permittivity specify the permittivity of a certain crystal and can be written in matrix form as follows.

$$\begin{bmatrix} \epsilon_{11} & \epsilon_{12} & \epsilon_{13} \\ \epsilon_{21} & \epsilon_{22} & \epsilon_{23} \\ \epsilon_{31} & \epsilon_{32} & \epsilon_{33} \end{bmatrix}$$

Equation (3) can also be written in the following summation form:

$$\begin{aligned} D_1 &= \sum_{j=1}^{3} \epsilon_{1j} E_j \\ D_2 &= \sum_{j=1}^{3} \epsilon_{2j} E_j \\ D_3 &= \sum_{j=1}^{3} \epsilon_{3j} E_j \end{aligned} \tag{4}$$

This summations can also be written as:

$$D_i = \sum_{j=1}^{3} \epsilon_{ij} E_j \quad (i = 1, 2, 3) \tag{5}$$

and by replacing the sum symbol by the Einstein summation rule one obtains:

$$D_i = \epsilon_{ij} E_j \quad (i, j = 1, 2, 3) \tag{6}$$

Einstein's summation rule states that any index appearing in more than one of the variables is being summed over, so Eq. (6) implicitly contains *one* sum sign, summing over j.

Even for the same symmetry of the crystal/system, nonlinearity will require further higher rank tensorial coefficients. In the same fashion as Eq. (3), tensors of higher rank are constructed. The excellent introduction by Nye (1985) will easily lead the reader through the basics how to do so.

Hysteresis on the other hand, is much harder to be captured Wadhawan (2001). It presently is a major field of research. Simple uniaxial forms will be introduced in Sect. 2.3. A comprehensive study is found in the seminal works by Devonshire (1949), Devonshire (1951), Devonshire (1954).

2.2 *The Thermodynamic Energy Approach*

Thermodynamics deals with energy. This energy can have many forms. These forms can be converted into each other, as long as the disorder, namely *entropy* (S) is reversibly exchanged. This fact has been discussed in detail in Sect. 3.2 of the chapter by Schröder (2016) included in this course volume. If entropy of the entire system increases, the processes become irreversible. In this context minimization of a thermodynamic potential allows to find the stable configuration under the imposed boundary conditions. Thus, the term $-TS$ will permit to search for a minimum of a thermodynamic energy potential, T being temperature.

At this point, we must briefly revisit the reasons why a certain thermodynamic energy potential is chosen. Fundamentally, the internal energy U^{tot} is all energy contained in a body. A change in internal energy dU^{tot} can occur by external work δW^{tot} performed on or heat exchange δQ^{tot} with the body under consideration:

$$dU^{tot} = \delta Q^{tot} + \delta W^{tot}. \tag{7}$$

We here exclude chemical energy which would unnecessarily complicate the issue. U^{tot} is the total energy of the system/body.

It is typical not to use the total energy, because the boundaries of a system/body must then be well defined. More commonly when one considers solids, normalized quantities are used. In the following, we will use normalized quantities throughout the text in order to facilitate the reading. Chemical literature normalizes with respect to the amount of chemical species, namely per Mole. Solid state physics and mechanics literature normalize per volume. This will be our common understanding for the remainder of the text unless explicitly stated otherwise:

$$U^{tot} = \int_V U \cdot dV \tag{8}$$

$$S^{tot} = \int_V S \cdot dV \tag{9}$$

$$H^{tot} = \int_V H \cdot dV \tag{10}$$

Thus, these three capital letters in the following are considered as normalized quantities namely *volume densities*. Their units are always per meter cubed (or equivalent).

For *solids* the change of internal energy density is given as:

$$dU = TdS + X_{ij}dx_{ij} + E_i dD_i + H_i dB_i. \tag{11}$$

already accounting for the tensorial components of each of the three contributions to work. As above, Einsteins summation rule applies. For solids the relevant variables are: stress X_{ij}, strain x_{ij}, electric field E_i, electric displacement D_i, magnetic field H_i, and magnetic flux density B_i. In order to avoid confusion with the dielectric constant ϵ_{ij} and the surface charge density σ, we will use the standard form used in ferroelectric literature and not the one used in mechanics for strain: $x_{ij} \equiv \epsilon_{ij}$ and $X_{ij} \equiv \sigma_{ij}$. Note that all contributions to work are normalized per volume by nature of the involved fields, which can be confirmed by a simple unit check.

As a note: for gases and liquids a shrinking volume increases internal energy, and thus a minus sign must be accounted for: $dU^{tot} = TdS^{tot} - pdV^{tot}$. This is the typical form used in the context of chemical reactions.

Due to the second law of thermodynamics, there is not an arbitrary choice of how much external work can be performed by the thermodynamic system. Entropy of the entire system (including the heat bath) can only increase, or in an idealized situation stay constant: $dS^{tot} \geq 0$. At the same time, the total internal energy is a constant $U^{tot} = const$. Thus, certain amounts of heat can be generated without providing performable work to the outside.

Now "free" quantities can be defined: the free energy F per volume or Gibbs' free enthalpy G per volume (the latter is also called Gibbs' energy). Both tend towards a minimum during the course of a change of state due to the fact that the term "$-T \cdot S$" is subtracted. A reasonable definition of the free energy is thus:

$$F = U - TS. \tag{12}$$

It is this fraction of internal energy, that can still be converted into useable work. $A \equiv F$ are both equivalently used in literature to denote free energy.

The (Helmholtz) enthalpy H^{tot} is used where the experimental boundary conditions limit the performable work. For *gases* or *liquids* H^{tot} is given by

$$H^{tot} = U^{tot} + p \cdot V^{tot}. \tag{13}$$

The plus sign accounts for the fact that a shrinkage of volume in a gas or liquid is representative of external work performed on the system and thus an energy gain inside (U^{tot} increases). The variation in H^{tot}, namely dH^{tot}, is then $dH^{tot} = TdS^{tot} + V^{tot}dp$. For typical *chemical* experiments, pressure is constant and the change in enthalpy is directly given by the exchanged heat $dH^{tot} = TdS^{tot} = \delta Q^{tot}$. So chemical binding energies are in general tabulated as chemical *binding enthalpy* of the chemical bond.

In *solids* several contributions to work must be considered: mechanical $dW_m = X_{ij}dx_{ij}$, electrical $dW_e = E_i dD_i$, and magnetic $dW_B = H_i dB_i$. Each of these three terms is already normalized per volume.

If experiment prescribes boundary conditions such that *NO* work is performed at all, a thermodynamic potential must be chosen that excludes variations in strain dx, electric displacement dD, and magnetic flux dB. In order to obtain this potential, a Legendre transform of the form $H = U - x_{ij}X_{ij} - D_iE_i - B_iH_i$ must be used.

Now in reality, certain experiments will not prescribe intensive variables (thermodynamic forces) alone, but a mix of boundary conditions may occur. We will later see that these mixed forms are particularly useful, when non-linearity has to be put into a formal form and explicit material functions are sought. The following set of equations summarizes all forms of energy potentials that could be used. We here added the magnetic parts to the story exceeding the form given in Lines and Glass (1977). We intentionally use the superscripts (not used so far in literature) in order to clarify that these quantities are scalars (energies) by nature. The notation commonly used for ferroelectrics is somewhat misleading for the newcomer. So in our notation G^1 is used while G_1 is the common notation in ferroelectric literature, Lines and Glass (1977), ($G_1 \equiv G^1$, $G_2 \equiv G^2$, $H_1 \equiv H^1$, $H_2 \equiv H^2$). A comprehensive list including all tensor indices is thus:

$$
\begin{aligned}
F &= U - TS && (14)\\
H &= U \quad - x_{ij}X_{ij} - D_iE_i - B_iH_i && (15)\\
H^1 &= U \quad - x_{ij}X_{ij} && (16)\\
H^2 &= U \quad - D_iE_i && (17)\\
H^3 &= U \quad - B_iH_i && (18)\\
G^1 &= U - TS - x_{ij}X_{ij} && (19)\\
G^2 &= U - TS \quad - D_iE_i && (20)\\
G^3 &= U - TS \quad - B_iH_i && (21)\\
G &= U - TS - x_{ij}X_{ij} - D_iE_i - B_iH_i, && (22)
\end{aligned}
$$

where the quantities H^3 and G^3 are newly introduced here. Again Einstein's summation rule over all tensorial indices applies. For clarity to the newcomer, Eq. (22) contains a total of seventeen summands. In order to account for all possible experimental boundary conditions, one could also use yet another set of variables, which are given by

$$H^{12} = U \qquad - x_{ij}X_{ij} - D_iE_i \tag{23}$$
$$H^{13} = U \qquad - x_{ij}X_{ij} \qquad - B_iH_i \tag{24}$$
$$H^{23} = U \qquad - D_iE_i - B_iH_i \tag{25}$$
$$G^{12} = U - TS - x_{ij}X_{ij} - D_iE_i \tag{26}$$
$$G^{13} = U - TS - x_{ij}X_{ij} \qquad - B_iH_i \tag{27}$$
$$G^{23} = U - TS \qquad - D_iE_i - B_iH_i \tag{28}$$

each containing different sets of contributions from mechanical, electric and magnetic fields. In each line of Eqs. (14) through (28) a thermodynamic potential is defined with a unique set of eigenvariables.

In the following, Gibbs' free energy $G(T, \bar{\bar{X}}, \mathbf{E}, \mathbf{H})$ is used to derive the relations between the different field variables in equilibrium, because experiment can typically prescribe temperature, elastic stress, and electric and magnetic fields. It has eigenvariables T, $\bar{\bar{X}}$, **E**, and **H**. G follows from internal energy by subtracting all product terms encountered so far also turning it into a *free* form. Thus in absolute terms:

$$G = U - TS - X_{ij}x_{ij} - E_iD_i - H_iB_i, \tag{29}$$

where X_{ij} is stress, x_{ij} strain, E_i electric field, P_i polarization, H_i magnetic field, B_i magnetic flux density, T temperature, and S entropy. Again, Einstein's summation rule applies, i=1,2,3; j=1,2,3. G is the normalized quantity, here the free enthalpy *density* per volume.

For chemical systems one can arrive at the chemical potential via $\int GdV = G^{tot}$. Then, μ is the chemical potential $\mu = \partial G^{tot}/\partial N$ with N being the number of particles in the respective system. Thus, the chemical potential μ normalizes G^{tot} with respect to particle number and not with respect to volume. In thermal equilibrium, μ is constant across the thermodynamic system like pressure or temperature. It is highly relevant for chemical considerations. For a single constituent, $G^{tot} = \mu \cdot N$ and one could change between both normalizations using the particle density ρ^*, $\rho^* = \rho/M$ with M being the mass of the atom or molecule and ρ the standard mass density. In the context of chemical reactions or alloy formation, several chemical potentials μ_k must be considered for each constituent and Eq. (29) is then valid for each μ_k as well as for the sum of all individual chemical potentials $\mu N = \sum \mu_k N_k$, with k being the number of constituents and N_k the number of particles of type k. It directly follows that chemical potentials are interrelated through concentrations $c_k = N_k/N$: $\mu = \sum \mu_k c_k$.

In solids, one typically does not have to consider these changes, because solids tend to react very slowly or at very high temperature only. At high temperatures, where chemical issues must be discussed, magnetic and electric ordering typically vanish and the field terms in Eq. (29) become obsolete. Chemical reactions will thus be neglected for now and "low temperature" effects will be discussed. Consideration of particle numbers will re-appear when free charge carriers are discussed later in the semiconductor Sects. 3 and 4.

The variation of G, dG, can be calculated from Eq. (29) by applying the mathematical product rule to each of the product terms. Inserting the variation of internal energy dU from Eq. (11) one arrives at:

$$dG = -SdT - x_{ij}dX_{ij} - D_i dE_i - B_i dH_i \tag{30}$$

We again see that Gibbs' potential is a function of T, $\bar{\bar{X}}$, $\vec{E}$, and $\vec{H}$. On the other hand, because Gibbs' function is a potential, it can mathematically be treated as a total differential. For a total differential the partial derivatives of the independent variables are then given by:

$$\begin{aligned} dG = &\left(\frac{\partial G}{\partial T}\right)_{\bar{\bar{X}},\vec{E},\vec{H}} dT + \left(\frac{\partial G}{\partial X_{ij}}\right)_{T,X_{k\neq i,l\neq j},\vec{E},\vec{H}} dX_{ij} + \\ &\left(\frac{\partial G}{\partial E_i}\right)_{T,\bar{\bar{X}},E_{k\neq i},\vec{H}} dE_i + \left(\frac{\partial G}{\partial H_i}\right)_{T,\bar{\bar{X}},\vec{E},H_{k\neq i}} dH_i \end{aligned} \tag{31}$$

The subscripts denote those variables (in all their index components), that must be kept constant. It is now easy to compare Eqs. (30) and (31). We see that the complementary variables are first partial derivatives of G:

$$-S = \left(\frac{\partial G}{\partial T}\right)_{\bar{\bar{X}},\vec{E},\vec{H}} \tag{32}$$

$$-x_{ij} = \left(\frac{\partial G}{\partial X_{ij}}\right)_{T,X_{k,l;\ k\neq i \text{ or } l\neq j},\vec{E},\vec{H}} \tag{33}$$

$$-D_i = \left(\frac{\partial G}{\partial E_i}\right)_{T,\bar{\bar{X}},E_{j\neq i},\vec{H}} \tag{34}$$

$$-B_i = \left(\frac{\partial G}{\partial H_i}\right)_{T,\bar{\bar{X}},\vec{E},H_{j\neq i}} \tag{35}$$

In order to render the concept of partial derivatives valid in this context, also all other components of e.g. stress must be kept constant if *one* of them is varied. This is explicitly denoted in the subscripts (and often leisurely omitted in literature).

At this point it is valuable to look at an example where it becomes immediately clear, why we have to be so careful about the details in boundary conditions. Effectively, Line (33) contains nine explicit equations. For each, eight other stress variables must be kept constant in order to be able to define the partial derivatives. This in mind, it is actually very difficult to assure Eqs. (15), (16), (19), (22)-(24), (26), (27) experimentally. Let us first consider a solid in hydrostatic compressive environment which is easily achieved experimentally in a pressurized liquid. Under these circumstances everything is clear and G is a valid potential. Let us now consider a thin film on a very rigid substrate. The substrate suppresses (or prescribes) all in-plane strains (e.g. $x - y-$plane). The stresses in the plane are generated through the elastic properties of the film itself and are an indirect consequence of the boundary conditions. The perpendicular direction z is free to displace and thus constant stress is prescribed. Strain is here the dependent variable. For that matter, one would actually have to define yet another potential $G^{thin\ film}$ that would explicitly use these boundary conditions. Fibres and other geometries will yield their own set of explicit boundary settings. In the context of this scenario, the set of equations in Line (33) would have to be separated into some equations yielding strain and other yielding stress. Thus, many more potentials can be thought of. In the following we will not make use of these details but go back to the general approach.

In order to link the fields by physical *material coefficients*, the second derivatives of the thermodynamic potentials must be used. One set of constants is derived from Gibbs' potential. These coefficients are given by:

$$-\left(\frac{\partial^2 G}{\partial X_{kl}\partial X_{ij}}\right)_{T,\vec{E},\vec{H}} = \left(\frac{\partial x_{ij}}{\partial X_{kl}}\right)_{T,\vec{E},\vec{H}} = c_{ijkl}^{T,\vec{E},\vec{H}} \quad (Elasticity) \tag{36}$$

$$-\left(\frac{\partial^2 G}{\partial E_j \partial E_k}\right)_{T,\bar{\bar{X}},\vec{H}} = \left(\frac{\partial P_k}{\partial E_j}\right)_{T,\bar{\bar{X}},\vec{H}} = \epsilon_{kj}^{T,\bar{\bar{X}},\vec{H}} \quad (Permittivity) \tag{37}$$

$$-\left(\frac{\partial^2 G}{\partial H_i \partial H_j}\right)_{T,\bar{\bar{X}},\vec{E}} = \mu_0 \left(\frac{\partial M_j}{\partial H_i}\right)_{T,\bar{\bar{X}},\vec{E}} = \chi_{ji}^{T,\bar{\bar{X}},\vec{E}} \quad (Susceptibility) \tag{38}$$

$$-\left(\frac{\partial^2 G}{\partial X_{ij}\partial E_k}\right)_{T,\vec{H}} = \left(\frac{\partial D_k}{\partial X_{ij}}\right)_{T,\vec{E},\vec{H}} = \left(\frac{\partial x_{ij}}{\partial E_k}\right)_{T,\bar{\bar{X}},\vec{H}} = d_{kij} \quad (Piezoelectric) \tag{39}$$

$$-\left(\frac{\partial^2 G}{\partial X_{ij}\partial H_l}\right)_{T,\vec{E}} = \left(\frac{\partial M_l}{\partial X_{ij}}\right)_{T,\vec{E},\vec{H}} = \left(\frac{\partial x_{ij}}{\partial H_l}\right)_{T,\bar{\bar{X}},\vec{E}} = q_{lij} \quad (Piezomagnetic) \tag{40}$$

$$-\left(\frac{\partial^2 G}{\partial E_k \partial H_l}\right)_{T,\vec{E},\vec{H}} = \left(\frac{\partial M_l}{\partial E_k}\right)_{T,\bar{\bar{X}},\vec{E}} = \alpha_{lk} \quad (Magnetoelectric) \tag{41}$$

$$-\left(\frac{\partial^2 G}{\partial X_{ij}\partial T}\right)_{T,\vec{E},\vec{H}} = \left(\frac{\partial S}{\partial X_{ij}}\right)_{T,\bar{\bar{X}},\vec{E}} = \alpha_{(\lambda)ij} \quad (Piezocaloric) \tag{42}$$

$$-\left(\frac{\partial^2 G}{\partial T \partial E_k}\right)_{T,\vec{E},\vec{H}} = \left(\frac{\partial P_k}{\partial T}\right)_{T,\bar{\bar{X}},\vec{E}} = p_k \quad (Pyroelectric) \tag{43}$$

$$-\left(\frac{\partial^2 G}{\partial T \partial H_l}\right)_{T,\vec{E},\vec{H}} = \left(\frac{\partial M_l}{\partial T}\right)_{T,\bar{\bar{X}},\vec{E}} \qquad = \quad \gamma_l \tag{44}$$

$$(Pyromagnetic)$$

where it is understood that in Eqs. (36), (39), (40) and (42) again all other stress components are kept constant: $X_{k \neq i, l \neq j}$ =const. In a similar fashion, also free energy F, enthalpy H, or internal energy U each yield a set of equations of similar nature. Each set is given for a different set of boundary conditions. For G, all first partial derivatives are derivatives with respect to thermodynamic forces. For U this is exactly the other way around, namely, U is differentiated with respect to matter response. For all other cases the partials mix. Again in the book of Nye (1985), the differences in *value* between the material *coefficients* determined from either U, F, H, or G are compared. For most material properties the differences between the coefficients derived from the different potentials are small and often literature does not differentiate between them. Formally, it always has to be specified, under which experimental boundary conditions the relevant parameters are measured thus determining from which energy potential it must be calculated in a rigorous fashion. Certain differences may actually become sizable, particularly the caloric variables Nye (1985).

Table 1 shows some physical properties, their tensor ranks, and physical parameter definitions.

The thermodynamic relations can be visualized in the Heckmann diagram Heckmann (1924). Its triangular form is depicted in Nye (1985) and reused in Spaldin and Fiebig (2005). In Fig. 1 we use an extended version. The trapezoidal prism represents the relations between the thermodynamic fields (prism base) and the material responses (prism top).

All potential *order parameters* of a material/system reside within the top plane. Order parameters are those thermodynamic variables in which matter can order upon a temperature dependent phase transition without application of an external force/field. Volume integrals of the material response represent extensive variables of the material, namely, total magnetic moment, total dipole moment, and total displacement of the body. For a detailed approach to ferroelectrics in the context of order parameters see Strukov and Levanyuk (1998). The links between any of the different quantities are the material *coupling coefficients*. They are here given in linear form. The same form is true for any *linearized* state of the system. So also highly non-linear properties yield a set of tangents which represent the linearized material coupling coefficients for certain finite values of the fields. The partial derivatives are the slopes of the tangents. Mathematically, the tangents define a tangent space for this particular state (i.e. a set of given values of e.g. T^0, X^0_{ij}, E^0_i, and H^0_i) providing a tangent space to this point in thermodynamic variable space. The thermodynamic variable space is *not* the physical 3D space, but a mathematical space of variables.

Table 1 Physical properties and coefficients in tensor notation. We explicitly denote the changes in variables by Δ in order to make apparent the differential nature of the coefficients. All coefficients are function of other variables and subject to change for changes in any of the thermodynamic potentials. All coefficients in equations containing "Δ-s" are tangent slopes

Tensor	Rank	Meaning	Equation	SI unit
C_p	Zero	Heat capacity	$\Delta S = \frac{C_p}{T}\Delta T$	J/K
p_i	1st	Pyroelectric effect	$\Delta P = p_i \Delta T$	$\mathrm{Cm^{-2}K^{-1}}$
p_i^S	1st	Electrocaloric effect	$\Delta S = p_i^S \Delta E_i$	$\mathrm{VKJ^{-1}m^{-1}}$
t_i	1st	Electrothermal effect	$\Delta E_i = t_i \Delta T$	$\mathrm{V\,K^{-1}m^{-1}}$
$S_{(H)i}$	1st	Magnetocaloric effect	$\Delta S = S_{(H)i} \Delta H_i$	$\mathrm{JK^{-1}A^{-1}m^{-2}}$
γ_i	1st	Pyromagnetic effect	$\Delta M_i = \gamma_i \Delta T$	$\mathrm{Am^2kg^{-1}K^{-1}}$
ϵ_{ij}	2nd	Permittivity	$\Delta D_i = \epsilon_{ij} \Delta E_j$	$\mathrm{Fm^{-1}}$
f_{ij}	2nd	Heat of deformation	$\Delta X_{ij} = f_{ij} \Delta T$	$\mathrm{Nm^2K^{-1}}$
$\chi_{(M)ij}$	2nd	Susceptibility	$\Delta M_i = \chi_{(M)ij} \Delta H_j$	$\mathrm{m^3kg^{-1}}$
α_{ij}	2nd	Magnetoelectric effect	$\Delta P_i = \alpha_{ij} \Delta H_j$	$\mathrm{Vm^{-1}Oe^{-1}}$
$\alpha_{(\lambda)ij}$	2nd	Thermal expansion	$\Delta x_{ij} = \alpha_{(\lambda)ij} \Delta T$	$\mathrm{K^{-1}}$
d_{ijk}	3rd	Piezoelectric coefficient	$\Delta x_{jk} = d_{ijk} \Delta E_i$	$\mathrm{mV^{-1}}$
q_{ijk}	3rd	Piezomagnetic coefficient	$\Delta M_i = q_{ijk} \Delta X_{jk}$	$\mathrm{Am^4N^{-1}kg^{-1}}$
c_{ijkl}	4th	Elasticity	$\Delta X_{ij} = c_{ijkl} \Delta x_{kl}$	$\mathrm{Nm^{-2}}$
Q_{ijkl}	4th	Electrostriction	$x_{ij} = Q_{ijkl} P_k P_l$	$\mathrm{m^2C^{-2}}$
N_{ijkl}	4th	Magnetostriction	$x_{ij} = N_{ijkl} M_k M_l$	$\mathrm{A^{-2}m^{-4}kg^2}$

2.3 *The Landau-Devonshire Polynomial Approximation*

The differential form of constitutive relations described in the previous section represents the linear part of a phenomenon or its linearization. Particularly in systems with *ordering*, the key aspects are non-linear and need to be addressed by considering an extension of the linear approach. In ferroelectrics polarization hysteresis loops and electrostriction are parameters that are highly non-linear. In order to stick to the historical development, we will first consider a description of only the electric behavior of a ferroelectric in the vicinity of a phase transition. It was L.D. Landau who designed a very simple but enormously powerful theory for phase transitions in general, Landau (1937, 1965). His assumption was that across a phase transition the thermodynamic energy function should be expandable as a Taylor series. Initially considered as a simple approximation, it turned out that particularly for ferroelectrics this expansion maintains to be valid for very large temperature differences from the phase transition temperature (up to 100 K or more in certain cases).

Devonshire used this approach to then derive a general formalism to describe ferroelectrics, Devonshire (1949, 1951, 1954). He considered a prototype state where initially the cross couplings to other than electric parameters (e.g. elastic, thermal, magnetic etc.) are neglected. The electric order parameter in concern is the electric displacement $D_i = 0$ in the equilibrium state. Near equilibrium it only changes by

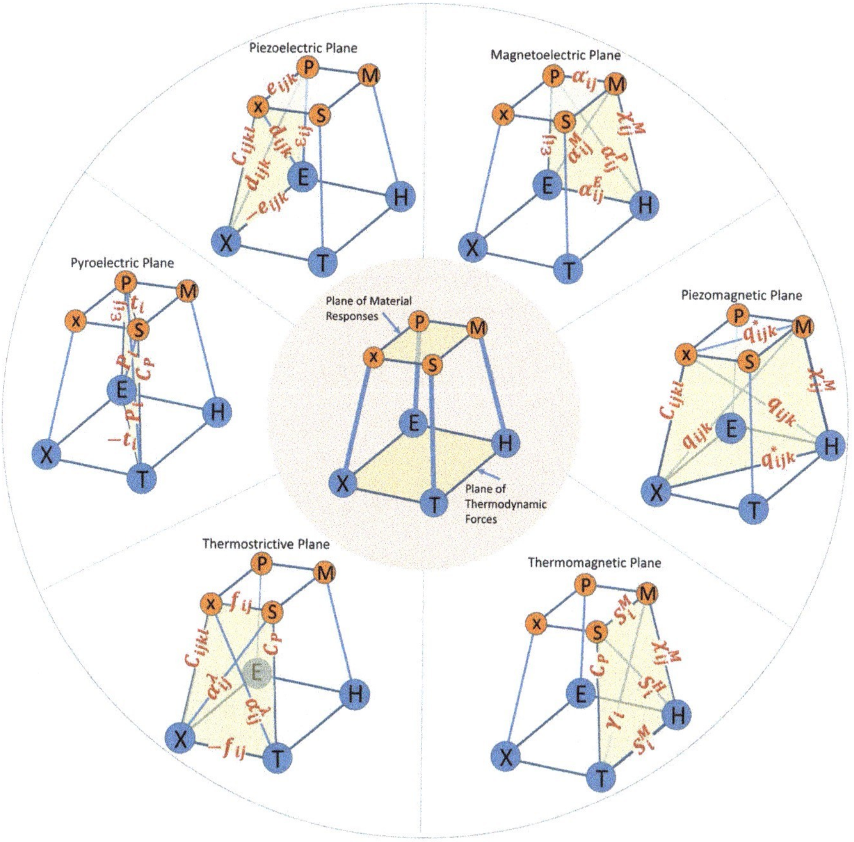

Fig. 1 The relationships between magnetic, thermal, electrical, and mechanical properties of crystals (© M. Etier)

a small value dD_i. The first assumption is based on the premise that any anomalies caused in the ferroelectric behavior due to a cross coupling with any non-electric parameters can in principle be neglected. The second assumption actually reflects the fact that we are considering a case very close to a dielectric-ferroelectric transition, since in a dielectric state the displacement D_i is zero (in absence of an externally applied electric field; $E = 0$, $D = 0$).

Let us now briefly consider, which of the potentials (Eqs. (14) through (28)) should be chosen. We want to explicitly use D. Thus, the potential should have D as a variable. But the remaining variables should be easily accessible to experiment, so we want T, X, and H as other variables. We neglect magnetism, so only the transforms G^1, Eq. (19), G^2, Eq. (20), or G, Eq. (22) apply. As we want D to be an explicit variable for the function, only G^1 is permissible (G^2 and G both depend on E). For the prototype phase the Taylor expansion of Gibbs' free energy function by considering D as the only independent parameter can thus be put forward as

$$G^1 = G_0 + \chi_i^{\sigma,T} D_i + k_{ij}^{\sigma,T} D_i D_j + \xi_{ijk}^{\sigma,T} D_i D_j D_k + \psi_{ijkl}^{\sigma,T} D_i D_j D_k D_l \ldots\ldots \quad (45)$$

where $\chi_i^{\sigma,T}$, $k_{ij}^{\sigma,T}$, $\xi_{ijk}^{\sigma,T}$, and $\psi_{ijkl}^{\sigma,T}$ are coefficient tensors of rank 1, 2, 3, and 4, respectively. It can be easily imagined that with increasing order of the polynomial, the difficulty in determining symmetry restrictions on the coefficient tensors also becomes more complicated. Hence physical inputs are needed to reduce these to a convenient approximation. For example, since the approximation is supposed to be valid for both the polar (below T_c) and non-polar (above T_c) phases, we consider only the even order terms such that for a polar phase a reversal of electric field would produce an equivalent polarization state. In that case, Eq. (45) reduces to

$$G^1 = G_0^1 + k_{ij}^{\sigma,T} D_i D_j + \xi_{ijk}^{\sigma,T} D_i D_j D_k + \psi_{ijkl}^{\sigma,T} D_i D_j D_k D_l \quad (46)$$

$$E_i - E_0 = \left(\frac{\partial G^1}{\partial D_i}\right)_{X,T} \quad (47)$$

$$= k_{ij}^{\sigma,T} D_j + \xi_{ijk}^{\sigma,T} D_j D_k + \psi_{ijkl}^{\sigma,T} D_j D_k D_l \quad (48)$$

Here it can be inferred that in the absence of spontaneous polarization the coefficient $k_{ij}^{\sigma,T}$ is the reciprocal dielectric permittivity $k_{ij}^{\sigma,T} = \left[\epsilon_{ij}^{\sigma,T}\right]^{-1}$.

From here onwards it is simpler to evaluate the electric response of a ferroelectric as a function of temperature by assuming that the displacement is directed along *one* of the crystallographic axes only ($D_i = D$) and that the non-polar phase is centrosymmetric. Eq. (46) can then be formulated in scalar form

$$G^1 = (\alpha/2)D^2 + (\gamma/4)D^4 + (\delta/6)D^6 \quad (49)$$

For a first order transition, which is often the case among the available ferroelectric materials, the constants γ, and δ are considered temperature independent with γ being negative and δ being positive. The dielectric equation of state for such a system derived by differentiating Gibbs' function w.r.t. D is:

$$E = \frac{\partial G^1}{\partial D} = \alpha D + \gamma D^3 + \delta D^5. \quad (50)$$

Here E is parallel to D. Now the basic assumption of the conventional phenomenological theory Devonshire (1949, 1951, 1954) is that α is nothing but the reciprocal isothermal permittivity (48) that follows Curie-Weiss' law close to the Curie temperature

$$\alpha = k_{ij}^{\sigma,T} = \beta(T - T_0). \quad (51)$$

β is a positive constant and T_0 is the Curie point which is a temperature not equal to the actual transition temperature T_c as we will see shortly. Substituting $\gamma' = -\gamma$ such that β, γ', and δ are positive constants, Eqs. (51) and (49) yield

$$G^1 = (\beta/2)(T - T_0)D^2 - (\gamma'/4)D^4 + (\delta/6)D^6 \quad (52)$$
$$E = \beta(T - T_0)D - \gamma' D^3 + \delta D^5 \quad (53)$$

The curves for G versus D at different temperatures are qualitatively shown in Fig. 2a. As can be seen from Fig. 2, a first order transition takes place at temperature $T = T_c$ since G as well as the first derivative of G w.r.t. D become zero leading to the following set of equations:

$$(\beta/2)(T - T_0) - (\gamma'/4)P_S^2 + (\delta/6)P_S^4 = 0 \quad (54)$$
$$\beta(T - T_0) - \gamma' P_S^2 + \delta P_S^4 = 0 \quad (55)$$

In the absence of an external electric field the displacement D equals the spontaneous polarization P_S that appears at two of the free energy minima. The solution for the set of Eqs. (54) and (55) comes out to be in the form of the actual transition temperature

$$T = T_c = T_0 + (3\gamma'^2/16\beta\delta) \quad (56)$$

Thus, T_c is *not* equal to T_0. Substituting this back into Eq. (54) the spontaneous polarization can be expressed as

$$P_S^2 = 3\gamma'/4\delta \ (T = T_c) \quad (57)$$

The reciprocal permittivity in the absence of any electrically induced displacement above and below the transition temperature can now be written as:

$$k_{ij}^{\sigma,T} = \beta(T - T_0) - 3\gamma' P_S^2 + 5\delta P_S^4 \ (T < T_c) \quad (58)$$
$$= \beta(T - T_0) \quad (T > T_c) \quad (59)$$

The spontaneous polarization thus contributes to the reciprocal permittivity below T_c. Solving Eq. (53) for spontaneous polarization ($D = P_S$; $E = 0$) and substituting in Eq. (58) yields

$$k_{ij}^{\sigma,T} = \frac{3\gamma'^2}{16\delta} + \beta(T - T_c) \quad (60)$$

with $T_0 = T_c - 3\gamma'^2/16\beta\delta$.

Approaching T_c from the paraelectric side Eq. (59) can be written as

$$k_{ij}^{\sigma,T} = \frac{3\gamma'^2}{16\delta} + \beta(T - T_c) \quad (61)$$

It can be inferred that the slope of the permittivity curve above and below T_c has a ratio of 8 and that the reciprocal permittivity undergoes a discontinuity at $T = T_c$. This can be seen in Fig. 1c along with the discontinuity in polarization P_S (Fig. 2b). Such discontinuities are in the first place experimentally observed. As shown in Fig. 3

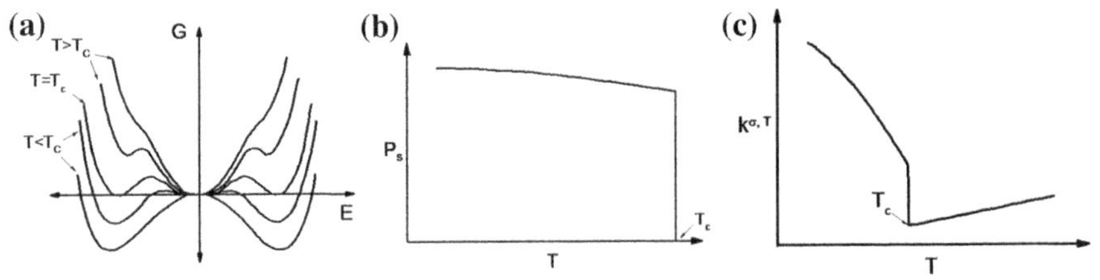

Fig. 2 Qualitative dependence of free energy versus electric displacement at different temperature (**a**), along with the temperature dependence of spontaneous polarization P_S (**b**) and the reciprocal permittivity k_σ^T (*c*), Lines and Glass (1977)

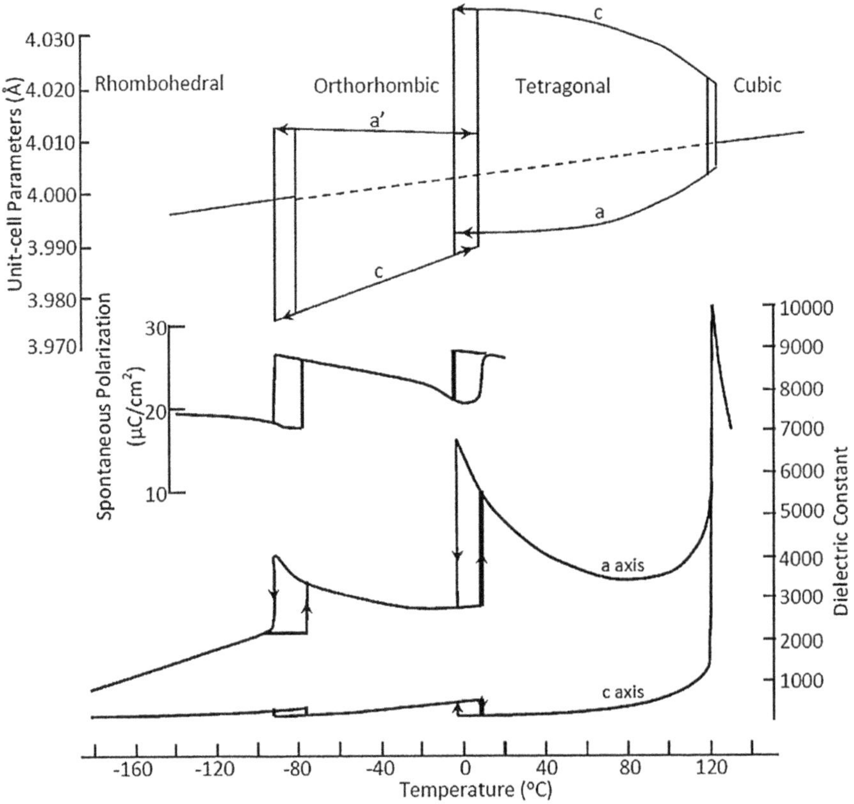

Fig. 3 Polymorphic phase transitions in barium titanate ($BaTiO_3$) single crystals observed through changes in the unit cell parameters, spontaneous polarization, and dielectric constant, Aksel and Jones (2010)

for a $BaTiO_3$ ferroelectric crystal the dielectric permittivity seems to be showing a highly positive slope while approaching T_c from lower temperatures whereas a smaller negative slope while moving away towards higher temperature. In experiment it is also found that the transition temperature marked by the discontinuity in dielectric permittivity lies at different values during heating as compared to cooling. This thermal hysteresis is due to the fact that some amount of non-polar metastable phase can exist below T_c and likewise some polar phase can still survive beyond T_c . This is typical for a first order phase transition.

Apart from discontinuity in dielectric constant, the discontinuity in polarization also induces an entropy change as the phase transforms from a non-polar to a polar state. As mentioned in the previous section entropy $S = -(\frac{\partial G}{\partial T})_{D,\sigma}$ is zero in the non-polar phase. The small change in entropy upon transition is given as

$$\Delta S = \frac{1}{2}\beta P_c^2 \tag{62}$$

The amount of ΔS is small but sufficient to have generated intensive research in solid state cooling devices, termed as ferroic cooling.

In the way we have constructed G, it does not contain contributions to heat from the phonon bath. Obviously in a real system entropy is never zero at finite temperature. But in order to work out the essentials, we will not carry along these extra contributions which would also necessitate some quantum mechanical considerations.

The equation of state, Eq. (49) ca be rewritten in terms of reduced variables as follow

$$d = (\frac{2\delta}{\gamma'})^{\frac{1}{2}} D \quad e = (\frac{2}{\gamma'})^{\frac{5}{2}} \delta^{\frac{3}{2}} E \quad t = 4\beta \frac{\delta}{\gamma'^2}(T - T_0) \tag{63}$$

yielding

$$e = td - 2d^3 + d^3. \tag{64}$$

where the reduced temperature t is the only variable. Figure 4a qualitatively shows the family of curves defined by Eq. (64). As t approaches zero the system transforms from a linear dielectric into a non-linear ferroelectric. The relevant observation with respect to hysteresis now is that Eq. (64) is a function of d. But in reality measurement determines $d(e)$. Thus, in a certain range of e-field values two or more solutions exist. This means two or more metastable states may coexist. This becomes the field range of multiple possible states. Macroscopically, we typically observe a hysteresis scanning the outskirts of this range, while microscopically, domains form. In each domain a state on either direction is stable spontaneously. The states shown by dotted lines in Fig. 4a are inaccessible to the system (dotted lines) which would here be the domain. The permittivity would become negative making this an impossible state.

Macroscopically one can reach a state apparently within the hysteresis, which has become the issue of Rayleigh- and switching distribution theory Damjanovic and Demartin (1996). Such data suggest that one can really enter the inside of the

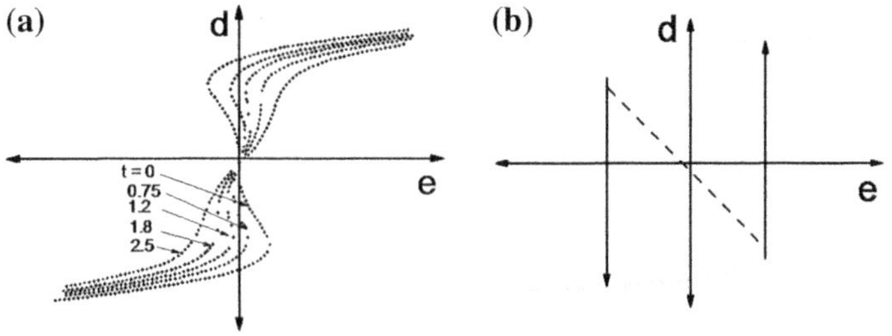

Fig. 4 d vs. e isotherms based on Eq. (64) for the case with absolute states (**a**) and by considering metastable state (**b**). The arrow in **b** indicates the path along which switching occurs as the system approaches instability. The dotted line in **b** indicates states that are inaccessible to system, Lines and Glass (1977)

hysteresis. But this is an effect of a macroscopic ensemble of microscopic domains. Partial cancellation of the polarization directions of the different domains yields an apparent intermediate state. The single loop that exists below the phase transition, Fig. 4b, resembles the experimentally observed $P - E$ hysteresis loop, but is is a macroscopic average of an ensemble while the dotted lines Fig. 4a are the instable thermodynamic polynomial approximations.

In close proximity to the real phase transition several loop configurations including single, double and broken loops are possible Jona and Shirane (1993), for which certain combinations of the coefficients α, β and γ arise, have to match, and they do near the phase transitions.

The fact just developed involves domain walls in between regions of well defined thermodynamic state. Upon macroscopic polarization reversal, these domains will move. This is a dissipative process and generates heat because the stress and electric fields of the wall itself interact with defects, dislocations and grain boundaries. This can lead to considerable heat-up rates for fast reversal processes.

2.4 *The Depolarizing Field*

At this point two terminologies must be differentiated: the *depolarizing field* which is a physical property of the crystal and the *depolarization* of a poled ceramic or single crystal ferroelectric. We will first deal with the depolarizing field and then briefly discuss de-poling of a poled ceramic.

The depolarizing field: The spontaneous polarization P_S is the primary order parameter in pyroelectrics. It typically arises along a well defined polar axis in the crystal which becomes the polar axis by the formation of $\mathbf{P}_s$. It is thus an intrinsic property of pyroelectrics at the microscopic level for a given thermodynamic condition. The onset of spontaneous polarization P_S is occurring at the Curie temperature upon cooling. For crystals with the perovskite structure an atom is displaced to an off-center position of the elementary cell. The collective shift of these charges across the entire domain/crystal yields a charge imbalance of *bound charges* between the opposing faces of the crystal. The charges only shift, they are *not* itinerant. If one applies electrodes onto the two polar faces of the crystal connecting them via a set of wires enables the flow of an amount of charge carriers ΔQ from one electrode to the other. The *dielectric displacement* **D** reflects the area density of these charges on the outer surfaces: $\Delta Q = \int_{\vec{A}} \mathbf{D} \cdot d\vec{A}$. It is thus the itinerant counterpart to polarization. On the other hand, it can be shown from Maxwell's equations of electrodynamics that the component of **D** normal to a surface must be continuous. Thus, in the pyroelectric $\mathbf{D} = \varepsilon_0 \mathbf{E} + \mathbf{P}$. **P** itself contains two contributions, one is the spontaneous polarization and the other one is the response of the material to an external field **E** *without* polarization switching:

$$\mathbf{D} = \varepsilon_0 \mathbf{E} + \epsilon_0 \left(\bar{\bar{\varepsilon}}_r - 1 \right) \mathbf{E} + \mathbf{P}_s \tag{65}$$

This leads to the generation of surface charge density $\sigma = q/A$. In a polarized crystal the uncompensated charges ($\mathbf{D} = 0$) induce an electric field which is oppositely oriented to P_S (see Fig. 5). It reduces the net value for P_s. This field is called the *depolarizing field* E_d. It is essential for the behavior of a ferroelectric in the nano regime. At this level depending on the energy state of the material, formation of domains takes place according to the value of the depolarizing field and the domain wall energy density σ. Due to the depolarizing field one big domain is splitting into two or more domains separated by 180° domain walls (see Fig. 5). In this case the free enthalpy $G^1(P, T)$ Eq. (19) must be used in its total form, namely the integration over the relevant volume like in Eqs. (8) to (10) must be considered. It needs to be extend by two terms taking into account the depolarizing field energy and the domain wall energy:

$$G^{1\ total}(P, T) = G_0^{1\ total} + \int \left(\frac{A}{2} P^2 + \frac{B}{4} P^4 \right) dV + W_w + W_E \tag{66}$$

The coefficients A and B in general are temperature dependent. For the simplest description of a ferroelectric 1st and 2nd order phase transitions B is assumed to be independent of temperature. W_w is the energy that is consumed to generate the domain wall. It is proportional to the domain wall area

$$W_w = \sigma \left(\frac{V}{d} \right) \tag{67}$$

where σ is the energy density of the domain wall energy per unit area. The volume V divided by the average distance d between domain walls (typically along the normal vector of the wall) yields the average domain wall area in the crystal. W_E is the energy due to the depolarizing field caused by the non-compensated bound charges on the crystal surface:

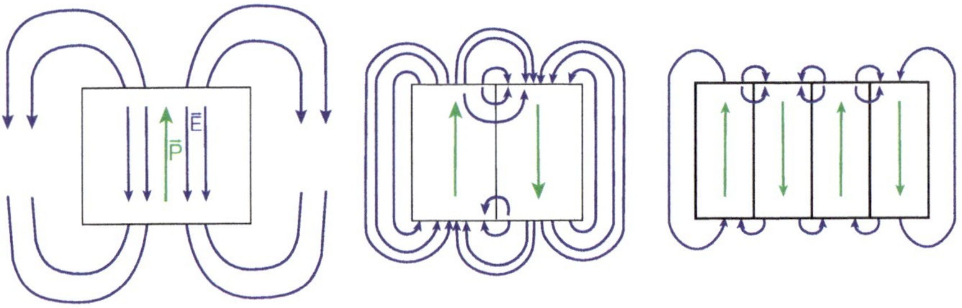

Fig. 5 The depolarizing field E_d develops as a response to the spontaneous polarization. E_d is opposite to P_s. The more domain walls exist, the less volume is filled with high fields reducing the overall electrostatic energy of the system, according to Sonin and Strukow (1974)

$$W_E = \frac{1}{2}\int \mathbf{D}\cdot\mathbf{E}dV = \frac{1}{2}\int \frac{\varepsilon}{\epsilon_0}L^2P^2dV \tag{68}$$

$\mathbf{D} = \varepsilon_0\mathbf{E} + \mathbf{P}(\mathbf{E}) = \varepsilon_r\varepsilon_0\mathbf{E}$ is the dielectric displacement. The first equality holds in any dielectric, the second for linear material behaviour.

The depolarizing field is given by $\mathbf{E} = -L/\varepsilon_0 \cdot \mathbf{P}$, where L is the depolarizing factor. L primarily contains geometrical factors. At equilibrium, the polarization charge density at the crystal surfaces is compensated by free carriers, so that no depolarizing field emerges. For insulators in an isolated environment, this balance is achieved only very slowly by defect charge carriers. In a mono-domain case, the energy W_E can increase to very large values. To avoid this, the crystal is forming domains with different polarization directions. If the equilibrium is reached and all depolarizing fields are compensated, the mono-domain state represents the energetically most favorable configuration. In general, this equilibrium state is rarely achieved. This depends on many factors, such as crystal symmetry, electrical conductivity, the defect structure and the pre-treatment course as well as the geometry of the sample.

Upon fast cooling, no free charge carriers are available. Then, $\mathbf{D} = 0$ and the system strives to generate as many domain walls as possible in order to reduce the absolute volume affected by large field values. Figure 6 shows that in a 180° domain structure only the edges where domain walls interact with an outer surface experience high fields. The majority of the volume is field free. An outer surface can be a free surface or grain boundary.

One fundamental question is why the formation of additional domain walls does not occur infinitely. The answer is very simple, each domain wall locally represents a discontinuity in crystal order. This does not occur for an infinitely fine extension of the wall. The discontinuity of polarization and spontaneous strain generates a certain amount of elastic/electrostatic energy W_w within the wall itself, which decays into the bulk of the domain in a small but finite extension. The total wall energy is again the integral of the wall energy density over the whole wall area A.

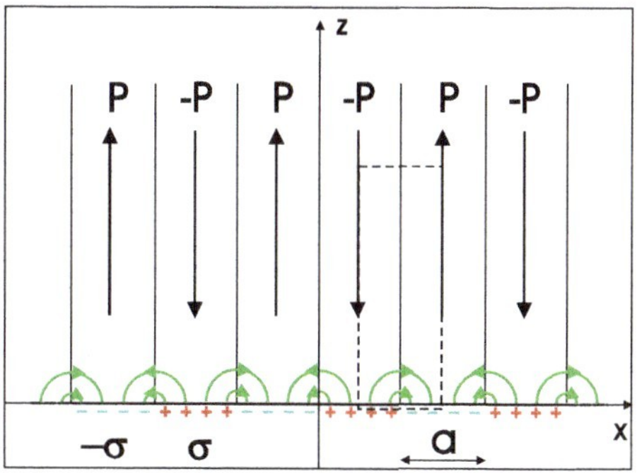

Fig. 6 Local field distribution of a multi-domain state. The volume experiencing large fields becomes very small around the edges where the walls intersect with the outer surface/grain boundary, Genenko and Lupascu (2007)

$$W_w = \int_A w_w dA \tag{69}$$

The local fields generate additional local changes in semiconductor properties, e.g. higher conductivity due to the existence of a free electron gas in the vicinity of domain walls. This will be discussed later, see Sect. 5.1, Kittel (1946), Sonin and Strukow (1974), Jackson (1998).

All ferroelectric materials contain 180° domain walls unless an order-disorder transition is present, as e.g. in certain liquid crystals. 90° domain walls electrically generate similar boundary conditions. Additionally, due to the associated spontaneous strain in the material, they are capable to compensate mechanical stresses. Without electric field, they represent ferroelastic domains. Along with their spontaneous polarization they are ferroelectric ferroelastic. Like for 180° domain walls, the reduction of electrostatic field energy is the main reason for their formation.

2.4.1 Depoling a Poled Ceramic

Poling a macroscopic ferroelectric turns it into a useful piezoelectric. This poling process requires prior application of a sufficiently large electric field. The ceramic itself must contain many ferroelectric domains and a low amount of hard dopants Hagemann (1980), if poling is done at low temperature. Hard dopants are advantageous, when the material is poled at high temperature, because the dopants do not inhibit domain wall motion at high temperature. Upon cooling, this polar state is imprinted into the ceramic microstructure and stabilized by the hard dopants.

Depoling denotes the fact that external temperature, pressure, or reverse electric fields can reduce the degree of poling of a poled ceramic. Typically, this also reduces the overall piezoelectric constant of the device, because, e.g. $d_{eff} = \frac{\partial x}{\partial E} = 2QP(E)\varepsilon(E)$ where $P(E)$ is the macroscopic polarization of the sample, see e.g. Lupascu and Rödel (2005). Thus, absolute polarization linearly enters the piezoelectric performance of the material. Depolarization is thus a very much unwanted effect.

Hard dopants, which are typical in ultrasonic materials, stabilize a poled state at the cost of loosing some dielectric constant entering the piezoelectric coefficient linearly. But nevertheless the long term stability is fundamental for ultrasonic applications.

3 Energetics of a Semiconductor

3.1 The Band Structure

Physical Background There is a large number of excellent introductory books to semiconductors, e.g. Blakemore (1962), Ashcroft and Mermin (1988), Sze (1981), Hellwege (1988). This text cannot replace these detailed works. It is merely intended to review the main features relevant with respect to the semiconductor effects in ferroelectrics. The nomenclature used here is mostly adopted from Möschwitzer (1992).

Semiconductors differ from metals in their electronic structure. Fundamentally, they are not different from insulators. The electronic shell of an atom consists of *discrete* energy levels. At this point we leave the entirely continuous world of thermodynamics including mechanics. Quantum mechanics is able to describe the detailed features of the discrete nature of the energy landscape of the electron structure for each atom, at least in principle. It is the basis for all chemical reactions and solid state formation. One discrete electronic state of an atom interacts with the electronic states of the neighboring atoms. Depending on the electron energy level (with respect to vacuum) and the geometrical structure of the electronic state, namely its quantum mechanical *wave function*, this interaction may lead to *joint* electronic states which are distributed between both atoms. This is the classical covalent chemical bond. If multiple atoms of the same kind join to form a solid, the joint electronic state extends over the entire space that the solid is occupying. Each atom contributes one electron into this joint state. Due to Pauli's principle, two electrons cannot occupy the absolutely identical electronic state at the same location. So even though we talk about the same electronic state of all atoms involved, very small differences in energy level are generated in order to still differentiate between the different electrons. But, and this is often hard to understand to a newcomer, the electronic state of *each* electron extends over the entire solid. So quantum physics also talks about *delocalized states*. The atomic electron state that provides electrons into the band determines the degree to which the electron is rather localized near the atomic core or widely distributed. While complete inner shells hardly contribute to the effective electron bands it is rather the outer shell that determines the electronic properties of the crystal.

As all electrons occupy nearly the same electron energy, but with a certain small finite energy width, the term *electron band* has become common. It describes the extension of electronic states to the same primary atomic state across a certain

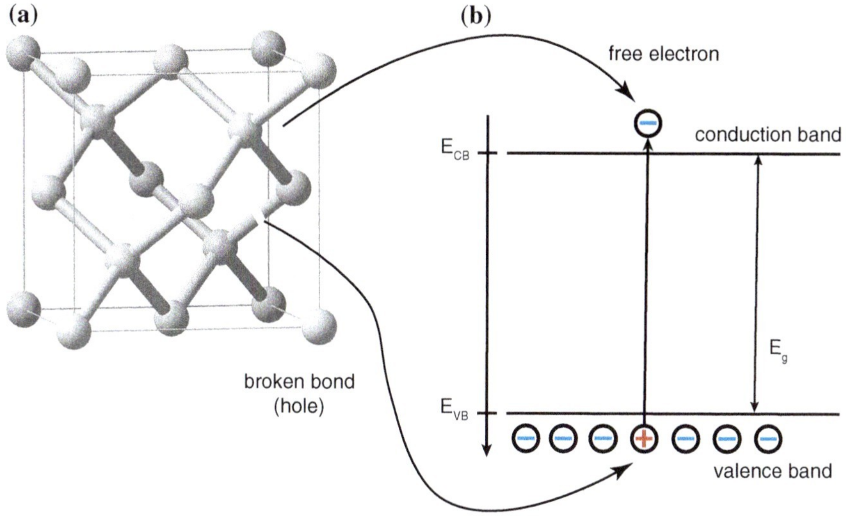

Fig. 7 Diamond crystal structure semiconductors (silicon, germanium). **a** Atoms in the crystal lattice, **b** energetic states of intrinsic charge carriers in the electronic bands

energy range. Energy bands are fully filled, if the electronic shell of the involved atom becomes completely filled due to chemical bonding. One more issue must be addressed. In Fig. 7a the crystal structure of the most common semiconductor silicon is shown. The quantum mechanical wave functions extending over the entire crystal must match the periodicity of the crystal! Otherwise the notion of a unit cell would completely loose its sense in quantum mechanics. It is a very simple thought experiment that every direction in the crystal implies a different periodicity of this overall wave function. Thus, the wave function *and* the associated energy of the electronic state become direction dependent. For every direction, only one electronic energy (with the previously discussed certain width) applies for a given direction of the wave function in reciprocal space. For the scope of this article it is too difficult to discuss the reciprocal lattice in depth. Many solid state physics books have taken care of this. But, everyone must remember that the energy is direction dependent and the typical horizontal depiction as given in Fig. 7b sums up energies of *all* permissible directions in space.

Let us again consider silicon. It can form four bonds. If they are fully bonded, each silicon sees eight electrons in its electronic shell and this shell then is that of a noble gas, namely completely filled. Completely filled electron bands *cannot* provide electron transport. It is somewhat like a fully occupied parking garage deck. No motion is possible, unless one car leaves the deck. For higher energies there are (many) possible electronic states, but they are not occupied. Only if an electron gains sufficient energy to be launched into such an enhanced energy state, it becomes available to the transportable electronic entities. In the parking deck picture, one car moves up onto the next empty deck and now has plenty of space to move. The fully occupied deck is the valence band, the empty deck is the conduction band. Already from this simple picture it is also clear that electron mobility will be different from hole mobility (the empty space on the lower deck that moves across the crystal as a reaction to electron motion in the reverse direction).

Now we come to differentiate between metals and semiconductors / insulators. In metals, due to the nature of the metallic chemical bond, the individual bands are *not* fully occupied. Thus, electrons can move freely across the solid. In the picture of the parking deck, it is only partly filled and movement of cars is not difficult. In a semiconductor/insulator the valence band (lower parking deck) is fully occupied, and the conduction band (upper parking deck) is completely empty. The difference between insulators and semiconductors is the difference between energy levels (the energy to get one car onto the higher deck), namely the *energy band gap*. In semiconductors the band gap is small and typical temperatures enable the transition of a sufficient number of charge carriers into the conduction band by thermal excitation. In insulators the band gap is so large that typical temperatures cannot provide enough energy to lift charge carriers into the conduction band. Conduction does not occur due to lacking charge carriers. The differentiation between a semiconductor and an insulator thus depends on temperature. In daily life, we always refer to the room temperature state and materials are typically denoted according to their properties at room temperature.

Intrinsic Conduction Like all thermally activated processes, also the thermal occupancy of electronic states is determined by an energy term and temperature. The energy term in an un-doped semiconductor is exactly half the band gap. Unlike classical particles, electrons follow Fermi-Dirac statistics, see Ashcroft and Mermin (1988). For finite temperature, the whole lattice vibrates and provides energy to the electronic system. This vibration results in depopulating the valence band and filling the conduction band. Fundamentally, no extra charges can be generated due to the law of charge conservation. Thus, *every* electron in the conduction band is matched with a hole in the valence band (hole = positive charge carrier representing the missing electron). This process is the generation of an electron-hole pair. In the reverse process, electrons from the conduction band seek a hole in the valence band. When the electron retakes its state in the valence band, the electron in the conduction band disappears and the hole is compensated. This is called recombination. Thus: ***In an intrinsic semiconductor in thermal equilibrium the electron density***, n_0, ***and hole density***, p_0, ***are always of same value and called intrinsic charge carrier density***, n_i.

$$n_i = n_0 = p_0 \tag{70}$$

Energy Band Diagram The energy band diagram graphically illustrates the possible energetic states of the charge carriers for one particular crystal. A complete representation shows the possible energy states $E(\vec{k})$ of the electron with respect to their propagation directions in the crystal $\vec{k}$ with $|\vec{k}| = 2\pi/\lambda$. This is termed electron dispersion. The involved quantum mechanical state of the electron must match the periodicity of the crystal. As the atomic positions are discrete, $\vec{k}$ can also only take on discrete values. For a given electronic state of the contributing atom(s), this entails one well determined energy value per direction. In 3D this yields the Fermi surface. One direction among all yields the highest energy value in the valence band $E = E_V$ or the lowest value in the conduction band $E = E_{CB}$ (Fig. 8).

For Germanium, both energies arise for the same direction in space (direct gap), in silicon the lowest energy in the valence band does not arise in the same direction as for the conduction band. This yields an indirect gap where the system needs an additional phonon in the excitation process that will take on the difference in impulse of the two states. The energy gap in both cases is $E_g = E_{CB} - E_V$. In a simplified discussion of semiconductors, this is all what people consider. In order to be excited, an electron from the valence band has to be provided with the band gap energy minimum and in the case of an indirect gap with the associated difference in impulse. Electrons can not occupy any energy state within the forbidden zone between valence and conduction band unless defect states are present.

A parameter that will be vital later is the density of states $g(E)$. It tells us how many possible electron states are found in a certain energy interval. For a strong dispersion, the density of states must be low (few states per energy interval), for shallow bands (small dispersion) the density is high. Numerically, it can be directly calculated from the band structure itself:

$$g(E) = \int \frac{d\vec{S}}{4\pi^3 \nabla_k E} \tag{71}$$

Whether it is feasible to compute $g(E)$ in practice, depends on the computation power of semiconductor theory.

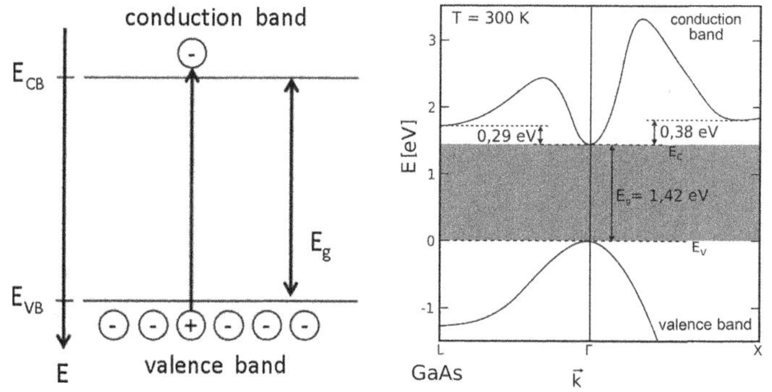

Fig. 8 Energy band diagram of a semiconductor averaging over all directions in the crystal. Real band structure of gallium arsenide incorporating the direction ($\vec{k}$-) dependence of the electronic states, right image © Wikipedia

For facilitating the understanding of electron energy dispersion, a semiclassical approach is chosen, namely the kinetic energy of the electron is given as $E = \frac{\vec{p}^2}{2m_{ij}}$ with $\vec{p}$ the momentum of the electron given in the context of quantum mechanics as $\vec{p} = \hbar\vec{k}$. m_{ij} is the effective mass tensor. This concept now puts all the physics of the band structure into the effective mass:

$$m_{ij} = \hbar^2 \left(\frac{\partial^2 E}{\partial k_i \partial k_j} \right)^{-1} \tag{72}$$

3.2 *Electron Statistics*

3.2.1 Fermi-Dirac Distribution Function

Due to Pauli's principle, no electronic state can be occupied more than once. This entails a different distribution function than for classical particles (Maxwell-Boltzmann distribution) or phonons and photons (Bose-Einstein distribution), Reif (1965). For electrons the Fermi-Dirac distribution applies, Blakemore (1962), Ashcroft and Mermin (1988):

$$f(E) = \frac{1}{1 + e^{(E-E_F)/(k_B T)}} \tag{73}$$

In a simple approach, the Fermi-Dirac distribution function can be considered as a melting ice-block. While at T=0 it is a Heaviside function, 1 up to E_F and zero above, it melts off into a finite occupancy of higher energies and a concurrent reduction near the edge for finite temperatures (Fig. 9). A detailed description of the subtleties in semiconductor statistics can be found in Blakemore (1962).

The *Fermi level* is a thermodynamic quantity. It equals the chemical potential μ_e of the electrons and is thus an equilibrium quantity. We will later see that this directly couples into the thermodynamics of the ferroelectric as a whole.

The newcomer must also observe that in certain cases *quasi Fermi levels* are introduced. This is an artificial construction to account for external electric potential gradients in a material. The quasi Fermi levels are not equilibrium quantities. The concept is as follows: We assume that the electrons stemming from the donor levels generate their own Fermi level as well as the acceptors above the valence band. In equilibrium the two will equalize and there will be a potential generated on the outside. If we now deviate from this built in potential externally or by introducing further charge carriers e.g. in a solar cell, the two quasi Fermi levels will separate exactly by the amount of externally applied additional voltage. For solar cells this would be the open circuit voltage V_{OC}.

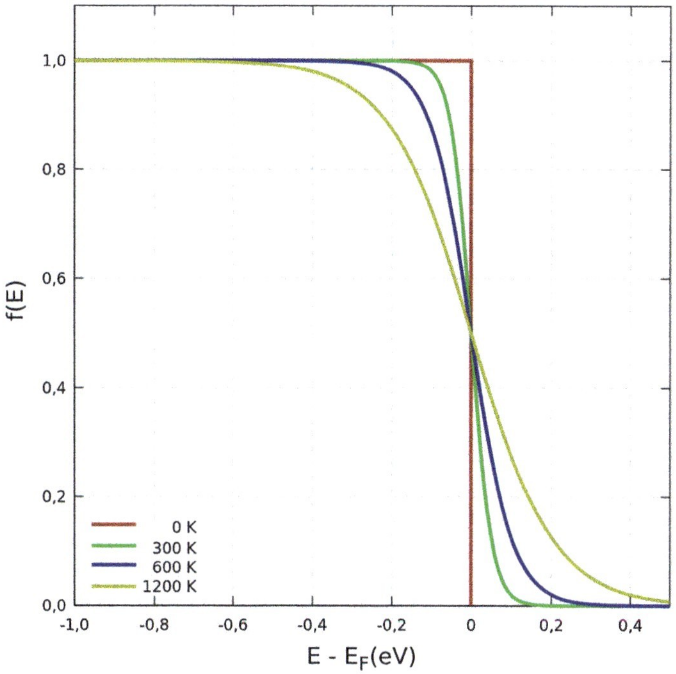

Fig. 9 Fermi-Dirac distribution function plotted for different temperatures, © Wikipedia

The intrinsic semiconductor For an intrinsic semiconductor the Fermi level is located in the middle of the band gap. At absolute zero temperature ($T = 0$K) all energy-terms of the conduction band are vacant and all energy-terms of the valence band are occupied. At $T > 0$K the same number of energy-levels vacant in the valence band becomes occupied in the conduction band. The Fermi level of an intrinsic semiconductor is thus:

$$E_F = \frac{E_{CB} - E_V}{2} \tag{74}$$

at $T = 0K$. If the densities of electronic states in the valence N_V as well as the conduction band N_{CB} are symmetric (with respect to energy values) and of equal number, this is independent of temperature. If both are different, and this is the relevant case for describing a *pn*-junction, then the Fermi energy level will slightly shift with rising temperature even for an intrinsic semiconductor.

3.3 Semiconductors with Impurities

3.3.1 Law of Mass Action

Following the law of mass action the product of electron and hole densities is an intrinsic material quantity for a semiconductor at thermal equilibrium. It depends on the density of states, the band gap, and temperature. It does *not* depend on the impurity density. The following applies for a p-type semiconductor:

$$n_p \cdot p_p = n_i^2 \tag{75}$$

And for an n-type semiconductor:

$$n_n \cdot p_n = n_i^2 \tag{76}$$

The n-type semiconductor has an excess of electrons representing the *majority charge carriers* in this case. The holes are then the *minority charge carriers*. Technically, the difference between majority and minority charge carrier densities is about ten orders of magnitude for typical donors in classical semiconductors. For a p-type semiconductor the entire setting is simply reverse.

The insertion of impurity atoms into the lattice (impurities) enables the specific generation of additional electrons and holes in the conduction band or valence band, respectively. Two types of impurities exist: donors and acceptors. Under particular circumstances the same defect can serve as donor or acceptor, but most classical semiconductors and typical impurities are either. A *donor* (in the context of silicon this is a group V element) generates an energy level, E_D, within the forbidden zone, which is energetically close to the conduction band, E_{CB}. Hence, small energies are needed to ionize these defect states. The impurity atom donates its fifth valence electron to the conduction band. This process generates a mobile electron in the conduction band and a *localized* ionized foreign atom in the lattice and hence, *no mobile hole* (Fig. 10). The *localization* of the remaining *impurity* will become very

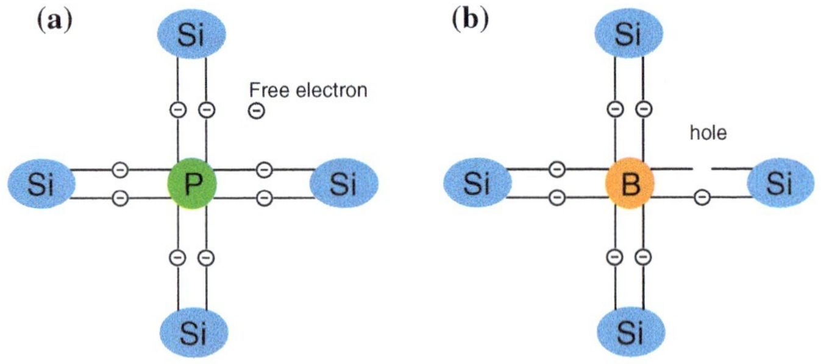

Fig. 10 Impurity atoms in silicon: **a** donor and **b** acceptor

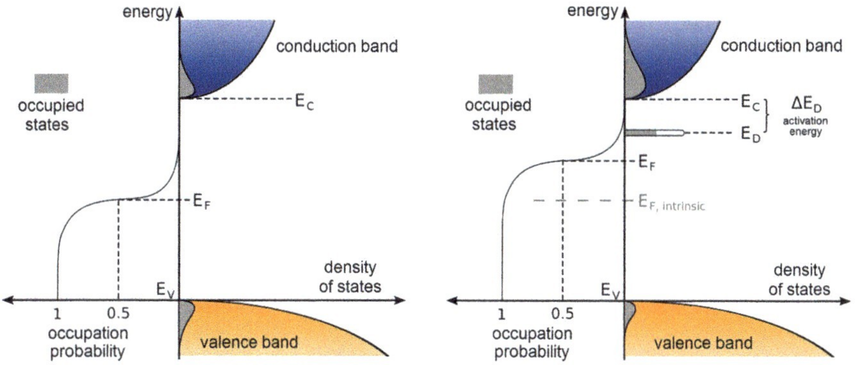

Fig. 11 Temperature dependence of the electron density in an intrinsic semiconductor (left) and for an n-doped semiconductor (right), © Wikipedia

important in the context of ferroelectrics later. Overall this semiconductor has an excess of electrons and is called an n-type semiconductor (Fig. 11). An *acceptor* (for the host semiconductor silicon this is a group III element) generates an energy level close to the valence band. A valence band electron can easily occupy this energy state, which generates a negatively charged impurity atom in the lattice. Simultaneously, a hole is generated in the valence band resulting in an excess of mobile holes turning the semiconductor into a p-type semiconductor.

Three different characteristic states can result for a doped semiconductor as a function of temperature. The first one is called impurity reserve, in which not all foreign atoms have been ionized yet. This is typical for low temperatures. The Fermi energy level is located in between the defect energy value and the corresponding band, e.g. the donor level and the conduction band for an n-semiconductor. The second range is called impurity depletion range meaning all foreign atoms are ionized (usually at room temperature). Hence, as the density of impurities, either donor density, N_D, or acceptor density, N_A, can be controlled technically, the density of free charge carriers (p or n) is determined as well. The following applies: for a p-type semiconductor $p = N_A$ and for an n-type semiconductor $n = N_D$ *in depletion*. Figure 12 displays the temperature dependence of the number of mobile charge carriers in an n-type semiconductor. At low temperature (right in the graph) thermal excitation of the electrons from the donor level determines the amount of carriers in the conduction band, the range of reserve of impurities. In this range the number of actually *ionizes impurities* $N_D^+ \neq N_D$ is different from the number of impurity atoms and temperature dependent through the Fermi-Dirac distribution. The nearly horizontal line represents charge depletion. In depletion the Fermi levels moves towards the center of the band-gap (Fig. 13). Only when the temperature has risen much (going to the left in the graph), intrinsic charge carriers contribute, because now the energy that has to be provided to the electron system is the entire band gap. The symmetry of the Fermi-Dirac distribution requires that, in depletion, the Fermi level will shift from in between the donor (acceptor) level at low temperature to the center of the band gap at high temperature in order to assure conservation of charge.

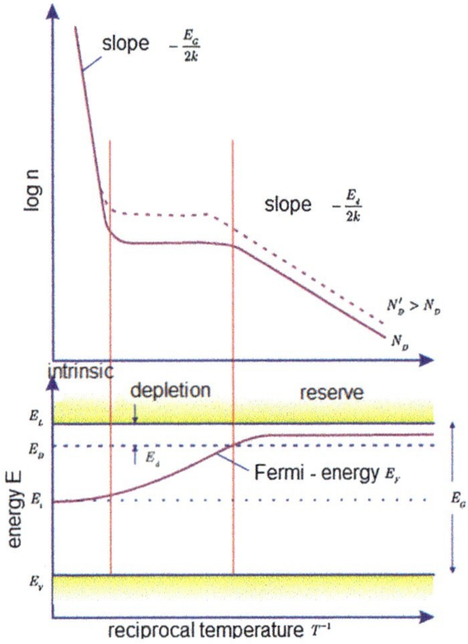

Fig. 12 Shift of the Fermi-level with temperature in the impurity depletion temperature range along with the change in carrier density (top), © Othmar Marti (2016)

3.4 Semiconductor with both Donors and Acceptors

If a semiconductor is doped with donors and acceptors, charge neutrality is obtained by the following equation:

$$p + N_D^+ = n + N_A^- \tag{77}$$

In depletion, all impurities are ionized and $N_D^+ = N_D$ or $N_A^+ = N_A$ for n- and p-type semiconductors, respectively. The number of impurities dominates the properties of the doped semiconductor. If $N_D > N_A$ an n-type semiconductor with an electron density of $n_n = N_D - N_A$ is obtained. If $N_A > N_D$ a p-type semiconductor with a hole density of $p_p = N_A - N_D$ is obtained.

Donors are atoms with a higher number of electrons than the host matrix, e.g. phosphorus in silicon. This generates an additional defect energy level E_D. In impurity reserve the first entity to provide a mobile charge carrier will be the associated electronic donor level, because the energy to ionize it is small. Two things are fundamentally different for a defect state from a band state: first it is *localized* while a band state is always delocalized. For a direct band gap semiconductor no extra impulse needs to be transferred. For an indirect band gap, a certain amount of impulse $\Delta\vec{p} = \hbar\Delta\vec{k}$ must be provided to the charge carrier in order to lift it into the relevant band state. Secondly, a singly ionized defect state will permit only one electron on the site. This entails a single spin state on the defect site. The underlying statistics in the grand canonical ensemble changes, because the number of countable states changes, Reif (1965). The resulting Fermi-Dirac distributions become, Blakemore

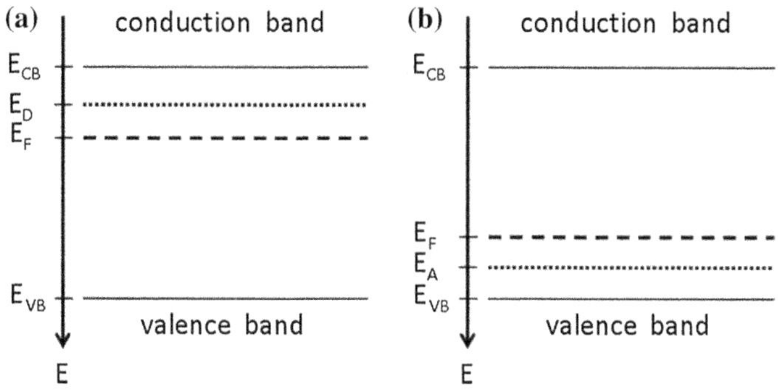

Fig. 13 Integral energy band diagrams of doped **a** n-type and **b** p-type semiconductors in depletion not resolving the directional dependence of the band energies themselves

(1962):

$$f_D(E) = \frac{1}{1 + g_D e^{(E_F - E_D)/(k_B T)}} \tag{78}$$

with the degeneracy factor $g_D = 2$ for a typical donor state.

$$f_A(E) = \frac{1}{1 + g_A e^{(E_A - E_F)/(k_B T)}} \tag{79}$$

with the degeneracy factor $g_D = 4$ for a typical acceptor state in Ge, Si, or GaAs, because the bands at $\vec{k} = 0$ are doubly degenerate, Sze (1981).

3.5 Transport of Charge Carriers

3.5.1 Conductivity of a Semiconductor

The electrical conductivity, κ, of a semiconductor (also denominated σ in much literature) is given by the product of mobility times charge carrier concentration for both electrons and holes:

$$\kappa = e\left(\mu_n \cdot n + \mu_p \cdot p\right), \tag{80}$$

$e = 1.610^{-19}$C is the elementary charge and μ_p and μ_n are the mobilities of holes and electrons, respectively. The latter two are material quantities. They are influenced by the quality of the crystal lattice, the density of impurities, and by the temperature dependent lattice vibrations. Figure 14 (left) shows the impact of these factors on the mobility of charge carriers for silicon. The mobility of electrons is typically twice

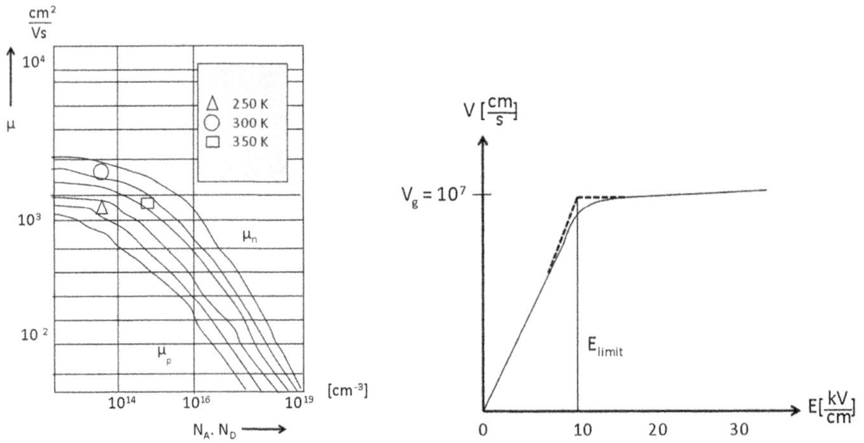

Fig. 14 (left) Mobility of electrons and holes in silicon as a function of defect density and temperature, (right) field dependence of the drift velocity of electrons in silicon, Möschwitzer (1992)

as large as that of holes in silicon. The correlation of the current density, $\vec{j}$, and the drift velocity, $\vec{v}$, with the electric field are:

$$\vec{j} = \kappa \vec{E}, \qquad \vec{v} = \mu \vec{E} \tag{81}$$

For weak electric fields the drift velocity, $\vec{v}$ is directly proportional to the electric field, $\vec{E}$. Based upon phonon interactions the charge carriers cannot be further accelerated above a specific limit of the electric field, E_{limit}, when the drift velocity saturates. This is typically not reached in silicon devices. This limit may more readily be reached in ferroelectrics due to the high fields applied or intrinsic to them. As most of them are not so well studied as silicon, these numbers are typically not available.

3.5.2 Charge Carrier Continuity

The balance equation for electrons and holes must be fulfilled. The general continuity equation, see e.g. Reif (1965), then becomes, Möschwitzer (1992):

$$\mathrm{div}\vec{j}_p = -e\left(R - G + \frac{\partial p}{\partial t}\right) \tag{82}$$

for holes and

$$\mathrm{div}\vec{j}_n = e\left(R - G + \frac{\partial n}{\partial t}\right) \tag{83}$$

for electrons. R is the recombination rate, G the generation, rate and $\vec{j}_p$ and $\vec{j}_n$ the respective current densities. As electrons and holes are able to recombine, this is one of the rare cases, where continuity must include source terms (G and R) in order to be complete. Continuity states that a quantity cannot appear or disappear unless sources exist.

3.5.3 Poisson's Equation in a Semiconductor

The following section is highly relevant in the context of ferroelectrics, because the distribution of defects in the crystal influences the field distribution within the material. In this section we will outline the most simple case of a linear geometry and a constant defect density. The correlation between charge distribution, course of electric field, and the course of the electric potential, ϕ, within a region of a semiconductor can be calculated from Poisson's equation:

$$div\ \vec{E} = \frac{\rho}{\epsilon_{HL}} \tag{84}$$

where $\rho = e(N_D^+ + p - N_A^- - n)$, E is the electric field component along the direction of integration, ρ is the space charge density and ϵ_{HL} is the dielectric constant of the semiconductor. ρ contains *all* charge carriers at the respective location in space, localized and mobile ones.

One-dimensional example

Typically, the following argument is given for the depletion region where all defects are ionized. In order to allow for partial ionization of the defects in the vicinity of grain boundaries later, we will directly utilize the ionized defects N_D^+ and N_A^- in our nomenclature not restricting ourselves to a particular case. We consider only the charged impurities and assume the associated free charge carriers to have been transported to the outside.

For the charge carriers in Fig. 15 the following applies: $\rho = (N_D^+ - N_A^-)$, because all free charge carriers are missing. Poisson's equation then reads:

$$\frac{dE}{dx} = \frac{e}{\epsilon_{HL}} \left(N_D^+ - N_A^-\right) \tag{85}$$

Integration yields the electric field:

$$E(x) = \frac{e}{\epsilon_{HL}} \left(N_D^+ - N_A^-\right) x + E_0 \tag{86}$$

As field is the gradient of potential:

$$E(x) = -\frac{d\phi(x)}{dx} \tag{87}$$

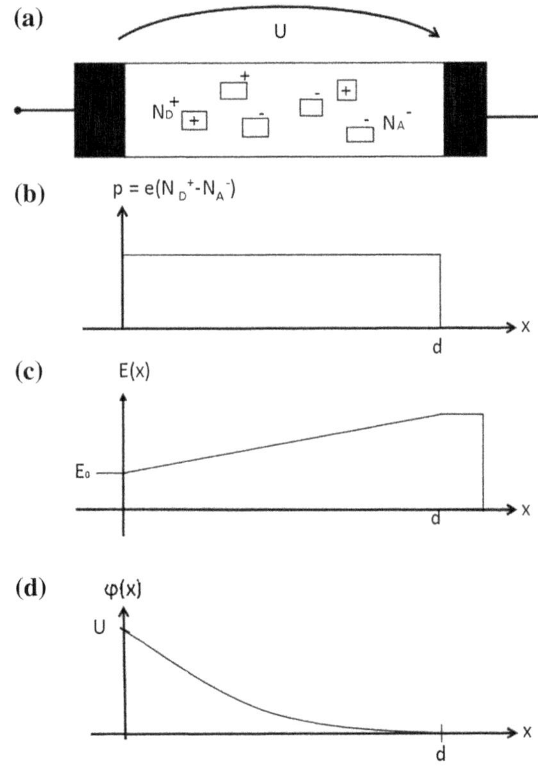

Fig. 15 **a** Semiconductor including space charges, **b** course of space charge density, **c** course of electric field, and **d** course of potential across the sample in between the electrodes. Adopted from Möschwitzer (1992)

integration and application of the boundary condition $\phi(d) = 0$, which determines the arbitrary integration constant, gives:

$$\phi(x) = -\frac{e}{2\epsilon_{HL}} (N_D - N_A) \left(d^2 - x^2\right) - E_0 (x - d) \tag{88}$$

This represents the course of the potential in dependence of location, d. Hence, the potential difference across the depletion region is given by:

$$U = \phi(0) - \phi(d) = -\frac{e}{2\epsilon_{HL}} \left(N_D^+ - N_A^-\right) \cdot d^2 + E_0 \cdot d \tag{89}$$

From this very simple example of a constant density of charged impurities, we see that the local fields differ much from the case when a field is applied to an impurity-free sample, when the electric field is constant throughout the material interior. A space impurity charge region will bend the associated electronic bands. This happens near ferroelectric grain boundaries, see Sect. 5.2.

3.5.4 Space Charge Limited Current (SCLC)

For low conductivities, the transport of charge carriers is not instantaneous across a piece of semiconductor, because current actually progresses through drift of charge carriers and not through their mutual interaction like in a metal. The associated charge density itself influences the field situation in the semiconductor. If the drift time of charge carriers through a segment of semiconductor is smaller than the characteristic relaxation time $\tau_{relax} = \epsilon_{HL}/\kappa$ of the semiconductor, with κ again the electrical conductivity, then a space charge region is formed, Rose (1955), Tredgold (1966). We integrate the (uniaxial) current density of the electrons:

$$j = e\mu_n n E \tag{90}$$

Space charge is $\rho = -en$ and Poisson's equation reads:

$$\frac{\mathrm{d}E}{\mathrm{d}x} = -e\frac{n}{\epsilon_{HL}} \tag{91}$$

Substituting en from Eq. (91) into (90), we obtain

$$j = -\mu_n \epsilon_{HL} E \frac{\mathrm{d}E}{\mathrm{d}x}. \tag{92}$$

Integrating this equation from $x = 0$ to $x = d$ with boundary conditions $E(0) = E_A$ and $E(d) = 0$ yields

$$j = \frac{1}{2}\mu_n \epsilon_{HL} \frac{E_A^2}{d}. \tag{93}$$

These boundary conditions mean that the electric field is zero on one side of the dielectric and of finite value E_A on the reverse side. Thus, in an insulator, the fields on both sides of a sample are significantly different. Externally one only sees the potential difference as a whole (integrated across the entire sample). This is very different from a "good" conductor when E is constant throughout the sample and the voltage drop is linear. Integrating the field distribution within the sample:

$$U = \int_0^d E\mathrm{d}x = \int_0^{E_A} \frac{\mu_n \epsilon_{HL}}{j} E^2 \mathrm{d}E = \frac{\mu_n \epsilon_{HL}}{3j} E_A^3 \tag{94}$$

Eliminating E_A from the last two equations, we generate

$$j = \frac{8}{9}\mu_n \epsilon_{HL} \frac{U^2}{d^3} \tag{95}$$

The behaviour is thus very different from the linear case in ohmic resistors. A number of systems with low conductivity shows space charge limited current, e.g. molecular electronics or crystalline dielectrics with low carrier concentration. It is typical for ferroelectrics with medium bandgap.

3.6 Metal-Semiconductor Junctions

As the *pn*-junction does not play a major role in ferroelectrics, we have transferred the issue to the Appendix.

A metal semiconductor interface is always present in an electroded and thus usable ferroelectric. The metal semiconductor interface can have several different characteristics, see Fig. 16. The most simple case is an ohmic contact. The top series of illustrations in Fig. 16 shows a p-conductor. Thus, the majority carriers are holes. The Fermi level of the aluminium adjusts to that of the semiconductor closely. Holes from the semiconductor can easily move up from the valence band into the conducting band states of the metal. For reverse polarity, the electrons from the metal fill the holes in the valence band and thus carry the current.

The Schottky junction is a junction between a metal and a semiconductor material. Some (most of) such interfaces will develop a thin insulating interface layer. Suppose that the metal and the semiconductor are both electrically neutral and separated from each other. The band diagram is shown in Fig. 17a for an n-type semiconductor ma-

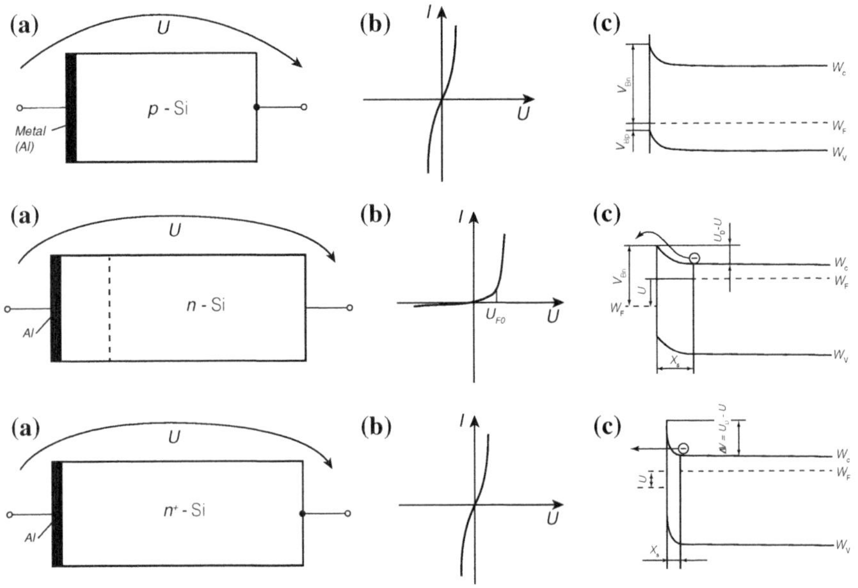

Fig. 16 Three types of metal semiconductor interfaces, (top) ohmic contact, (center) Schottky barrier, and (bottom) charge tunneling due to high charge carrier concentration and correspondingly narrow barrier layer, adopted from Möschwitzer (1992)

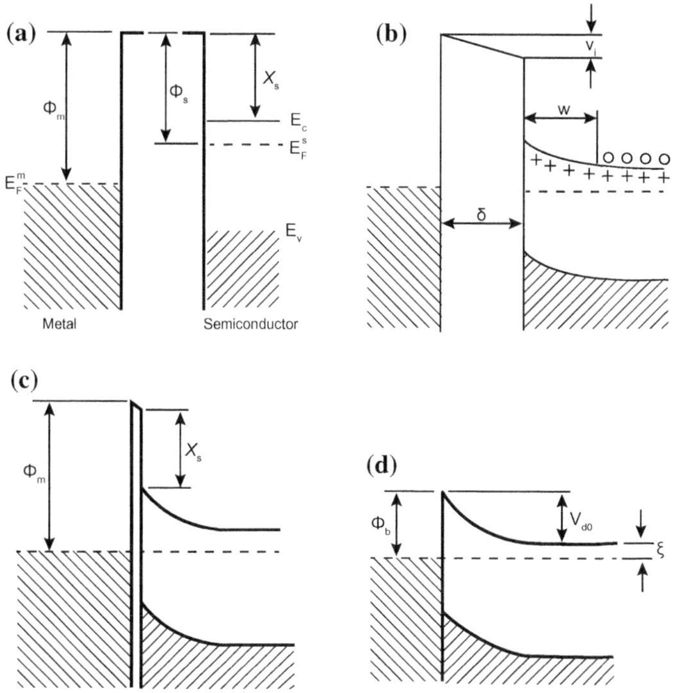

Fig. 17 Barrier formation between a metal and a semiconductor for different widths of an intermediate insulating layer. **a** Isolated, **b** In electric contact, **c** Separated by a narrow gap, and **d** In perfect contact. ∘ ⇒ electrons in conduction band; + ⇒ donor ions, adopted from Rhoderick and Williams (1988)

terial with a work function less than that of the metal. If the metal and semiconductor are in electrical contact by a wire the two Fermi levels are forced to be continuous across the junction at thermal equilibrium as shown in Fig. 17b.

An electric field develops in the gap directed from the right side to the left due to the different work functions (energy between the Fermi-level and vacuum) of each, the metal and the conduction band of the semiconductor. A negative charge must be on the surface of the metal balanced by a positive charge in the semiconductor. Since the semiconductor is n-type, the positive charge will be provided by conduction electrons receding from the surface, leaving uncompensated positive donor ions in a depletion region of electrons. Due to the many orders of magnitude lower donor concentration in comparison with the concentration of electrons in the metal, the uncompensated donors possess a layer of evident thickness; w, comparable to the width of the depletion region in a p-n junction, and the bands in the semiconductor are bent upwards as shown in Fig. 17b. The potential difference V_i between the outside surface potentials of the metal as well as the semiconductor is given by $V_i = \vec{\delta} \cdot \mathbf{E}_i$, where δ is the insulating gap between the two materials and $\mathbf{E}_i$ the electric field in the gap. If the metal comes closer to the semiconductor, V_i must tend to zero if E_i

is to remain finite as shown in Fig. 17c. When finally the metal comes into contact with the semiconductor, the barrier due to the vacuum disappears altogether and an ideal metal-semiconductor contact is obtained (Fig. 17d). The height of the barrier Φ_b relative to the Fermi level is given by

$$\Phi_b = \Phi_m - \chi \tag{96}$$

where Φ_m is the difference between the vacuum level and Fermi level of the metal and χ is the difference between the conduction band of the semiconductor and vacuum level.

In many real metal-semiconductor junctions, a thin insulating layer of oxide remains on the surface of the semiconductor and an ideal contact like shown in Fig. 17d is never reached. This insulating film is commonly referred to as the interfacial layer. A metal-semiconductor junction is therefore more like that shown in Fig. 17c. The insulating layer is usually small enough that electrons can tunnel through it quite easily. Hence, Figs. 17c, d present almost the same contact concerning (free) electrons. Furthermore, the potential difference, V_i, in the oxide layer is very small meaning Eq. 96 is still a very good approximation.

For ferroelectrics the thickness of the interface layer is not well investigated. It may represent an extra insulating layer or merge into a single material, namely no pronounced extra insulation at the interface. In the context of fatigue of ferroelectric memories (FeRAMs) much discussion has been circling around the relevance of the interface to the contact metal and/or an insulating interface, Scott (2000). Iridium metal electrodes turned out to be a good contact electrode to PZT-based ferroelectric memory elements. It was assumed that iridium forms an IrO_x oxidic layer in direct contact with the ferroelectric. It has never been fully dis-entangled, whether it was just the appropriate work function of the iridium and its oxidation enthalpy, or whether it was actually the insulating oxide layer that improved the fatigue resistance.

3.6.1 Junction Width and Capacity

In the p-n junction and other junctions we have talked / will talk about, a region of depleted charge carrier density arises. This means that we arrive at a situation where a certain charge density on one side is separated from a charge carrier density of opposite sign on the other side by a layer that appears effectively insulating. This is a classical capacitor. Thus every depletion region is associated with a certain capacitance. A very good introduction can be found at http://ecee.colorado.edu/ bart/book/book/chapter4/ ch4_3.htm and in numerous semiconductor books. At this point we only want to discuss the effect for a ferroelectric. As in ferroelectrics the number of charge carriers is generally still low, we deal with capacitors anyhow. Thus equivalent circuits for a semiconducting ferroelectric itself will always contain a conductive contribution (parallel resistor) as well as a series of capacitances. The equivalent circuit is similar to the one for the Maxwell-Wagner effect, only it stems from a slightly different origin, see Fig. 31. The interface depletion region adds another capacitor in series. If in practice such a situation is encountered, the interested reader is referred to the above link or the relevant sections in the introductory books on semiconductors.

3.7 Heterojunctions

The metal-semiconductor contact and the well-known p-n junction are the typical interfaces discussed in semiconductor textbooks. For heterogenous materials additional interfaces within the microstructure/ layer system can exist, namely adjacent semiconductors of different band widths and electron affinities, the *isotype heterojunction*, also called n-N (or p-P) junction. An isotype heterojunction is different from an anisotype heterojunction in that the dopants of the two sides are of the same type. The isotype heterojunction may also occur between two ferroelectric materials with different energy band gaps or between a ferroelectric and a non-ferroelectric semiconductor Zhan et al. (2015).

When materials with different work functions are brought into contact, charge redistribution occurs. In the case of a heterojunction a rectifying barrier may be produced. Figure 18 shows an example of a heterojunction between two semiconductor materials with n-type (Nd) and N-type (ND) doping. When these two materials are brought into contact, the charge redistribution produces a space charge region. There will be electrons that change sides from the N region to the n region in order to align the Fermi levels of both materials. The space charge distribution is shown in step 2 for the regions $x < 0$ and $0 < x < xN$. The electric field distribution (step 3) and the potential profile (step 4) result from this charge distribution. The corresponding band diagram of an n-N heterojunction is shown in step 5, which is very similar to that of the unbiased p-N heterojunction. It can be observed that the difference ΔE_C acts as an electron-blocking layer at the conduction band and the difference ΔE_V as hole-transport blocker at the valence band. Different alignments of the energy bands at the interface determine the effective behavior of the heterojunctions. Figure 19 shows the possible three different types of heterojunctions arrangement. They are defined as: straddling gap (type I), staggered gap (type II), or broken gap (type III). The equilibration of the fermi levels across the entire structure yields the corresponding band bendings.

Figure 20 shows the energy band diagrams for a P-n-N heterojunction. Here, it is assumed that the n region between the P- and N-doped semiconductor materials is wide enough such that the space charge regions near the two junctions (P-n and n-N) do not merge. If this region is narrow, the flat portion in the n-doped semiconductor will not be present, and the band diagram is further distorted. Many more scenarios can be envisaged, each yielding a different space charge, field and electrical characteristic.

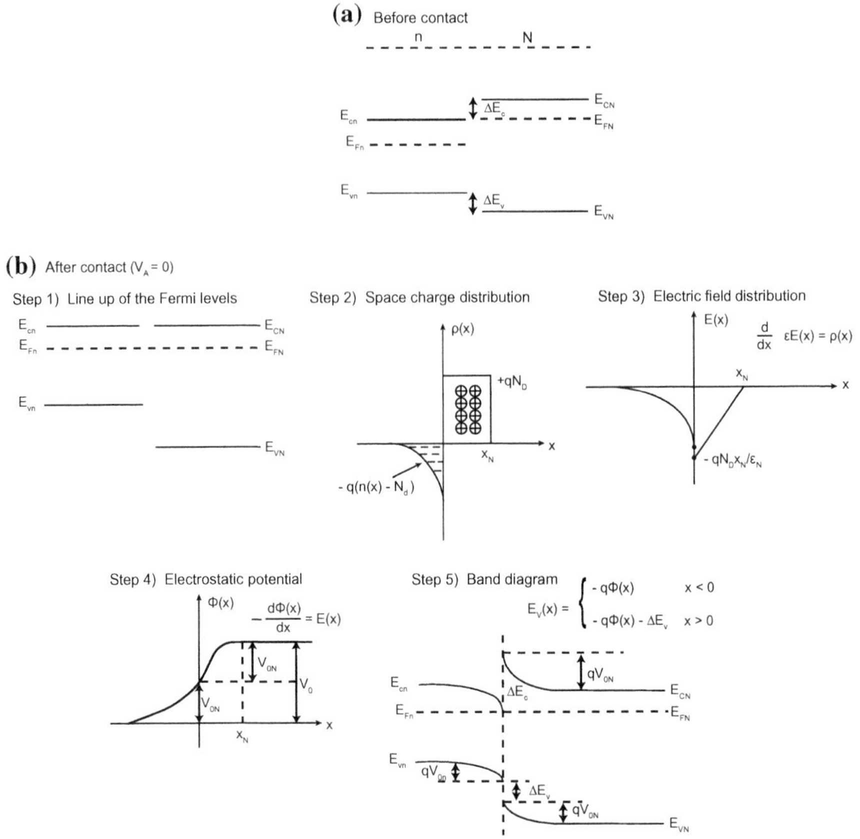

Fig. 18 A step-by-step clarification for the energy band diagram of an isotype n-N heterojunction (**a**) before contact and (**b**) after contact showing five steps to achieve the steady-state energy band diagram, Chuang (2009)

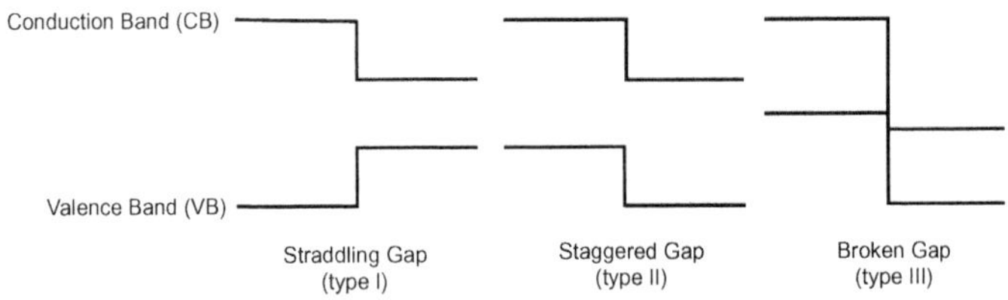

Fig. 19 Three types of heterojunctions arranged by band alignment, from © Wikipedia

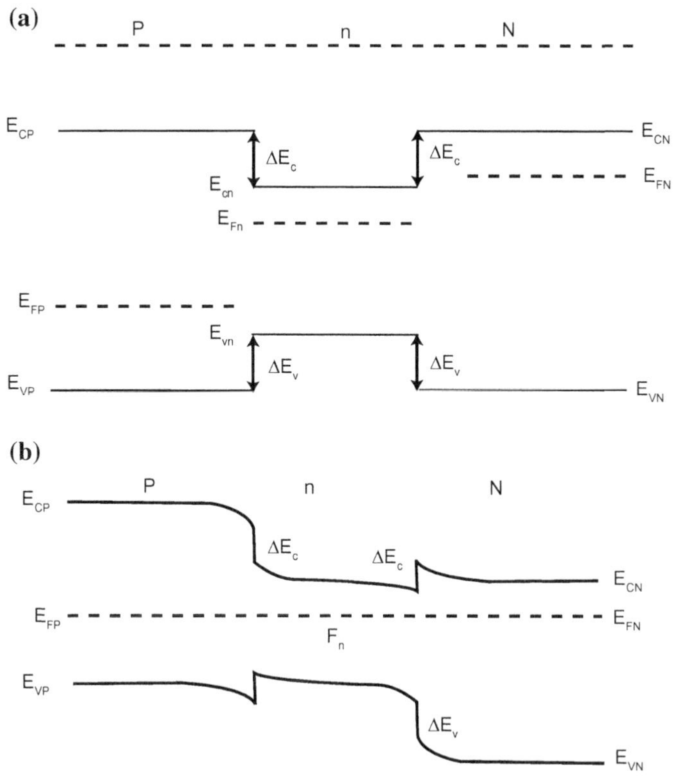

Fig. 20 Energy band diagrams for a P-n-N heterojunction structure: **a** before contact and **b** after contact. Assumption: thermal equilibrium Chuang (2009)

3.8 *Structural Defects*

„Crystals are like people, it is the defects in them which tend to make them interesting!", ***Colin Humphreys***

A perfect crystal with regular arrangement of atoms can not exist, it is an idealization. All crystals have some defects. In semiconductor materials the defects play a very important role in the understanding of their electrical and optical properties. Some properties, such as density and elastic constant are proportional to the concentration of defects, so the effect on such properties is small if the defect concentration is small. Other properties such as color of a material or the conductivity of a semiconductor crystal are much more sensitive to the presence of a small number of defects.

The knowledge of characteristics of defects to achieve the desired property of any semiconductor device is essential in design and fabrication of the device Henderson (1972), Hummel (2011), Wang et al. (2013). Based of their geometry, crystal lattice defects can be classified into the following categories:

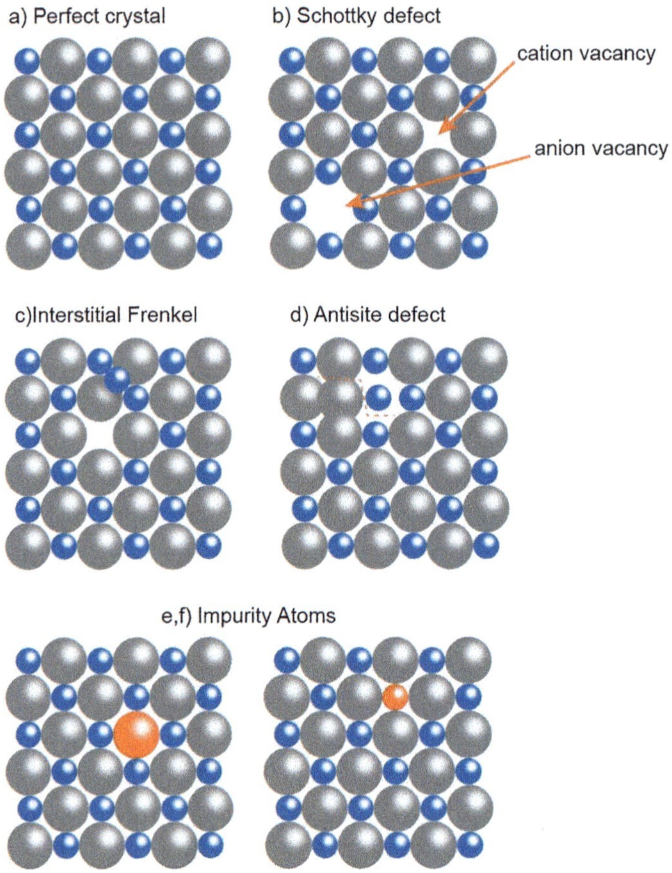

Fig. 21 Different types of point defects

(i) Zero-dimensional (0-D)
(ii) One-dimensional (1-D)
(iii) Two-dimensional (2-D)
(iv) Three-dimensional (3-D)

(i) Zero-Dimensional (0-D) Defects
Are also called point defects. They are places where an atom is missing or is irregularly placed within the lattice structure. Point defects exist as vacancies, interstitial defects, substitutional impurity atoms, and interstitial impurity atoms. Combinations hereof constitute combined or more complex point defect configurations.

To understand how these defects are formed we consider a perfect binary-crystal of type M^+A^-.

Vacancies Vacancies are atoms or ions missing from a crystal lattice. Vacancies are point defects. Their extension is roughly that of an atom. In metals it is hard to identify vacancies experimentally, because they are immediately compensated by free charge carriers. One of the most effective methods for their identification locally is positron annihilation lifetime analysis Keeble et al. (1998). A positron is a positively charged nuclear anti-particle. It is attracted by the higher density of electrons at the vacancy with respect to the perfect lattice and becomes localized around the vacancy forming a hydrogen like electronic orbital. When the quantum mechanical interaction between the wave function of the positron and an electron from the metal is sufficiently large, the positron and the electron do a matter-anti-matter recombination generating a large amount of energy emitted as a detectable photon.

In ionic compounds vacancies can have different charge states. Electron paramagnetic resonance is a method to identify them individually Keeble et al. (1996). For oxygen, the external partial pressure at high temperature can be used to set the amount of oxygen vacancies in the lattice, see Sect. 4.5. As the vacancies are charged, they much interact with polarisation *and* the electronic system of the semiconductor.

Vacancies have to be strictly discerned from *voids*. Voids are large defects containing hundreds, thousands or even millions of (grossly uncharged) vacancies. They are thus microstructural defects rather than point defects. The difference between voids and pores then only depends on the scientific society speaking. Inner surfaces of pores may be charged to a certain extent by chemical species or dangling bonds.

Schottky Defects This defect arises when a pair of positive and negative ions is missing from a crystal lattice as shown in Fig. 21. This type of defect maintains charge neutrality. The quasi chemical reaction of formation of Schottky defects can be written as

$$0 \rightarrow V_{\mathrm{M}} + V_{\mathrm{A}}. \tag{97}$$

The 0 (=zero) denotes a perfect lattice. Schottky defects typically arise in NaCl, BeO, CaO, or CsX, (X =Cl, Br, I) during crystal growth or at moderate temperatures when ionization is still weak but a certain gain in entropy due to the formation of the vacancy pair already yields a slight lowering of Gibbs' free enthalpy.

Interstitial Defects If an atom or ion is removed from its normal position in the lattice and is placed at an interstitial site, the point defect is called an interstitial. The most common interstitials are protons existing in many materials. In metals interstitial hydrogen is one of the most feared point defects, because it entails hydrogen embrittlement. At high temperature also oxides can contain a certain amount of proper atoms occupying interstitial sites. Due to the high energy involved, they are typically rare. One of the prerequisites to obtain larger amounts of interstitial atoms are open crystal structures. In such systems often interstitial sites are occupied leaving a regular lattice site unoccupied (Fig. 22). This easily maintains charge neutrality. The balance between entropy gain and energy cost of the disorder determines how many interstitial sites are taken. In certain crystals the amount of interstitial atoms/ions significantly alters e.g. magnetic properties. Close packed systems like perowskites rarely exhibit interstitials. Hydrogen H^+ is a very small ion. It can thus occupy many sites that are forbidden elastically for other ions. The role of hydrogen in oxides is complex and not yet understood for many systems.

Well known examples of ionic crystals showing anionic interstitials are: CaF_2, SrF_2, BaF_2, ZnO, CeO_2, ThO_2. Cationic interstitials have been well characterized for: AgCl, AgBr, $NaNO_3$, or KNO_3. Both types of defects are rare in perowskite ferroelectrics but may arise in other crystal systems. Lithium niobate, one of the ferroelectrics with the highes Curie point and much utilized in optical systems, is a crystal where lithium vacancies occur and lithium interstitials may form. For long, this crystal was believed not to be a ferroelectric, because the many point defects would electrically suppress domain switching. This form is called congruent lithium niobate according to its manufacturing route. Stoichiometric $LiNbO_3$ as well as $LiTaO_3$ then proved to yield well switchable ferroelectric domain systems much later, Shur et al. (2001, 2002).

Frenkel Defects A Frenkel defect is an atom that leaves its regular lattice site and occupies a nearby interstitial site. It gives rise to two defects i.e. one vacancy and a self interstitial. These two defects are called Frenkel defect or Frenkel pair, Fig. 21c, d. The formation of interstitial defects is symbolized by (Fig. 22):

$$\mathrm{M_M} \rightarrow V_\mathrm{M} + M_i \tag{98}$$

Antisite Defects Arise, when atoms of different types exchange the positions in lattice, Fig. 21e. The defect reaction for the anti-site is therefore:

$$\mathrm{A}_A + \mathrm{M}_M \rightarrow \mathrm{M}_A + \mathrm{A}_M$$

(ii) One-Dimensional (1-D) Defects

Are line defects. These defects form as and extend along a line in a crystal lattice. Dislocations are the most commonly encountered line defect. The only other line defect could be a breakdown path or chemical alignment of atoms along a line. This is a rare scenario.

There are two basic types of dislocations:

(a) Edge dislocation
(b) Screw dislocation

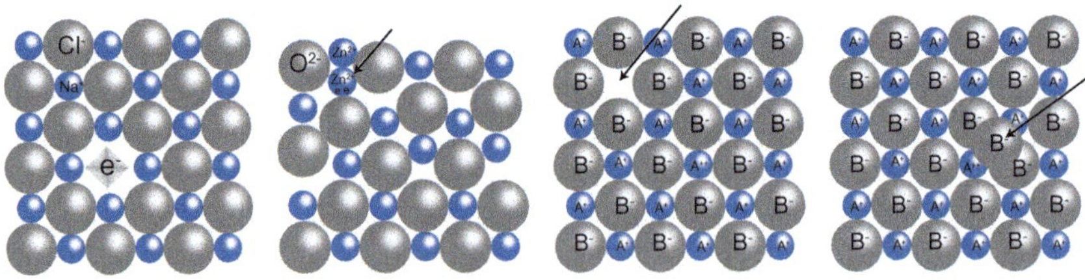

Fig. 22 Point defects in ionic crystals: color center, charged Frenkel defect, uncharged vacancy, and charged interstitial (untypical as oversized)

Along a closed dislocation loop, the character of a dislocation changes from edge to screw and back Hull and Bacon (1984). Dislocation loops have been visualized in non-stoichiometric $BaTiO_3$, Suzuki et al. (2001). They are very rare in the bulk of an ionic crystal.

Edge Dislocation Mechanical forces acting on a crystal can be partly compensated by the formation of dislocations plastically relaxing the stresses on a crystal. Dislocation motion is a dissipative process. Energy is consumed by interaction of dislocations with each other and by heat. When enough shear stress is applied plastic deformation sets in. In typical metals the motion of large numbers of dislocations arises in response to shear stress and plastic deformation may account for 30% strain or more.

Figure 23 shows an edge dislocation. It is evident that an extra plane of atoms is inserted from the top side. This side of the dislocation thus experiences compressive stress. The bottom side, where this extra plane is missing, experiences tensile load. In ferroelectrics dislocations are typical in thin films grown on substrates with non matching lattice constants. During epitaxial growth the new layers of atoms try to match the underlying substrate's periodicity. Beyond a few atomic layers, stresses amount to such high values that the system tries to escape this mechanical mismatch. The stresses in the films are partly compensated by dislocation formation directly at the interface to the substrate Chu et al. (2004), Yang et al. (2015). They are a source of mechanical domain fixation and extra charges.

The *Burgers vector* of a dislocation is a crystal vector, specified by Miller indices, that quantifies the difference between the distorted lattice around the dislocation and a perfect lattice, Hull and Bacon (1984). An edge dislocation has a Burgers vector perpendicular to its dislocation line (Fig. 23).

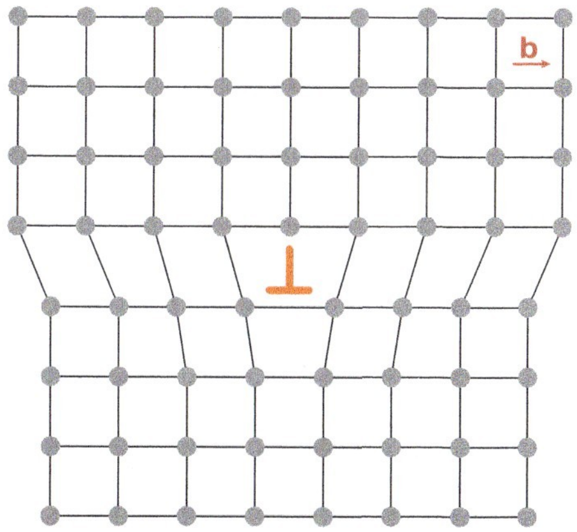

Fig. 23 Edge dislocation

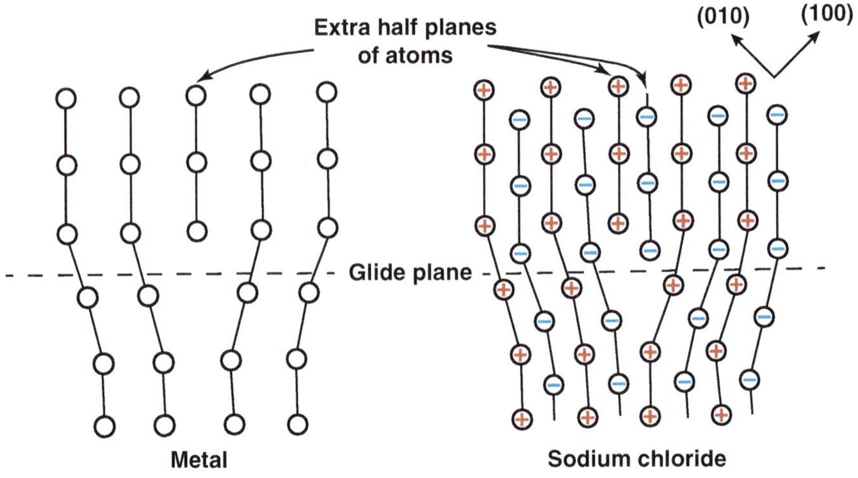

Fig. 24 Edge dislocation in an ionic crystal

In ionic crystals edge dislocations equally exist, Kingery et al. (1976). The major difference to mono-atomic metals is that layers of both polarities must be passed by the Burgers vector. Thus, the Burgers vector becomes at least twice longer. As the Peierls-Nabarro-Energy of the energy stored in the dislocation is proportional to the square of the Burgers vector, at least four times the energy of dislocation formation has to be applied in order to generate a dislocation in an ionic crystal compared to a metal (of equal elastic shear constant and domain width) (Figs. 24 and 25).

Screw Dislocation: A screw dislocation is more complex. The Burgers vector is parallel to the dislocation line. The motion of a screw dislocation is also a result of shear stress, but the defect line moves perpendicular direction of the atom displacement. To visualize a screw dislocation is much harder. You can imagine a block of metal that begins to rip by applying a shear stress to one end. The atoms represented by the

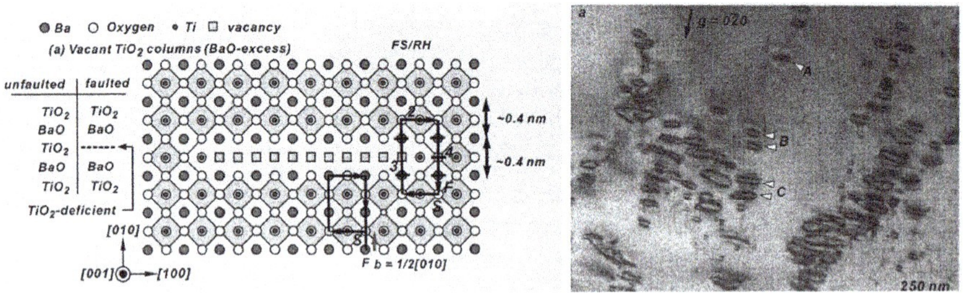

Fig. 25 (Left) Ionic arrangement in a dislocation loop in $BaTiO_3$. A similar picture can be drawn for TiO_2 excess. (right) Transmission electron microscopy image of such dislocation loops. Images taken from Suzuki et al. (2001)

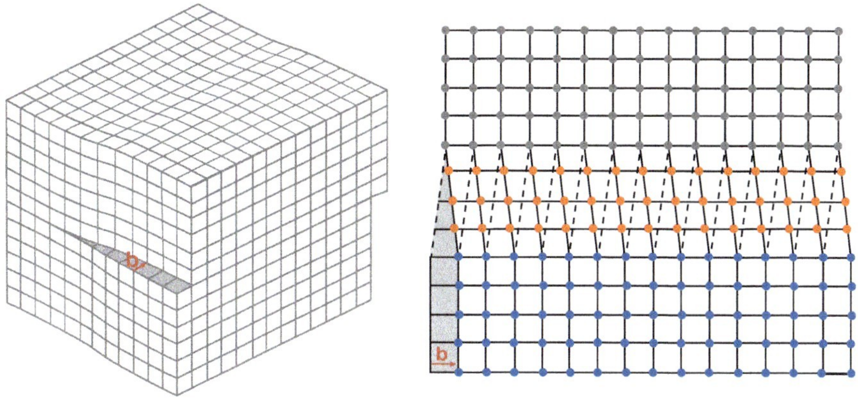

Fig. 26 Screw dislocation in a simple cubic lattice (metal)

gray balls will not move from their original position. The atoms represented by the orange balls are in the process of moving and only a small portion of the bonds are broken at a time. And the atoms represented by the blue balls have moved to their new position in the lattice and have re-established metallic bonds. If the shear force is increased, the atoms will continue the movement. This is shown in the Fig. 26. So a screw dislocation has its Burgers vector parallel to the dislocation line.

(iii) Two-Dimensional (2-D) Defects

Two-dimensional defects *in* crystals arise as stacking faults, or at the limit as twin boundaries. Twin boundaries in some context are also considered grain boundaries. All other planar defects are interfaces. These can be grain boundaries, sub-boundaries, and outer surfaces. The most important 2-D-defect in semiconductors is an interface between two different semiconductor materials, the classical p-n-junction. It is the working horse of a vast number of semiconductor devices.

Any type of 2-D-defect has a significant influence on the optical and electrical properties.

Stacking faults A stacking fault occurs wherever the regular layer pattern in a crystal structure is broken (Fig. 27). There are three kinds of stacking faults: intrinsic, extrinsic, and twin boundaries Weertman and Weertman (1992). In face-centered cubic lattices (fcc) the ideal stacking sequence of the close packed planes is ABCABCABC.

Extrinsic stacking faults can be created by inserting an extra plane of atoms into the structure. In the ABC*B*ABC pattern, the B plane in the middle is an extrinsic stacking fault.

Intrinsic stacking faults can be created by removing a plane of atoms. For example in the ABCABC pattern of a face centered cubic lattice ABC*BC*ABC is missing an A-plane which represents an intrinsic stacking fault.

Stacking faults are well known and easily described in metallic systems, where we have shown the most common representative here. We refer the interested reader to

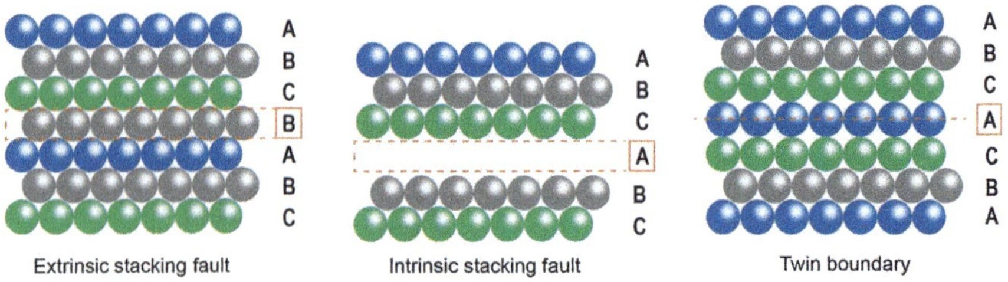

Fig. 27 Schematic representation of the intrinsic and extrinsic stacking faults and twin boundary

more details in the relevant textbooks, particularly on the three dimensional images involved. It is also easy to understand the layer sequence in an fcc lattice if one buys marbles of three colours and puts them into three layers. If the top level coincides with the bottom layer we have ABA which is the hexagonal layer sequence. If the third empty triangle is filled, we obtain the face centered cubic lattice. Please observe that the layers are all [111]-layers thus the vector perpendicular to the plane is along a <111>-direction.

In close packed ionic crystals stacking faults are again rare. If they occur, they represent a discontinuity in electronic wave function and will alter the local electronic states.

A twin boundary is a surface that separates two volumes of crystal that are mirror images of one another. In the ABCACBA pattern, the plane A is a twin boundary. But twin boundaries can have many forms. In ferroelectrics the most typical twin boundary is the ferroelectric domain wall. Typically, we are thus concerned with 180° and 90° twins in tetragonal systems and 110° and 70° twins in rhombohedral crystals, Eknadiosiants et al. (1990).

Grain boundaries A grain boundary is a general planar defect that separates regions of different crystalline orientation (Fig. 28). In the closer sense grain boundaries only arise in between crystals of same type. Ordered grain boundaries are twin boundaries and contain certain periodicity. In general, grain boundaries are not well ordered and contain many unmatched chemical bonds. This is the reason why electrical defects are found at grain boundaries in large amount. Poly-crystalline solar cells e.g. generate much less output than single crystal solar cell. Charge carriers scatter at the grain boundary and many of them are reflected back into the grain of their origin. Later we will see, that electrical properties in ferroelectrics can be immensely influenced by the grain boundary as well, see Sect. 5.2 (Figs. 37, 38 and 39).

Hetero Interfaces A grain boundary is a general planar defect that separates regions of different crystalline orientation. If these two crystals are of different kind, i.e. in real composites, they form semiconductor hetero interfaces. In semiconductor physics of solar cells these are termed heterojunctions. Two semiconductors of different work functions, electron affinities, and band gaps meet at the interface. If the interface is clean, i.e. without surface states, the classical heterojunction arises, Sze (1981),

Fig. 28 Grain boundaries

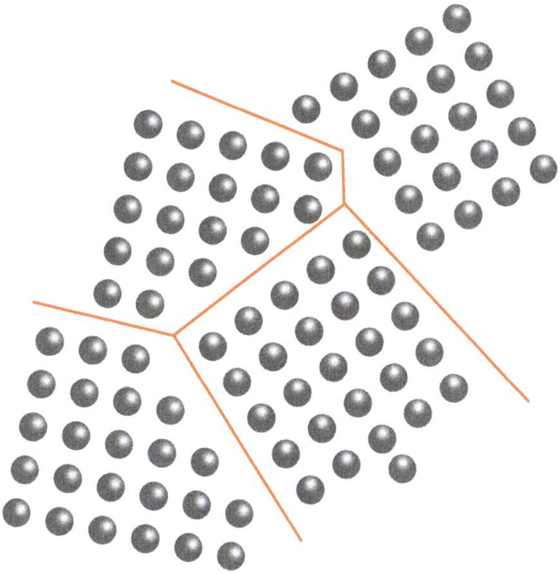

Möschwitzer (1992). We will later see that e.g. multiferroic composites contain such heterojunctions, see Sect. 5.3. Particularly if one of the two partners is conducting, pronounced effects like space charge layers and interface fields arise.

(iv) Three-Dimensional (3-D) Defects

There are numerous three dimensional defects: precipitates, second-phase particles or dispersants, inclusions, and voids. From a semiconductor point of view they all represent localized external interfaces to the semiconductor. Depending on the character of the constituting material, we may be concerned with any one of the previously discussed interfaces, namely, ohmic contacts, Schottky contacts, open surface to a gas (potentially humid), or a grain boundary to another semiconductor. All previously discussed electrical scenarios may thus occur at these interfaces. Material synthesis must avoid such 3-D-defects as much as possible, because the material behaviour becomes enormously complex and unpredictable.

In semiconductor physics, people have artificially designed particular precipitates, namely quantum dots or similar confined structures. These may yield wanted properties, but manufacturing is a challenge. Lithographic or self assembly techniques must be applied. In the context of our present manuscript, we will not address these effects and designs.

Precipitates are small particles (fraction of a micron) that are introduced into the matrix by solid state reactions. Their role is e.g. to increase the strength of structural alloys by acting as obstacles to the motion of dislocations. The size, the internal properties and the distribution through the lattice of the precipitates play a very important point.

Dispersants are larger particles ($10 - 100\mu$m). When a microstructure contains dispersants, the mechanical strength and electrical conductivity are some average of the properties of the dispersant phase and the parent. Inclusions vary in size from a few microns to macroscopic dimensions. They are undesirable particles and enter in the system as dirt. In microelectronic device they disturb the geometry of the device by interfering with manufacturing requirements. In ferroelectrics they arise as dirt from the raw materials or as segregated particles from sintering of non-stoichiometric green bodies or by chemical segregation. One recent example where they have been intentionally introduced is in the case of ZnO in $BaTiO_3$ where they serve to stabilize the microstructure and aid local conductivity at grain boundaries reducing fatigue effects Zhan et al. (2015).

Voids (or pores) are undesirable defects in the solid formed by trapped gases or by vacancy condensation. Their principal effect is to decrease mechanical strength and promote fracture at small loads. As the voids contain a gas (air or other) their dielectric constant is near 1. This means that much of the electric potential drop arises across the pore and does not reach the environing material of high dielectric constant. This means that effectively, the fields in the ferroelectric material drop considerably. This often represents a serious problem irrespective of any semiconductor effects.

4 The Ferroelectric Semiconductor

Semiconductor effects in ferroelectrics and their interfaces have become a highly investigated topic in recent years. Many integrated devices for data storage are envisaged. The basic physics was described many years ago. Fridkin summarized a number of Russian papers from the 1970ies into his book on semiconductor ferroelectrics, Fridkin (1980). Anyone who will work on the topic more intensively must read this summary and maybe also the second book on photoeffects, Fridkin (1979). Also a good read in the context of linking semiconductor to dielectric effects can be found in Bunget and Popescu (1984).

Fundamentally, we will show that energy is the quantity that links quantum mechanics and thermodynamics. This means that things that happen in the electron system and on charged defects will influence the thermodynamic behaviour. On the other hand, thermodynamics also influence the quantum mechanical response. This book is primarily written for those persons who want to understand ferroelectrics in their thermodynamic context. This also includes charge transport where it affects the overall response of a ferroelectric piezoelectric.

We will not tackle the fact that phase transitions in conducting ferroelectrics will alter the electronic system slightly. Furthermore, we will not go into detail about the optical properties, that entail refraction index modifications, Wechsler and Klein (1988), holographic polarization fixation, photovoltaic effects in ferroelectrics, and other processes associated with illumination. The principal basics are summarized again in Fridkin (1979, 1980). Descriptions of the optical effects can also be found in the book by Waser (2012), particularly in the nice chapter on electron holography in ferroelectrics contributed by Theo Woike.

4.1 Energy Value Considerations

Fridkin (1980) has elaborated on the fundamentals to calculate internal energy. It necessitates integration of heat capacity from zero Kelvin up to the temperature at which the system is being held. In doing so, he compared the energetics of the electronic system with that of the dynamic lattice, namely the fundamentals of Landau theory. The ratio of heat capacities

$$\frac{C_V^{el}}{C_V^L} = \frac{N_c}{n_0} \tag{99}$$

tells us, that, typically, the energy contribution of the electronic system can be neglected compared to lattice vibrations. $N_c = \left(\frac{2\pi m^* kT}{h^2}\right)^{3/2}$ is the density of states in the respective bands near the band edge (in parabolic approximation) with m^* the effective mass of the respective charge carrier and n_0 the density of unit cells per volume. Only at very low temperatures or *near phase transitions* its influence becomes relevant for classical ferroelectric systems. Once we encounter low energy band gaps though, e.g. in the sulfides or hexaferrites, the electronic contribution becomes significant. It has also to be understood that in the case of doping, which commonly occurs in ferroelectrics e.g. as vacancies, the effective energy band gap is that of the donor / acceptor state with respect to the band state. The defect states often lie close to the bands and then the discussion changes directly.

4.2 A Joint Energy Function

We now come to the point where the energy worlds of the semiconductor and that of the ferroelectric start interacting. For this purpose we will simply write down the energy function of the entire system, namely its free energy. Energy is an additive quantity, so we obtain

$$F = F_0 + F_1 + F_2 \tag{100}$$

with $F_0(T)$ being the free energy of the system experiencing neither an external electric field, $E = 0$, nor mechanical stress, $\sigma = 0$. Also the density of free charge carriers $N_i = 0$ is zero.

F_1 is the well established Landau expansion of the free energy function according to Ginzburg-Landau theory as before in Sect. 2.3.

$$F_1 = \frac{1}{2}\alpha P^2 + \frac{1}{4}\beta P^4 + \frac{1}{6}\gamma P^6 - \frac{1}{2}s_{ik}\sigma_i\sigma_k - P^2 Q_k\sigma_k \tag{101}$$

assuming Einstein's summation rule and Voigt's matrix notation for the tensors. α, β, and γ are the expansion coefficients of the Landau function. s_{ij} is the compliance matrix and ν_k the electrostriction coefficient. The latter ist reduced in nomenclature from tensor notation as P here is unidirectional and scalar. We do not use the notation according to Fridkin's book ($\nu_k = Q_k$), because the typical electrostriction coefficient is denoted by Q_k these days. The matrix notation used here assumes that strain only forms with respect to a single polarization direction P and neglects the fact that Q is actually a tensor of rank four. Complexities arising in the context of less symmetric crystal classes and coupling between polarization components in different directions within these crystals are neglected, because they would overload the issue.

The third contribution to free energy is the energy of thee free charge carriers

$$F_2 = \sum_i N_i E_i\,(T, P, \sigma_k)\,. \tag{102}$$

The sum sign is explicitly used here, because it sums over the large number of charge carriers and not over direction indices in space to avoid confusion. The roman E will denote the energy in terms of typical semiconductor literature while electric field will be expressed as $\mathbf{E}$.

Every real ferroelectric is inherently doped due to impurities in the raw materials when making ceramics or single crystals. Typically, the degree of doping is several orders of magnitude larger than in classical silicon based systems. Besides the quality of the raw materials, also the susceptibility of the system to forming oxygen vacancies is large (see Sect. 4.5).

Let us consider a doping concentration N_D ($= M$ in Fridkin 1980) and assume it to generate N_D^+ charge carriers ($= N$ in Fridkin 1980). This number is much larger than the number of intrinsic charge carriers n.

$$F_2 = nE_g + N_D^+\,(E_{CB} - E_D) - p\,(E_A - E_{VB}) \approx N_D^+\hat{E} \tag{103}$$

with $\hat{E} = E_D - E_A$ defining the reduced band-gap width.

Overall, the coupling between thermodynamics and electronic states now leads to a re-scaling of properties on either side, namely in thermodynamics as well as in the electronic states. The reduced band-gap becomes

$$\begin{aligned}\hat{E}(T, P\sigma_k) = {} & \hat{E}_0(T) + \frac{1}{2}aP^2 + \frac{1}{4}bP^4 + \frac{1}{6}cP^6 \\ & + \frac{\partial\hat{E}}{\partial\sigma_k}\sigma_k + \frac{1}{2}\frac{\partial^2\hat{E}}{\partial\sigma_k\partial\sigma_i}\sigma_k\sigma_i + P^2\frac{\partial^3\hat{E}}{\partial P^2\partial\sigma_k}\sigma_k\end{aligned} \tag{104}$$

with

$$a = \left(\frac{\partial^2 \hat{E}}{\partial P^2}\right)_0, \quad b = \left(\frac{\partial^4 \hat{E}}{\partial P^4}\right)_0, \quad c = \left(\frac{\partial^6 \hat{E}}{\partial P^6}\right)_0. \tag{105}$$

Again Einstein's summation rule applies. The free energy turns out to be

$$\begin{aligned} F(T, P, \sigma_k, N_D^+) &= F_0^N + \frac{1}{2}\alpha^N P^2 + \frac{1}{4}\beta^N P^4 + \frac{1}{6}\gamma^N P^6 \\ &+ N_D^+ \frac{\partial \hat{E}}{\partial \sigma_k}\sigma_k - \frac{1}{2}s_{ik}^N \sigma_i \sigma_k - P^2 Q_k^N \sigma_k \end{aligned} \tag{106}$$

now using a superscript to denote the re-scaled quantities. This entails a number of re-scaled coefficients:

$$\begin{aligned} F_0^N &= F_0 + N_D^+ \hat{E}_0, \qquad \alpha^N = \alpha + aN_D^+, \qquad \beta^N = \beta + bN_D^+ \\ \gamma^N &= \gamma + cN_D^+, \quad Q_k^N = Q_k - \frac{\partial^3 \hat{E}}{\partial P^2 \partial \sigma_k} N_D^+, \quad s_{ik}^N = s_{ik} - \frac{\partial^2 \hat{E}}{\partial \sigma_i \partial \sigma_k} N_D^+ \end{aligned} \tag{107}$$

The equations of state become

$$\mathbf{E}_{||P} = \frac{\partial F}{\partial P} = \alpha^N P + \beta^N P^3 + \gamma^N P^5 - 2PQ_k^N \sigma_k, \tag{108}$$

$$-x_k = \frac{\partial F}{\partial \sigma_k} = N_D^+ \left(\frac{\partial \hat{E}}{\partial \sigma_k}\right)_0 - \frac{1}{2}s_{ik}^N \sigma_i - P^2 Q_k^N \tag{109}$$

with $\mathbf{E}_{||P}$ being the electric field in direction of polarization and $-x_k$ strain.

4.2.1 Modification of Known Properties

A number of properties now change according to the semiconductor contribution.

- A *Shift of the Curie point* occurs with a reduced transition temperature

$$T_0^N - T_0 = -CaN_D^+ \tag{110}$$

- *Polarization* takes a more demanding form:

$$\left(P^N\right)^2 = P^2(T)\left[1 + \frac{bP^2(T)N_D^+}{\beta P^2(T) + 2\alpha} + \frac{c\alpha P^4(T)N_D^+}{\beta P^2(T) + 2\alpha} - \frac{cN_D^+}{\gamma}\right] \tag{111}$$

  This entails a modified jump in polarization at the transition temperature $P_0 = P(T_1)$:

$$\left(P^N\right)_0^2 = P_0^2\left[1 + \frac{bN_D^+}{\beta} - \frac{cN_D^+}{\gamma}\right] \tag{112}$$

- For first order phase transitions, the transition temperatures show temperature hysteresis. A *change in the temperature hysteresis* between rising an decreasing temperature arises due to the semiconductor contribution

$$\Delta T_h^N \approx \frac{3C}{16}\frac{\beta^2}{\gamma}\left[1+\frac{2bN_D^+}{\beta}-\frac{cN_D^+}{\gamma}\right] \tag{113}$$

- A change in *spontaneous deformation*

$$x_k^N = Q_k^N P_0^2 - N_D^+\left(\frac{\partial \hat{E}}{\partial \sigma_k}\right)_0 = Q_k P_0^2\left[1+\frac{(P_0^N-P_0)^2}{P_0^2}+\frac{N_D^2}{Q_k}\frac{\partial^3 \hat{E}}{\partial P^2 \partial \sigma_k}\right] \tag{114}$$

- A change in *dielectric properties*

$$\epsilon(T_1)\Delta T_h^N \approx C, \tag{115}$$

if it is assumed that the Curie constant does not change with the electron concentration then $\epsilon^N(T_1) > \epsilon(T_1)$.
- The *piezoelectric properties* are given by the equation of state, Eq. (109), for finite electric fields $\mathbf{E} \neq 0$:

$$d_k^N = \left(\frac{\partial x_k}{\partial \|\mathbf{E}\|}\right)^N = \epsilon^N Q_k^N P^N \tag{116}$$

- And finally a change in *latent heat* and *heat capacity*

$$\Delta S^N = \frac{1}{C}P_0^2\left[1+\frac{bN_D^+}{1+\beta}-\frac{cN_D^+}{\gamma}\right] \tag{117}$$

$$\Delta C^N = \frac{T_0 1}{C^2\beta}\left[1-\frac{bN_D^+}{\beta}\right] \tag{118}$$

This impressive list of formulae should not overwhelm us. As the density of charge carriers is small, the overall contribution of free electrons to the free energy remains marginal. Most terms modify the principal effects near the phase transition only by a small amount and farther away from the phase transition hardly at all.

4.3 Screening

The biggest effect of the charge carriers arises as screening. Screening denotes the fact that the bound charge of polarization is matched by free charge carriers. As such, all electrons in a metallic electrode screen the field effects of polarization in the ferroelectric. But as these electrons can be removed to the outside of the system and contribute to the measurable electric charge, typically, they are not included in a discussion of screening. Thus, the screening of polarization that we are concerned with in this subsection are those charge carriers that will reduce the field effect of polarization reaching the electrode. Screening thus happens *within* the ferroelectric material. Sources of screening can be electronic or ionic. Screening electrons are those that will reduce the polarization response within the electrode at the time scale that is relevant for the removal of charges from the electrode. All these processes are thus time/frequency dependent.

We will again follow Fridkin (1980) in that for screening the use of D instead of P is more appropriate. We will assume polarization along the z-axis only. The free energy then reads

$$F = F_0 + \frac{g}{4}D_0^4 + \frac{\kappa}{2}\left(\frac{dD}{dz}\right)^2 + \frac{a}{2}D^2 + \frac{b}{4}D^4 + \frac{c}{6}D^6 \tag{119}$$

yielding the equation of state through

$$E = \frac{dF}{dD} \tag{120}$$

Poisson's equation must be integrated for the actual charge carrier distribution in the crystal

$$\frac{dD}{dz} = \rho \quad \rho = e\left[p_0(\phi) - n_0(\phi)\right] \tag{121}$$

The boundary conditions are

$$\frac{\partial \sigma_k}{\partial z} = 0 \quad D|_{z=\pm 1/2} = 0 \quad D|_{z=0} = D_0 \quad \phi|_{z=0} = 0 \tag{122}$$

The term $\frac{\kappa}{2}\left(\frac{dD}{dz}\right)^2$ is the energy gradient term particularly necessary to describe boundaries, e.g. domain walls. It is now necessary to be incorporated, because we encounter an outer boundary to the finite crystal, which has been considered to be infinite so far. Under the boundary conditions (122) integrating yields $D = D(z)$, $E = E(z)$, $\rho = \rho(z)$ and $\phi = \phi(z)$. A situation similar to the space charge limited regime in Sect. 3.5.4 arises. D_0 is the dielectric displacement in the center of the crystal where surface effects are negligible. The coefficient g is determined by the elastic properties of the ferroelectric crystal.

We now consider the intrinsic case first, see Fig. 29. It is assumed that the band curvature is not too big, namely $E_{CB} - E_F - e\phi \gg kT$, and $E_F - E_{VB} + e\phi \gg kT$. The free charge carriers try to compensate polarization, thus $\rho = \rho(z)$ and accordingly, $D = D(z)$ and $\phi = \phi(z)$. The charge density takes the following form:

$$\rho = eN_i \exp\left(\frac{e\phi}{kT}\right) - eN_i \exp\left(\frac{-e\phi}{kT}\right) = 2en_i \sinh\left(\frac{e\phi}{kT}\right) \tag{123}$$

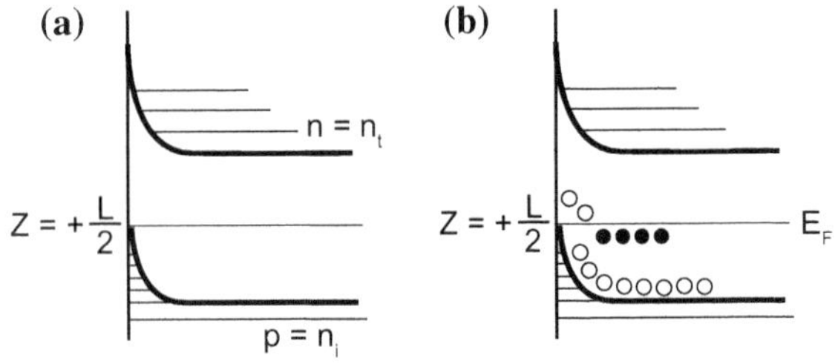

Fig. 29 The band bending of (left) an un-doped and (**b**) an acceptor-doped (p-type) ferroelectric in the screening regime, Fridkin (1980)

The hyperbolic sine only applies in the intrinsic case, obviously.

Guro et al. (1967) showed that the degeneracy of kT in the surface region still permits to use Eq. (123). They arrive at the implicit equation

$$\left[\kappa + \frac{kT}{32\pi^2 e^2 n_i}\left\{1 + \left(\frac{1}{8\pi e n_i}\frac{dD}{dz}\right)^2\right\}^{-1/2}\right]\frac{d^2 D}{dz^2} = aD + bD^3 + cD^5 \quad (124)$$

to determine the value of the dielectric displacement. Successive integration yields the relation between charge density ρ and dielectric displacement D:

$$\rho = \frac{D_0}{4\pi}\frac{1}{\sqrt{\kappa\zeta}}\left[1 + 2\xi + \zeta Q(t) - \sqrt{(1+2\xi)^2 + 2\zeta Q(t)}\right]^{1/2} \quad (125)$$

with t denoting normalized D: $t = D/D_0$. Implicitly, the dielectric displacement is given by:

$$z(D/D_0) = \sqrt{\kappa\zeta}\int_0^t \left[1 + 2\xi + \zeta Q(t') - \sqrt{(1+2\xi)^2 + 2\zeta Q(t')}\right]^{-1/2} dt' \quad (126)$$

using the following variables:

$$\xi = \frac{\kappa}{L_D^2} \quad \text{and} \quad \zeta = \frac{8\pi^2 D_0^2 \kappa e^2}{(kT)^2} \quad \text{and} \quad (127)$$

$$Q(t) = (1-t^2)\left[\frac{1}{4\pi\epsilon_0}\right] + (1-t^2)\left(\frac{bD_0^2}{2} + \frac{cD_0^4}{3} + (1-t^4)\frac{cD_0^4}{4}\right). \quad (128)$$

D_0 is given through

$$\frac{1}{4\pi\epsilon_0} = -\left(a + bD_0^2 + cD_0^4\right), \quad \frac{L}{2} = z(1) \quad \text{and} \quad l_D = \frac{1}{4\pi}\sqrt{\frac{kT}{e^2 n_i}} \quad (129)$$

l_D being the Debye screening length. It is large for small concentrations of carriers n_i or high temperatures and is typically large in ferroelectrics due to the small concen-

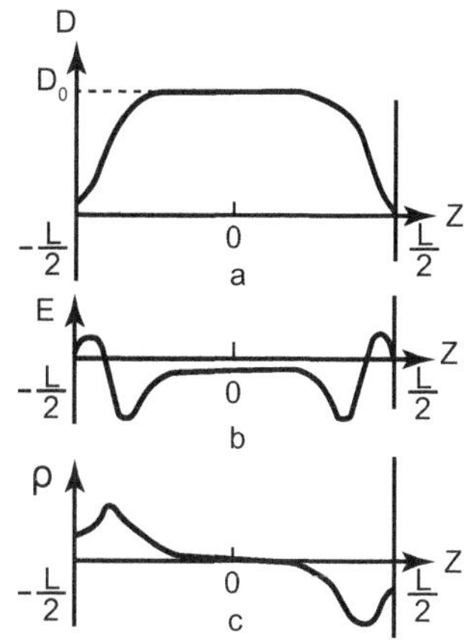

Fig. 30 Distribution of the electric displacement, electric field, and charge density for a sheet with single domain polarization at $T \simeq T_1$, Chenskii (1970)

tration of free carriers. Thus, Guro et al. (1967) assumed that $l_D \gg L$ for deriving the above equations.

Yet another length scale matters in ferroelectric semiconductors: the correlation length l_κ which is the width of the 180° domain wall. Using Eq. (125) for determining the surface charge ρ_S, corresponding to $D \simeq 0$ and $t \simeq 0$ along with knowing that $\zeta Q(0) \gg 1$

$$\rho_S \simeq \frac{D_0 1}{4\pi}\left[\frac{Q(0)}{\kappa}\right]^{1/2}, \quad n_S = \frac{\rho_S}{e} \simeq \frac{D_0}{4\pi e l_\kappa} \quad \text{with} \quad l_\kappa = \left[\frac{\kappa}{Q(0)}\right]^{1/2} \tag{130}$$

We are now at the point where we can determine numbers. Again Guro et al. (1967) determined the surface carrier density from $l_\kappa \approx 2\text{nm}$, $N_S = \rho_S/e \simeq 5 \cdot 10^{20}\text{cm}^{-3}$ using

$$n_i = N_c \exp\left[-\frac{E_c - E_v}{kT}\right], \quad \text{and} \quad N_c \simeq 5 \cdot 10^{19}\text{cm}^{-3}. \tag{131}$$

This yields the curvature of the bands to be $e\Delta\phi \simeq E_c - E_v$ thus a *curvature of the bands in the order of the band width*! Also the effective charge carrier density in this volume at the outer polar faces of a single domain ferroelectric crystal is large! Fig. 29 illustrates the band curvature.

As long as the Debye length l_D does not drop below l_κ, polarization can still arise. If the concentration of carriers is sufficient to fully screen polarization within one unit cell, then the polarization breaks down in the sense that it cannot be measured any longer electrically and is only potentially accessible by XRD or similar techniques. It will not function electrically.

Nevertheless, charge carrier concentrations much larger than assumed so far can be exploited without endangering the ferroelectric state itself. Fridkin (1980) also summarizes the results by Chenskii and Sandomirskii (1969) when the Debye screening length becomes the relevant length scale at higher carrier concentrations

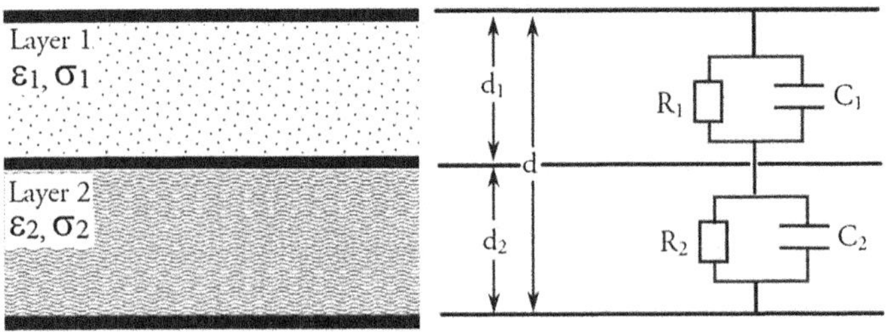

Fig. 31 Two-layer capacitor and its equivalent circuit. Taken from Prodromakis and Papavassiliou (2009)

($n \simeq 10^{17}\text{cm}^{-3}$). Figure 30 illustrates the resulting distribution of the electric displacement, electric field, and charge density for a single domain sheet of a ferroelectric in the case of the Debye length as the relevant length scale. It is obvious that the field situation at / near the electrodes becomes highly crucial. To go further into details would mean that we re-write the existing textbook, which is not the intention of this lecture. It was merely the intention to draw the attention to many subtleties one can encounter when dealing with ferroelectrics for which the band gaps are not so large, or the material contains significant amounts of defects.

All of this section is explicitly a rewrite from Chapter 3.1 in Fridkin (1980). Throughout this section, some parts have been used in cgs units, others in SI. This is due to the fact that the book by Fridkin (1980) is an assembly of many independent papers which have not been put in a unified notation. Unfortunately, this should have been corrected here, but is missing due to time constraints in finalizing the manuscript.

4.4 *Maxwell-Wagner-Relaxation*

To understand what its meant by the Maxwell-Wagner effect, the meaning of interfaces must be understood first, which play an important role in electronic devices. For a simple parallel plate capacitor with a structure of metal-insulator-metal, the metal insulator interface is employed to prevent charge injection from metals into insulator and accumulates charges on the metals, in proportion to the applied voltage. In p-n junctions (see appendix), the purpose of the p-n interface is to rectify carrier transport across the interface which results in the formation of charge distribution that is asymmetric with respect to the interface. These representative examples show that an interface is a meeting place of two dissimilar materials, where charges are accumulated. The Maxwell-Wagner (MW) effect accounts for buildup of charges at these two material interfaces, such as metal insulator interfaces, semiconductor semiconductor interfaces, and insulator semiconductor interfaces, Iwamoto (2012).

The simplest model to explain the Maxwell-Wagner effect is a double layer capacitor arrangement where each layer is characterized by its permittivity ε_i and by its relative conductivity σ_i. It is illustrated in Fig. 31 together with the equivalent circuit. It can be shown that the complex dielectric function for this circuit is given by, Schönhals and Kremer (2003):

$$\varepsilon^* (\omega) = \varepsilon_\infty + \frac{\Delta\varepsilon}{1 + i\omega\tau_{MW}}. \tag{132}$$

This equation is similar to the Debye equation for relaxation describing a single relaxation in dielectrics. However, the parameters have a completely different meaning and physical background. For $d_1 = d_2$ it holds that

$$\varepsilon_\infty = \frac{\varepsilon_1 \varepsilon_2}{\varepsilon_1 + \varepsilon_2} \tag{133}$$

$$\Delta\varepsilon = \varepsilon_0 - \varepsilon_\infty = \frac{\varepsilon_2\sigma_1 + \varepsilon_1\sigma_2}{(\sigma_1 + \sigma_2)^2 (\varepsilon_1 + \varepsilon_2)} \tag{134}$$

For the relaxation time τ_{MW} of the interfacial polarization, we have

$$\tau_{MW} = \varepsilon_0 \frac{\varepsilon_1 + \varepsilon_2}{\sigma_1 + \sigma_2}. \tag{135}$$

The relaxation time scales inversely with the conductivity of the system which means that Maxwell-Wagner effects are more pronounced for conductive materials. Thus, with the quite simple model, it is demonstrated that the dielectric response of an inhomogeneous medium can be frequency dependent although none of the individual components has frequency dependent dielectric properties. The frequency dependence can be similar to an orientational polarization. Certain dielectric systems have been reported to have giant dielectric constants. Representative examples are $CaCu_3Ti_4O_{12}$, Liu et al. (2004), Li or Ti doped NiO, Wu et al. (2002), $AFe_{0.5}B_{0.5}O_3$ (A= Ba, Ca, Sr; B = Na, Ta, Sb), Raevski et al. (2003), and $LaMnO_3$ Lunkenheimer et al. (2002). All these materials have similar dielectric behavior. The real part of electric permittivity ε' when plotted as a function of frequency, reveals an almost constant low-frequency value and shows a step-like decrease of the dielectric constant towards higher frequencies. This step-like decrease in ε' is accompanied by a loss peak in the imaginary part of the permittivity ε'' as illustrated in Fig. 32. It was realized that some extrinsic effects in materials play a more crucial role for the large values of dielectric constant $\epsilon' > 10^3$ for materials as compared with the intrinsic effects. These extrinsic effects can be attributed to different forms of the Maxwell-Wagner effect, Liu et al. (2004).

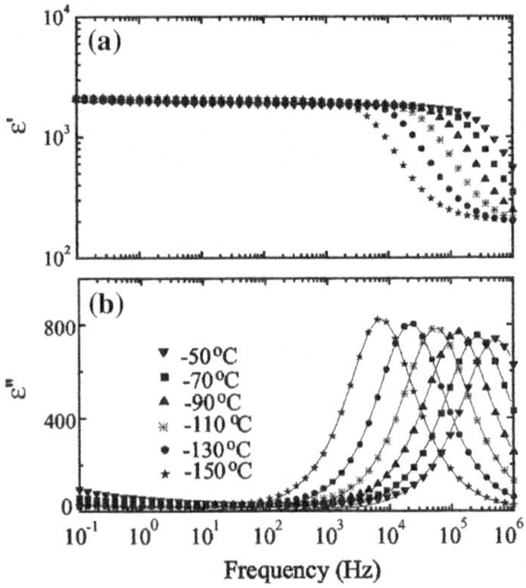

Fig. 32 Frequency dependence of permittivity $\varepsilon*$ at different temperatures. **a** Real part ε', **b** imaginary part ε''. Taken from Liu et al. (2004)

4.5 *The Electronic Impact of Defects*

The crystal structure of a ferroelectric can encounter numerous structural defects as discussed in Sect. 3.8. Let us first consider point defects in the bulk. For barium titanate, which is the drosophila of ferroelectric systems, interstitial defects are rare due to the dense packing of the ions in the perovskite crystal structure. Only hydrogen has been reported to enter in very small amount, Scott et al. (2001). Usually, $BaTiO_3$ is rather stoichiometric. Defects are thus vacancies or substitutional atoms, Baumard and Abelard (1984), Lewis and Catlow (1986). Sintered under reducing conditions either in a CO/CO_2-equilibrium- or in H_2-atmosphere, oxygen vacancies are introduced at high temperature according to the following defect reaction (see Moulson and Herbert 1989; Hench and West 1990; Waser and Smyth 1996);

$$O_O \rightleftarrows V_O^{\bullet\bullet} + 2e' + \frac{1}{2}O_2(g) \quad (136)$$

We use the Kröger-Vinck-Notation. The Kröger-Vinck-Notation was developed to describe defect equilibria in ionic solids. Instead of using the absolute charge state of an ion, it is rather specified with respect to its natural charge state in that particular compound. Thus, e.g. oxygen is doubly ionized in $BaTiO_3$, namely O^{2-} which is the classical notation in chemistry. In solid state chemistry it is mostly more convenient to express the difference from the natural charge state in the compound. The default charge state of oxygen at a regular lattice site is doubly negatively charged. The Kröger-Vinck-Notation uses O_O, because the ion does not differ in charge from its lattice site default value. When oxygen is put into its gaseous state, it is chemically reduced. The two remaining electrons are liberated and typically enter a delocalized band state in the $BaTiO_3$ lattice "$2e'$".

Fig. 33 Defect levels of the oxygen vacancy in the band gap of $BaTiO_3$, Moulson and Herbert (1989)

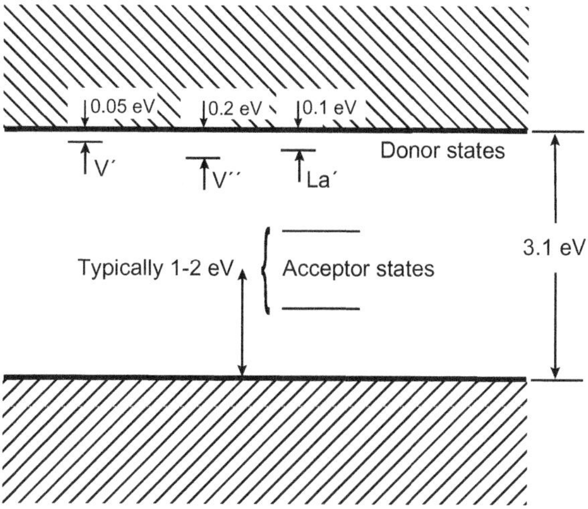

One has to observe that the reaction Eq. (136) is a thermally activated equilibrium reaction obeying to the classical chemical equilibrium equation (law of mass action). The equilibrium constant K exponentially depends on the inverse of temperature.

$$K = K' \exp \frac{-E_A}{kT} \tag{137}$$

with an activation energy E_A for the creation of the oxygen vacancy.

The subtlety for the electronic structure now enters by the fact that the vacancy itself is not necessarily ionized. It rather represents a double donor site with ionization energies of 0.05 eV and 0.2 eV beneath the conduction band minimum, respectively. Figure 33 shows the defect levels underneath the conduction band within the band gap of $BaTiO_3$. Thus, for chemically reduced $BaTiO_3$, we are concerned with a classical n-doped semiconductor. High doping levels by oxygen vacancies turn $BaTiO_3$ black at room temperature. A similarly well characterized donor is Lanthanum.

Due to the dense packing of the oxidic perovskites, chemical dopants will occupy lattice sites determined by their ionic radii. Iron is a classical dopant for $BaTiO_3$ Hagemann (1980). According to the defect reaction

$$\mathrm{Fe_2O_3} \rightleftarrows 2\mathrm{Fe}'_{\mathrm{Ti}} + 3\mathrm{O_O} + V_{\mathrm{O}}^{\bullet\bullet}. \tag{138}$$

iron substitutes titanium as an acceptor and generates an additional oxygen vacancy. This is necessary, because the iron now occupies the site of titanium in the crystal lattice requiring two oxygen *sites* per ion, one of which is taken by the vacancy (which may seem a bit strange to a beginner in this field) entailing its particular electronic properties.

The substitutional iron itself can easily attract holes into a trapped state at the titanium site:

$$\mathrm{Fe}'_{\mathrm{Ti}} + \mathrm{h}^{\bullet} \rightleftarrows \mathrm{Fe}_{\mathrm{Ti}}. \tag{139}$$

A deep acceptor level of around 1 eV above the valence band is formed.

An oxygen vacancy itself can also migrate. This is important to be mentioned, because all oxides exhibit a certain contribution of ion conduction to the overall conductivity of the material. For certain ion conductors this is even the dominant mechanism. The activation energy of this process for $BaTiO_3$ is around 1 eV Zafar et al. (1998a, b, 1999). While the vacancy is rather mobile, ionic motion is nevertheless typically much slower than electron motion. When the vacancy traverses a crystal, a continuous chain of ions along its path has moved, thus much ion movement has taken place. But each individual ion has only moved by a single lattice site. This is also the reason, why tracer experiments (radioactively marked ions) yield a completely different diffusion coefficient than measurement of the vacancy drift itself, which can only be determined macroscopically from average effects. The same argument also holds for substitutional dopant ions. Thus, while the oxygen vacancy is rather mobile, the substitutional iron is not. Obviously, at typical sintering temperatures, the lattice is so much in motion that also iron will diffuse, but the process is slow, because it necessitates the passage of the much rarer titanium vacancies. Also the jump rate of titanium is much smaller than for oxygen, because it is body centered in the unit cell and has to jump a full unit cell length across an oxygen site while the face centered oxygens can jump to a next neighbor site directly.

Donors for $BaTiO_3$ are either trivalent A-site (barium) or pentavalent B-site (titanium) substitutes. A typical A-site donor is lanthanum.

$$\mathrm{La_2O_3} \rightleftarrows 2\mathrm{La}^{\bullet}_{\mathrm{Ba}} + V''_{\mathrm{Ba}} + 3\mathrm{O}_{\mathrm{O}}. \tag{140}$$

As only two lattice oxygens match the two BaO which are substituted, this necessitates the addition of oxygen onto the crystal surface hence generating a barium vacancy.

$$\mathrm{La}_{\mathrm{Ba}} \rightleftarrows \mathrm{La}^{\bullet}_{\mathrm{Ba}} + e' \tag{141}$$

Figures 34, 35, and 36 show the dependence of charge carrier densities on the external oxygen partial pressure and the types of defects involved. Figure 34 show the general behaviour, Fig. 35 a detailed theoretical study and Fig. 36 shows the electrical conductivity of barium titanate as function of partial pressure and temperature. The lowest temperature that one can find such effects in $BaTiO_3$ (around 550°C) below which all ionic motion ceases and thus partial pressure dependencies lose their meaning, because a prior defect state (established at higher temperature) is kinetically frozen in. For more detailed discussion we refer the reader to the numerous excellent books on solid state ionics available in literature.

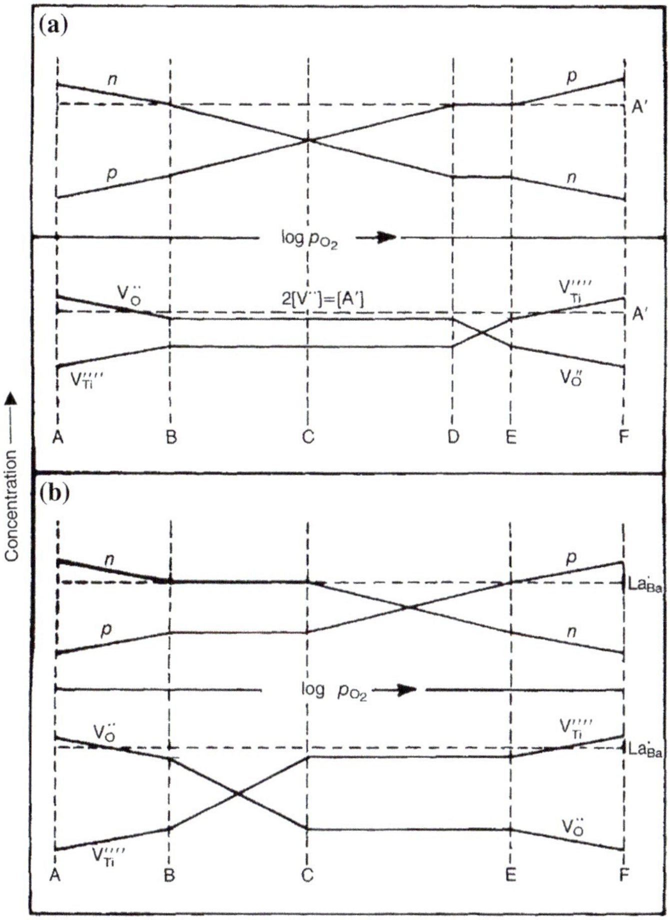

Fig. 34 Shift of electron density, hole density and vacancy densities as a function of oxygen partial pressure for donor as well as acceptor doped $BaTiO_3$, Waser and Smyth (1996)

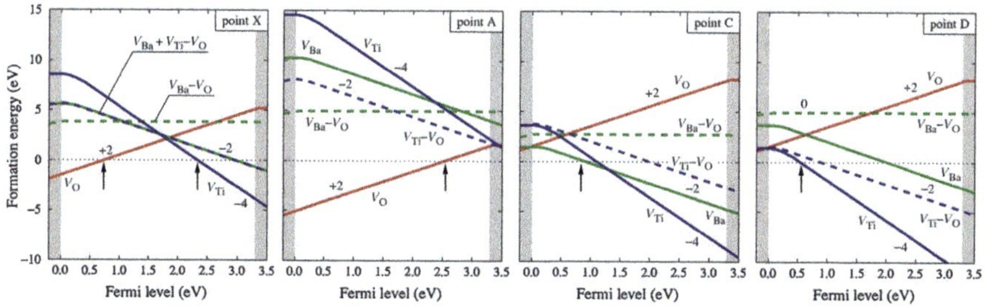

Fig. 35 Shift of the Fermi-Level around different defects in $BaTiO_3$, Erhart and Albe (2007)

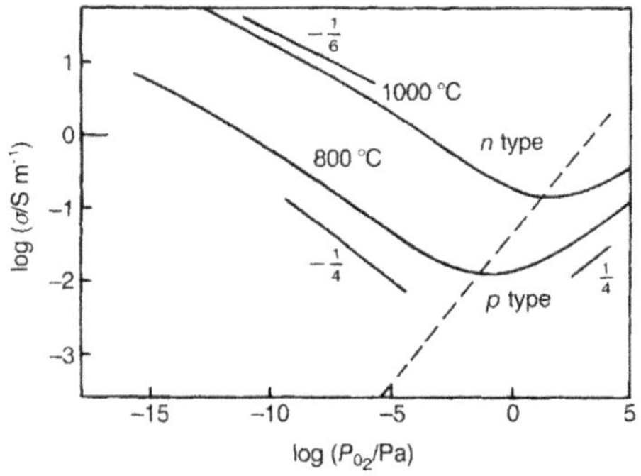

Fig. 36 Conductivity of $BaTiO_3$ as a function of oxygen partial pressure, Moulson and Herbert (1989)

5 Case Studies

5.1 The Domain Wall

The recent interest in the abnormally high electrical conductivity along charged domain walls (CDW) in ferroelectrics is due to the possible applications of this effect in high-density memory storage devices and reconfigurable nanoscale electronic circuits, Seidel et al. (2009), Seidel (2012), Catalan et al. (2012), Eliseev et al. (2011), Sluka et al. (2012), Sluka et al. (2013), Seidel et al. (2010), Balke et al. (2012), Maksymovych et al. (2012), Schroeder et al. (2012). The strong increase of bulk conductivity in the vicinity of CDW was revealed in the seventies in single crystals of the semiconductor-ferroelectric SbSI, Vul et al. (1973). Recently, the effect has been studied intensively both in thin film multiferroics and ferroelectrics, Seidel et al. (2009), Maksymovych et al. (2012), as well as in bulk single crystals of $BaTiO_3$, Sluka et al. (2012), and $LiNbO_3$ (LN), Schroeder et al. (2012), Mizuuchi et al. (2004). The conduction current measurements were usually made using macroscopic electrodes, Grekov et al. (1976), Mizuuchi et al. (2004), or locally by a conductive tip of an atomic force microscope, Seidel et al. (2009), Sluka et al. (2013), Seidel et al. (2010), Schroeder et al. (2012).

A strong increase of the bulk conductivity during the poling process induced by formation of CDW was revealed in MgO doped LN (MgOLN), Mizuuchi et al. (2004). Recently, the strong increase of the conductivity along CDW during illumination of MgO-LN with photoactive light has been demonstrated using high resolution conduction AFM, Schroeder et al. (2012). It was demonstrated, that the formation of CDW also leads to the improvement of dielectric and piezoelectric crystal properties, Xu et al. (2015), Mitsui et al. (2013).

Domain wall conductivity in ferroelectric-semiconductors is caused by screening of the depolarization field by free carriers, Seidel et al. (2009), Vul et al. (1973). Modern studies have proven that nominally uncharged domain walls, Seidel et al. (2010), Vasudevan et al. (2012), Farokhipoor and Noheda (2011), as well as vortex structures in $BiFeO_3$, Balke et al. (2012), exhibit strongly enhanced room-temperature conductivity correlating with changes in the electrostatic potential and the local electronic structure, which show a decrease in the band gap at the domain wall. Nominally uncharged fabricated vortex structures in $BiFeO_3$ show an order of magnitude increase in conductivity over single domain regions.

Domain wall conduction in insulating $Pb(Zr_{0.2}Ti_{0.8})O_3$ thin films was measured experimentally and unambiguously differentiated from displacement currents associated with ferroelectric polarization switching, Guyonnet et al. (2011). The domain wall conduction, which is nonlinear and highly asymmetric due to the specific local probe measurement geometry, shows thermal activation at high temperatures and high stability over time, Guyonnet et al. (2011). Metallic conductivity of 180° domain walls was reported in the same material, Maksymovych et al. (2012). This effect was traced to ferroelectric nanodomains with tilted (charged) walls, which act as interfaces with inherently tunable carrier density.

CDW of ferroelectric micro- and nanodomains are unique model objects for fundamental investigations of polarization reversal processes in nano-scale areas. Experimental and theoretical studies in this field represent an important branch of nanophysics and nanoelectronics, Seidel et al. (2009), Seidel (2012), Catalan et al. (2012). At the same time, detailed analysis of publications related to this problem has shown that neither the CDW formation processes in ferroelectrics nor time dependencies of CDW-induced conductivity and dielectric permittivity have been studied systematically, Zuo et al. (2014), Gureev et al. (2011), Eliseev et al. (2012a), Eliseev et al. (2012b), Morozovska (2012), Eliseev et al. (2013), Morozovska et al. (2012).

Macroscopic polarization reversal in ferroelectrics has been investigated in detail. However, the kinetics of the wall motion of micro- and nanodomains as well as the emergence of electrical conductivity need complex experimental and theoretical studies. The local scanning probe microscopy (SPM) methods not only supplement optical microscopy, but are ideal for experimental investigations of domain wall dynamics, Shur et al. (2011), because of high resolution, electrical and topological characteristics, and numerous specific advantages. The three-dimensional structure of CDWs is presently attracting much interest, Kaempfe et al. (2014).

SPM allows visualization of domains, enables quantitative measurements of local electric conductivity and manipulation of electro-physical properties of domain walls with nanometer-scale spatial resolution by inhomogeneous electric field produced by a conductive tip of SPM. Though SPM methods are widely used for domains visualization and generation of micro- and nanodomain structures, the possibilities for simultaneous manipulation of domain shape, size, and conductivity of the domain walls, as well as quantitative measurements of conductivity and its correlation with topological properties of the walls have not been investigated thus far both theoretically and experimentally. Experimental and theoretical studies of CDW conductivity in micro- and nanodomain structures represents a challenge for nanophysics, nanotechnology, and nanoelectronics Waser (2012).

As this field is presently developing rapidly and the basic physics is still vaguely understood, we refer the interested reader to the current literature. To go into a detailed discussion goes beyond the scope of an introductory course.

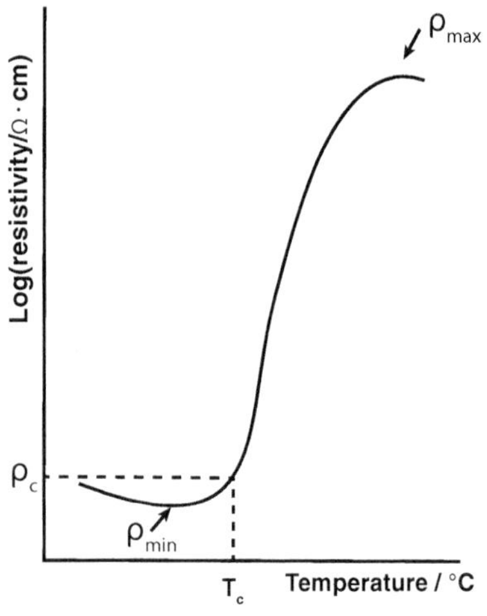

Fig. 37 Typical resistivity versus temperature curve of a $BaTiO_3$-based PTCR material (According to: Huybrechts et al. (1995))

5.2 *The PTCR-Effect at the Grain Boundary*

The positive temperature coefficient of resistance (PTCR) effect is a phenomenon which is observed in donor doped semiconducting polycrystalline $BaTiO_3$ ceramics. A transition from semiconducting to insulating behavior occurs at the T_c of the material. The resistivity jumps by several orders of magnitude as depicted in Fig. 37.

Conduction in undoped $BaTiO_3$
$BaTiO_3$ is an insulator at room temperature with a band gap of around 3.3 eV which is quite large compared with the typical semiconductors e.g. Si at 1.1 eV, and Ge at 0.67 eV. The phase diagram of $BaTiO_3$ is well known as depicted in Fig. 3. Before going into detail of the semiconducting behavior of doped $BaTiO_3$, let's first analyze the type of lattice defects which may exist in pure $BaTiO_3$. The intrinsic defects can be of two types: Schottky defects which arise when a pair of ions leaves the lattice in stoichiometric amounts simultaneously, creating a cation and an anion vacancy represented by the equation: $O_o = V_o^{\bullet\bullet} + \frac{1}{2}O_2 + 2e'$ at low oxygen partial pressure, pO_2, using the Kröger-Vink notation, Kröger and Vink (1956).

This sort of reaction leads to an n-type conduction mechanism in $BaTiO_3$. However, if $BaTiO_3$ is sintered in abundant oxygen, the following mechanism may hold resulting in p-type conduction: $\frac{1}{2}O_2 = O_o + V_{Ba}^{\prime\prime} + 2h^{\bullet}$ at high pO_2.

Hence conduction in undoped $BaTiO_3$ arises due to a combination of oxygen vacancies, cation vacancies, and associated electron holes.

Conduction in doped $BaTiO_3$
n-type semiconduction may be induced in ceramic and single crystal barium titanate in one of two ways: the first is by the reduction of some of the Ti^{4+} ions to Ti^{3+}, thereby releasing an electron into donor levels close to the conduction band, by heat-treating the material in a reducing atmosphere. The alternative method is to add

a small amount of donor ions which, upon dissolution in the lattice, have valency higher than that of the host ion. Commonly used dopants include La^{3+}, Ho^{3+}, Y^{3+}, which substitute for Ba^{2+} and Ta^{5+} and Nb^{5+} occupying Ti^{4+} sites.

Keeping in mind the charge neutrality condition of the overall lattice, a charge compensation mechanism follows which acts as a decisive factor for the type of conduction mechanism. In case of donor doping, charge compensation occurs either by electronic compensation or the vacancy compensation which are represented in Kröger-Vink notation, Kröger and Vink (1956), by:

$$M_2O_3 + BaTiO_3 = 2M^{\bullet}_{Ba} + 2e' + BaO + \frac{1}{2}O_2\,(\uparrow) \tag{142}$$

$$M_2O_3 + BaTiO_3 = 2M^{\bullet}_{Ba} + V''_{Ba} + BaO \tag{143}$$

The electronic compensation leads to conduction while vacancy compensation results in insulating behavior. Calculations have shown that electronic compensation is favored over cation compensation for donor doping at Ba sites. Cation vacancy compensation mechanisms are likely only at high (> 0.25 mol%) dopant concentrations and high pO_2.

The role of grain boundaries

Grain boundaries play a pivotal role in the electrical properties of a wide range of dielectrics. Grain boundaries act as obstacles/barriers for the movement of the charge carriers in certain ceramics which show ionic conduction, mixed ionic-electronic conduction, or electronic conduction. Mostly perovskite-type titanates are employed as high-permittivity dielectrics for capacitor applications. Due to the perpetual trend of miniaturization of electronic devices, the situation changes. In ceramic multilayers, the thickness of capacitors is much decreased. Hence, under the same operating voltages in the electronic circuit, the dielectric is exposed to significantly higher fields. For example, an applied operating voltage of 15 V at a capacitor with an inner dielectric thickness of 5 μm leads to a field of 3 MV/m. Consequently, the grain boundary (GB) barriers are reduced under such high fields. Hence it gives rise to a field-enhanced leakage current through the component and to an acceleration of resistance degradation. Vollman and Waser (1994) investigated the grain boundary space charge depletion layers in acceptor-doped $SrTiO_3$, and $BaTiO_3$, ceramics. Waser (1991) investigated the grain bulk conductivity of acceptor-doped $SrTiO_3$ ceramics by impedance analysis. For more literature about grain boundaries in classical semiconductors, please see Greuter and Blatter (1990).

The Heywang-Jonker Model

Various models have been proposed to explain the PTCR effect in semiconducting $BaTiO_3$, Saburi (1961), Heywang (1961), Jonker (1964), Peria et al. (1961). The most accepted models include the Heywang and the Jonker models. The Heywang

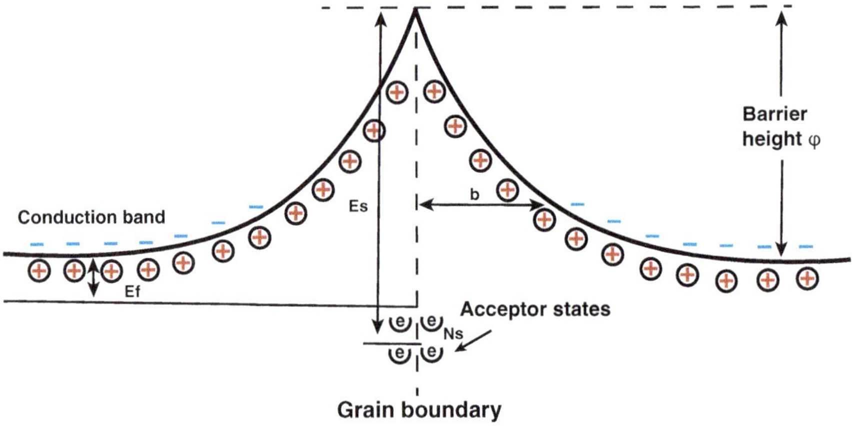

Fig. 38 Heywang PTCR model: a two dimensional potential barrier at grain boundaries, Heywang (1961)

model is based on a grain boundary potential barrier for the interpretation of the grain boundary mechanism of the phenomenon. The Jonker model, also called Heywang-Jonker model, is based on the influence of ferroelectric polarization on the resistivity below the Curie temperature, T_C.

As shown in Fig. 38, there exists a two-dimensional layer of electron traps, i.e. acceptor states, along the grain boundaries of $BaTiO_3$ due to which the grain boundaries exhibit different electrical properties from the bulk phase. The potential barrier ϕ_o is caused by a two-dimensional electron trap along the grain boundary where acceptor states attract electrons from the bulk resulting in an electron depletion layer having thickness b. The relation between the density of trapped electrons at the grain boundaries and the thickness of the depletion layer can be expressed as:

$$b = \frac{N_s}{2N_d} \tag{144}$$

where N_s is the concentration of trapped electrons and N_d is the charge carrier concentration. This depletion layer results in a grain boundary barrier, Φ_0,

$$\Phi_0 = \frac{e^2 N_s}{8\varepsilon_0 \varepsilon_{gb} N_d} \tag{145}$$

e is the electron charge, ε_0 the permittivity of free space and ε_{gb} the relative permittivity of the grain boundary region.

The overall resistivity, ρ, is related to the height of the potential barrier by

$$\rho = Aexp\frac{\Phi_0}{kT} \tag{146}$$

A is a geometrical factor and k the Boltzmann constant. Because $BaTiO_3$ is ferroelectric, the dielectric constant, ε, obeys the Curie-Weiss law above its Curie temperature. It is given by

$$\chi = \frac{C}{T - T_c} \tag{147}$$

where $\chi = \varepsilon - 1$, C is the Curie constant and T the absolute temperature. Incorporating Eqs. (145) and (147) into (146) and rearrangement yields:

$$\rho = A \exp\left[\frac{e^2 N_s^2}{8\varepsilon_0 N_d k C}\left(1 - \frac{T_c}{T}\right)\right] \tag{148}$$

Above the Curie point when doped $BaTiO_3$ is in the paraelectric phase, the grain-boundary permittivity, which follows the Curie-Weiss law, decreases with increasing temperature. The corresponding potential barrier increases proportionally and results in steeply increasing resistivity which depends exponentially on the potential barrier as denoted by Eq. (146). The energy of the trapped electrons in the grain boundary rises with temperature along with the potential barrier. When the energy of the electron traps reaches the Fermi level, trapped electrons start to jump to the conduction band, which can depress the increase in Φ_0 and ρ, thus ultimately enhance conductivity. This also explains the negative temperature coefficient resistivity (NTCR) effect when passing the point ρ_{max} in the high temperature range, see Fig. 37.

Heywang's model could not accurately explain the PTCR behavior below T_c. Hence Jonker's model was developed as an improvement. Below the Curie point, $BaTiO_3$ is ferroelectric with its polarization along the tetragonal crystal axis as shown in Fig. 3. Jonker introduced a refinement of Heywang's model. The Jonker model is based on the ferroelectric behavior of $BaTiO_3$ below T_c. The spontaneous polarization in domains near the grain boundary regions neutralize built up charge at the boundaries allowing for charge transport across the grains. The influence of domains at the grain boundary regions is shown in Fig. 39. The normal component of the spontaneous polarization in the domain acts to compensate space charge buildup at the boundary affecting a decrease in the potential barrier height. The barrier height is now given by the equation:

$$\phi = \frac{e^2 N_S^2 - \chi P_N^2}{8 s N_d} \tag{149}$$

where P_N is the spontaneous polarization. The importance of the Jonker refinement is the fact that Eq. (149) is of a more general nature. It not only takes into consideration the behavior of the material below T_c based on the spontaneous polarization in the ferroelectric state, but it also predicts the behavior above T_c where P_S becomes 0 in the non-ferroelectric state.

There are a number of the other improvements which have been made to these existing theories on the PTCR effect. Daniels and Wernicke (1976a, b) proposed an

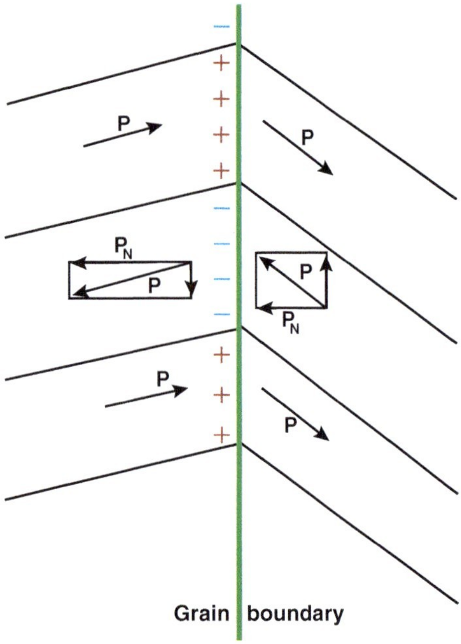

Fig. 39 Jonker PTCR model-domains at a grain boundary between two crystallites of different orientation. Compensating surface charges match the discontinuity of the normal component of polarization across the grain boundary, Jonker (1964)

excess of barium cation vacancies to be responsible for acceptor states generated at the grain boundaries. These acceptor states are thought to extend in a three dimensional manner, extending inward from the grain boundary, unlike the original two dimensional Hewyang model. Desu and Payne (1990a, b, c, d) have suggested a change in the compensation mechanism at the grain boundaries caused by donor segregation at the boundaries. Roseman (1993) showed the existence of fine domains and regions without them near the grain boundary in annealed samples, suggesting that electronic charge transport occurs across the grain boundaries through the fine domain regions. The abrupt rise in resistivity in annealed samples is attributed to the breakup of the fine domain regions above T_c.

5.3 *Magnetoelectric Composites*

Magnetoelectric composites are a class of heterogeneous materials where often inorganic piezoelectric and piezomagnetic phases are combined to achieve indirect magnetoelectricity at room temperature with a free choice on the type of the constituent phases. Such materials have been increasingly studied for the last two decades mainly because they offer the opportunity to control magnetic parameters in sensors, filters, and data-storage via an electrical voltage instead of current, Van Run et al. (1974), Bichurin et al. (2007), Srinivasan (2010), Vaz et al. (2010), Jahns et al. (2013), Duong and Groessinger (2007). The most frequently chosen materials are

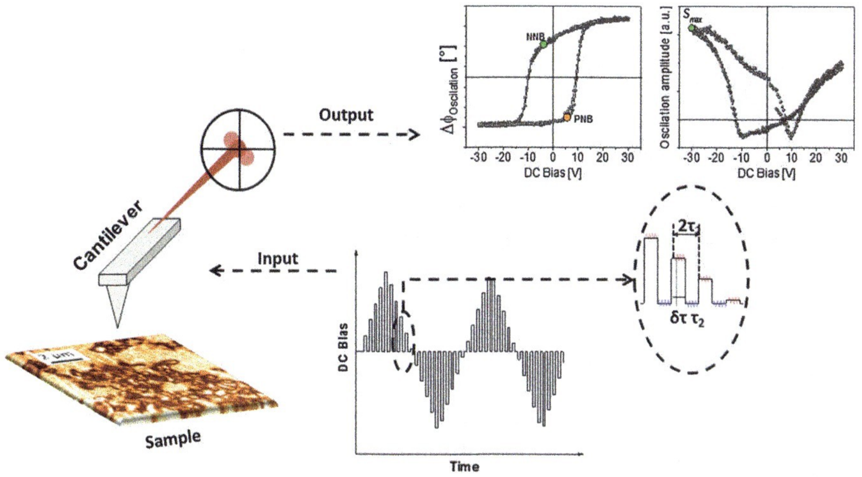

Fig. 40 Schematic operation of Switching Spectroscopy Piezoresponse Force Microscopy (SSPFM). The regions enclosed within the dark boundaries in the sample image represent the ferroelectric phase ($BaTiO_3$) whereas the surrounding matrix comprises the magnetic phase ($BaFe_{12}O_{19}$). Data partly from Trivedi et al. (2015)

PZT, $BaTiO_3$, PMN-PT as excellent piezoelectric (ferroelectric) components, and a ferrite ($CoFe_2O_4$, $NiFe_2O_4$, $BaFe_{12}O_{19}$) as a highly piezomagnetic constituent.

Now it is known that ferrites are small band gap (0.1–1 eV Smit and Wijn (1959)) semiconductors with one varying from other in terms of resistivity. The main conduction mechanism is due to electron hopping between mixed valent ions e.g. from Fe^{+2} to Fe^{+3} or Ni^{+2} to Ni^{+3} which are known to be present even in a perfectly stoichiometric composition mostly due to the crystal chemistry induced by the presence of Fe-ions. On the other hand, $BaTiO_3$ is known to possess an optical band gap of >3.0 eV and shows highly modified electrical properties especially at interfaces (PTCR effect, see Sect. 5.2). Owing to this large difference in the electrical properties of $BaTiO_3$ and the ferrites, it is convenient to presume a hypothesis where an interface between the two can undergo charge accumulation owing to the highly conducting nature of the ferrites. This scenario, being analogous to the commonly encountered case of the Si/SiO_2 interfaces in field-effect based electronic devices, is hence not uncommon.

Recently, experimental investigations on a $BaTiO_3$–$BaFe_{12}O_{19}$ ($E_g = 0.8$ eV) composite system using Switching-Spectroscopy Piezoresponse Force Microscopy (SSPFM) revealed that indeed some form of space charge exists at such heterojunctions, Trivedi et al. (2015). The technique (SSPFM) used is an Atomic Force Microscopy (AFM) based technique (Fig. 40) wherein a sharp tip moves over the sample surface, applying highly concentrated electrical bias, and realizes ferroelectric switching within a quasi-local region of around 100 nm in diameter, Jesse et al. (2006). The switching process is monitored through-out by means of changes in the piezoelectric displacement (butterfly loop) reflected by the tip in the form of

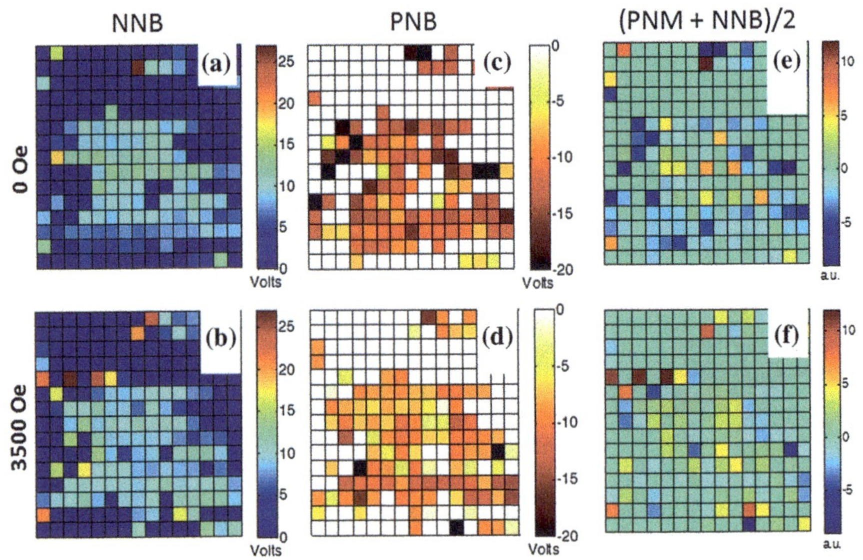

Fig. 41 Spatial maps of NNB (**a**)–(**b**), PNB (**c**)–(**d**), and (PNB + NNB)/2 (**e**)–(**f**) as a function of magnetic field, derived from the loop acquisition on the sample as shown in Fig. 40. The maps **e** and **f** reflect the existence of accumulated charges at the ferroelectric-magnetic interface, Trivedi et al. (2015)

mechanical deflection. The obtained loops are not equivalent to the macroscopically measured PE or strain-field (butterfly) loops. For that matter it is worthwhile to refer to the detailed clarifications provided by Lupascu et al. (2015). The most informative parameters in an SSPFM experiment as well as in the present context is the negative nucleation bias (NNB), which represents the onset of the appearance of an opposite polarity nanodomain under the tip with reversing electric field, and likewise a positive nucleation bias (PNB). The key results of this report are shown in Fig. 41 showing a spatial distribution of the NNB and PNB along with a quantity (PNB+NNB)/2 which reflects the horizontal asymmetry (offset due to random fields) of the phase-field loop. It can be seen that in the absence of magnetic field (0 Oe) there already exists an asymmetry in the loops quite close to the interface (Fig. 41) which is slightly lifted on application of magnetic field (3500 Oe). This can be understood in the realm of the magnetostrictive stress generated by the magnetic ($BaFe_{12}O_{19}$) phase residing next to the ferroelectric $BaTiO_3$. The presence of the preexisting charges can be associated to the semiconductor effect, which is believed to be modulated by the magnetostrictive stress. The magnetostrictively generated stress sets up an internal field (offset field, see also Lupascu and Rödel (2005)) in the ferroelectric matrix surrounding the switched nanodomain, which adds up to the pre-existing field created by the interfacial charges. The resultant field is finally responsible for lifting the asymmetry of the local switching loops.

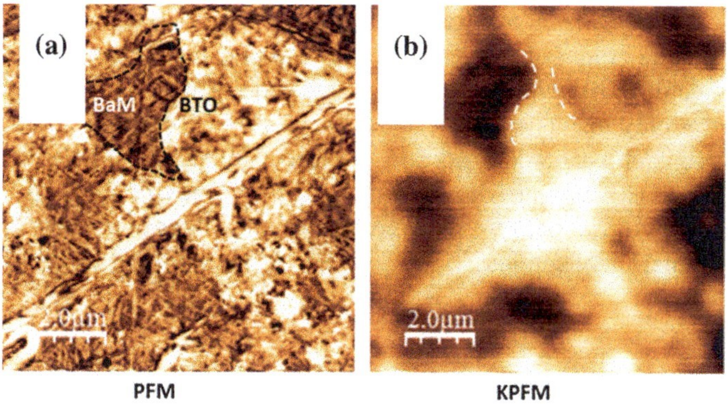

Fig. 42 PFM (**a**) and KPFM (**b**) images of an identical spot on the sample surface. The dashed line in (**a**) schematically differentiates between $BaTiO_3$ and $BaFe_{12}O_{19}$ phases. The dashed line in (**b**) highlights the gradient of contrast within the $BaTiO_3$ region

The existence of such interfacial space charges in the $BaTiO_3$ - $BaFe_{12}O_{19}$ composite system was later-on probed using another AFM variant called Kelvin Probe Force Microscopy (KPFM). KPFM spatially maps out any potential difference that exists between the AFM tip and the sample surface below it, Jacobs et al. (1999). Although a bifurcation of the observed potential between various possible origins (e.g. work function difference, space charge etc.) cannot be made in KPFM, a qualitative comparison of the signal intensity (image contrast) between different spatial regions can indeed provide useful information. Figure 42 shows the PFM and a KPFM images corresponding to an identical spot on the sample surface. The first set of information that these images provide is that there exists a strong correlation between the PFM and the KPFM contrasts, hence it could be concluded that the observed potential in KPFM is a material dependent quantity. Apart from this, the interface (marked by dashed white lines) shows a diffuse contrast i.e. the contrast changes continuously from the interface towards the center of each region while staying almost constant within.

This observed gradient in contrast at the interface can be associated with the proposed existence of space charges at interface which additionally modulates any pre-existing potential difference between the tip and the sample. Although an accurate quantitative estimation of these charges is extremely complex, attempts can be made experimentally as well as theoretically to improve the spatial resolution of KPFM, leading to a moderately accurate estimation of the width of the space charge regions.

5.4 *Polarization Stability in Heterostructures*

A very interesting recent finding shows the imminent impact of semiconducting properties on ferroelectric states. Zhan et al. (2015) used ZnO inclusions in

$Bi_{0.5}Na_{0.5}TiO_3$-based lead free ceramics to stabilize polarization. The typically strong thermal de-poling of this lead free compound reduced its temperature range of use considerably. Apparently, the free charge carriers at the interface to the semiconductor not only maintain the piezoelectric properties of the material, they also help to overcompensate de-poling due to pyroelectric reduction of polarization when approaching the Curie point. This very new development will likely receive much attention in technology, because these compositions will turn out to be much more viable in technical applications.

6 Conclusion and Outlook

Typically, semiconductor effects can be neglected in large band gap ferroelectrics when one wants to understand the basic ferroelectric, piezoelectric and dielectric properties. Fridkin (1980) already outlined this fact by a very straight forward energetic argument. Only for small band gaps the number of free charge carriers is sufficiently large to generate semiconductor effects. On a second view though, defect levels within the band gap will alter the overall charge balance in the system. They can provide free *and* localized charge carriers to the system. When changing the charge state of a defect atom or atomic/ionic vacancy this defect stays localized at temperatures when ferroelectricity is still present. As such it highly interacts with the polarization of the ferroelectric domains. Typical results of this are slowly changing states of charge imbalance within the microstructure.

Acknowledgements Discussions with Vladimir V. Shvartsman are greatly acknowledged. The authors highly appreciate financial support by the "Deutsche Forschungsgemeinschaft" (DFG), research grants Lu 729/12-2 and Schr 570/12-2, within the research group FOR 1509 on "Ferroic Functional Materials, Multiscale Modeling and Experimental Characterization".

Appendix

pn-Junction

A pn-junction is a device, in which a p-type and a n-type semiconductor meet each other. In this region the different semiconductors influence each other resulting in a characteristic electrical behaviour.

A pn-junction determines the functional characteristics of a semiconducting diode. Further, it is an important functional component of many other electronic devices.

The fabrication of a pn-junction is done by spatially doping of a p-type or n-type semiconductor with the opposite dopant.

Physical Operating Principle

The gradients of the hole or electron density at the interface of a p-type and n-type semiconductor cause a diffusion of majority charge carriers to the opposite side of the interface. Hence, spatially the density of free charge carriers decreases by several orders of magnitude, which lowers the conductivity in that region. As most of free charge carriers are missing in this region, it is called depletion region. The remaining ionized impurity atoms generate a space charge layer. This space charge generates an electric field, which causes a current in opposite direction to the diffusion (and hence, a drift current). Between these two currents equilibrium is obtained, which is correlated with a defined course of space charge and of electric field. The distribution of hole density, electron density, space charge density ρ, electric field, and potential ϕ are determined by the system of differential equations of the Poisson-equation, the transport mechanisms and charge carrier continuity (see Fig. 43).

Starting the calculation with Poisson's equation. (Eq. 84)

$$\frac{d(\epsilon_{HL} \cdot E)}{dx} = \rho = e\,(N_D - N_A + p - n) \tag{150}$$

The solution of the system of differential equations together with the charge neutrality condition at the interface

$$N_A \cdot \xi_p = N_D \cdot \xi_n \tag{151}$$

gives the extension of the space charge region, d_s;

$$d_s = \sqrt{\frac{2\epsilon_{HL} U_d\,(N_A + N_D)}{e N_A \cdot N_D}} \tag{152}$$

$$U_D = \frac{k_B T}{e} \ln\left(\frac{N_A \cdot N_D}{n_i^2}\right) \tag{153}$$

The regions outside of the space charge region are electrically neutral and the outer electric contacts are deposited in these outer regions.

1. In thermodynamic equilibrium the Fermi level is horizontal throughout the whole device.
2. Marking the conduction and valence band edge as well as the macro potential within the neutral regions of the pn-junction.
3. Connecting the macro potential at the transition region according to Poisson's equation

    (a) No surface charge at the interface ⇒ no sharp bend within the course of the macro potential
    (b) Poisson-equation is a differential equation of second-order ⇒ quadratic course of the potential with space charge density being constant

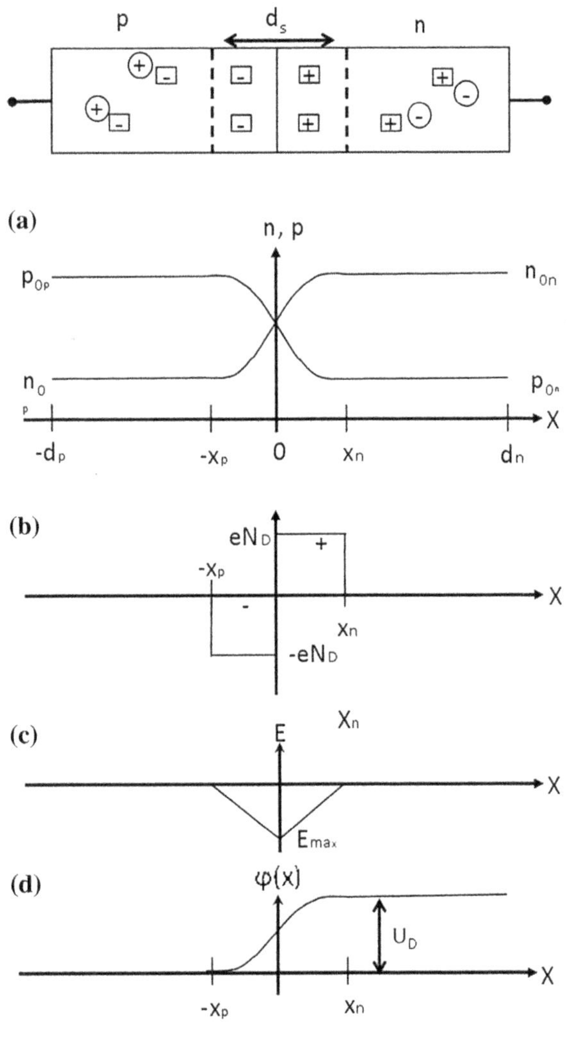

Fig. 43 **a** charge carrier densities, **b** space charge density, **c** electric field, and **d** course of potential at the pn-junction

(c) The change of sign of bending of the course of the potential takes place at the interface caused by the change of charge of the ionized impurity atoms (Fig. 44)

4. Conduction and valence band edges have to be parallel to the macro potential within the transition region.

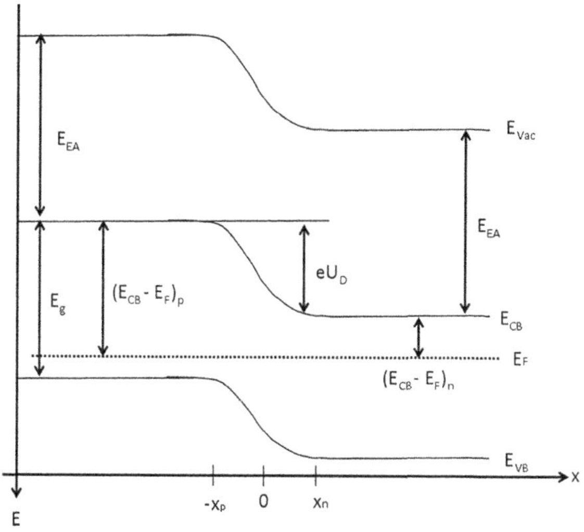

Fig. 44 Energy band diagram of a pn-junction

References

Aksel, E., & Jones, J. L. (2010). Advances in lead-free piezoelectric materials for sensors and actuators. *Sensors, 10*(3), 1935–1954.

Anderson, D. J. (2012). *An introduction to the principles of thermodynamics—Including 35 problems to solve*. U.K.: Alcester.

Ashcroft, N. W., & Mermin, N. D. (1988). *Solid Sate Physics*. Rinehart & Winston International Edition, Sounders College, Philadelphia: Holt.

Balke, N., Winchester, B., Ren, W., Chu, Y. H., Morozovska, A. N., Eliseev, E. A., et al. (2012). Enhanced electric conductivity at ferroelectric vortex cores in $BiFeO_3$. *Nature Physics, 8*(1), 81–88.

Baumard, J. F., & Abelard, P. (1984). An approach to the defect chemistry of barium titanate. *Solid State Ionics, 12*, 47–51.

Bichurin, M. I., Viehland, D., & Srinivasan, G. (2007). Magnetoelectric interactions in ferromagnetic-piezoelectric layered structures: Phenomena and devices. *Journal of Electroceramics, 19*, 243–250.

Blakemore, J. S. (1962). *Semiconductor statistics*. Oxford, UK: Pergamon Press.

Bunget, I., & Popescu, M. (1984). *Physics of solid dielectrics*. Amsterdam: Elsevier.

Burfoot, J. C. (1967). *Ferroelectrics: An introduction to the physical priciples*. London, Princeton, Toronto: D. Van Nostrand.

Callen, H. B. (1985). *Thermodynamics and an introduction to thermostatistics*. NY: Wiley.

Catalan, G., Seidel, J., Ramesh, R., & Scott, J. F. (2012). Domain wall nanoelectronics. *Reviews of Modern Physics, 84*(1), 119–156.

Chenskii, E. V. (1970). Single-domain polarization of a ferroelectric having a second order phase transition. *Fiz. Tverd. Tela (Sov. Phys. Solid State, USSR), 12*, 586 (446).

Chenskii, E. V., & Sandomirskii, V. B. (1969). Field effect in a ferroelectric semiconductor above Curie point. *Fiz. Tekh. Poluprovodn. (Sov. Phys. Semiconductor, 3*, 857 (724).

Chiang, Y.-M., Birnie, D, I. I. I., & Kingery, W. D. (1997). *Physical ceramics*. New York: Wiley.

Chu, M. W., Szafraniak, I., Scholz, R., Harnagea, C., Hesse, D., Alexe, M., et al. (2004). Impact of misfit dislocations on the polarization instability of epitaxial nanostructured ferroelectric perovskites. *Nature Materials, 3*, 87–90.

Chuang, S. L. (2009). *Physics of photonic devices*. (2nd ed.). Hoboken, N.J: Wiley.
Damjanovic, D., & Demartin, M. (1996). The Rayleigh law in piezoelctric ceramics. *Journal of Physics D: Applied Physics, 29*, 2057–2060.
Daniels, J., & Wernicke, R. (1976a). Part V. New aspects of an improved PTC model. *Philips Research Reports, 31*, 544–559.
Daniels, J., Wernicke, R. (1976b). Part V. New aspects of an improved PTC model. *Philips Research Reports, 31*, 526–543.
Desu, S. B., & Payne, D. A. (1990a). Interfacial segregation in perovskite: I, theory. *Journal of the American Ceramic Society, 73*, 3391–3397.
Desu, S. B., & Payne, D. A. (1990b). Interfacial segregation in perovskite: II, experimental evidence. *Journal of the American Ceramic Society, 73*, 3398–3406.
Desu, S. B., & Payne, D. A. (1990c). Interfacial segregation in perovskite: III, mirostrucure and electrical properties. *Journal of the American Ceramic Society, 73*, 3407–3415.
Desu, S. B., & Payne, D. A. (1990d). Interfacial segregation in perovskite: IV, internal boundary layer devices. *Journal of the American Ceramic Society, 73*, 3416–421.
Devonshire, A. F. (1949). Theory of barium titanate, part I. *Philosophical Magazine, 40*, 1040–1063.
Devonshire, A. F. (1951). Theory of barium titanate, part II. *Philosophical Magazine, 42*, 1065–1079.
Devonshire, A. F. (1954). Theory of ferroelectrics. *Advances in Physics, 3*, 85–129.
Duong, G. V., & Groessinger, R. (2007). Effect of preparation conditions on magnetoelectric properties of $CoFe_2O_4$ $BaTiO_3$ magnetoelectric composites. *Journal of Magnetism and Magnetic Materials, 316*, e624–e627.
Eknadiosiants, E. I., Borodin, V. Z., Smotrakov, V. G., Erekin, V. V., & Pinskaya, A. N. (1990). Domain structure of rhombohedral $PbTi_xZr_{1-x}O_3$ crystals. *Ferroelectrics, 111*, 283–289.
Eliseev, E. A., Morozovska, A. N., Svechnikov, G. S., Gopalan, V., & Shur, V. Ya. (2011). Static conductivity of charged domain walls in uniaxial ferroelectric semiconductors. *Physical Review B, 83*(23), 235313.
Eliseev, E. A., Morozovska, A. N., Gu, Y., Borisevich, A. Y., Chen, L.-Q., Gopalan, V., et al. (2012a). Conductivity of twin-domain-wall/surface junctions in ferroelastics: Interplay of deformation potential, octahedral rotations, improper ferroelectricity, and flexoelectric coupling. *Physical Review B, 86*, 085416.
Eliseev, E. A., Morozovska, A. N., Svechnikov, G. S., Maksymovych, P., & Kalinin, S. V. (2012b). Domain wall conduction in multiaxial ferroelectrics. *Physical Review B, 85*, 045312.
Eliseev, E. A., Yudin, P. V., Kalinin, S. V., Setter, N., Tagantsev, A. K., & Morozovska, A. N. (2013). Structural phase transitions and electronic phenomena at 180-degree domain walls in rhombohedral $BaTiO_3$. *Physical Review B, 87*, 054111.
Erhart, P., & Albe, K. (2007). Thermodynamics of mono- and di-vacancies in barium titanate. *Journal of Applied Physics, 102*, 084111.
Farokhipoor, S., & Noheda, B. (2011). Conduction through 71° domain walls in $BiFeO_3$ thin films. *Physical Review Letters, 107*, 127601.
Fridkin, V. M. (1979). *Photoferroelectrics*. Berlin: Springer.
Fridkin, V. M. (1980). *Ferroelectric semiconductors*. New York, Consultants Bureau: Plenum Publishing Corporation.
Genenko, Y., & Lupascu, D. C. (2007). Drift of charged defects in local fields as aging mechanism in ferroelectrics. *Physical Review B, 75*, 184107.
Grekov, A. A., Adonin, A. A., & Protsenko, N. P. (1976). Encountering domains in SbSI. *Ferroelectrics, 13*(1), 483–485.
Greuter, F., & Blatter, G. (1990). Electrical properties of grain boundaries in polycrystalline compound semiconductors. *Semiconductor Science and Technology, 5*, 111–137.
Gureev, M. Y., Tagantsev, A. K., & Setter, N. (2011). Head-to-head and tail-to-tail 180° domain walls in an isolated ferroelectric. *Physical Review B, 83*, 184104.
Guro, G. M., Ivanchik, I. I., & Kovtonyuk, N. F. (1967). Semiconducting properties of ferroelectrics. *Pis'ma Zh. Eksp. Teor. Fiz. (JETP Letters-USSR, 1*(5(1)), 9 (5).

Guyonnet, J., Gaponenko, I., Gariglio, S., & Paruch, P. (2011). Conduction at domain walls in insulating $Pb(Zr_{0.2}Ti_{0.8})O_3$ thin films. *Advanced Materials, 23*(45), 5377–5382.
Hagemann, H.-J. (1980). *Akzeptorionen in $BaTiO_3$ und $SrTiO_3$ und ihre Auswirkung auf die Eigenschaften von Titanatkeramiken*. Dissertation, Aachen University of Technology, Germany.
Heckmann, G. (1924). *Über die Elastizitätskonstanten der Kristalle*. PhD thesis, Universität Göttingen, Phil. Diss.
Hellwege, K.-H. (1988). *Einführung in die Festkörperphysik*. Berlin: Springer.
Hench, L. L., & West, J. K. (1990). *Principles of electronic ceramics*. New York: Wiley.
Henderson, B. (1972). *Defects in crystalline solids*. The Structure And Properties Of Solids 1. Edward Arnold, London.
Heywang, W. (1961). Resistivity anomaly in doped barium titanate. *Solid State Electronics, 3*, 51.
Hong, Seungbum. (2004). *Nanoscale phenomena in ferroelectric thin films*. Boston: Kluwer.
Hull, D., & Bacon, D. J. (1984). *Introduction to dislocations*. Oxford: Pergamon Press.
Hummel, R. E. (2011). *Electronic properties of materials* (4th ed.). New York: Springer.
Huybrechts, B., Ishizaki, K., & Takata, M. (1995). The positive temperature-coefficient of resistivity in barium-titanate. *Journal of Materials Science, 30*(10), 2463–2474.
Ikeda, T. (1990). *Fundamentals of piezoelectricity*. Oxford, UK: Oxford Science Publishers.
Iwamoto, M. (2012). Maxwell-Wagner effect. In B. Bhushan (Ed.), *Encyclopedia of nanotechnology* (pp. 1276–1285). Netherlands: Springer.
Jackson, J. D. (1998). *Classical electrodynamics*. Routledge.
Jacobs, H. O., Knapp, H. F., & Stemmer, A. (1999). Practical aspects of Kelvin probe force microscopy. *Review of Scientific Instruments, 70*, 1756–1760.
Jaffe, B., Cook, W. R, Jr., & Jaffe, H. (1971). *Piezoelectric ceramics*. Marietta, OH: Academic Press.
Jahns, R., Piorra, A., Lage, E., Kirchhof, C., Meyners, D., Gugat, J. L., et al. (2013). Giant magnetoelectric effect in thin-film composites. *Journal of the American Ceramic Society, 96*, 1673–1681.
Jesse, S., Baddorf, A. P., & Kalinin, S. V. (2006). Switching spectroscopy piezoresponse force microscopy of ferroelectric materials. *Applied Physics Letters, 88*, 62908.
Jona, F., & Shirane, G. (1993). *Ferroelectric crystals*. Oxford England: Dover Publications, reprint from Pergamon Press.
Jonker, G. H. (1964). Some aspects of semiconducting barium titanate. *Solid State Electronics, 7*, 895–903.
Kaempfe, T., Reichenbach, P., Schröder, M., Haußmann, A., Eng, L. M., Woike, T., et al. (2014). Optical three-dimensional profiling of charged domain walls in ferroelectrics by cherenkov second-harmonic generation. *Physical Review B, 89*, 035314.
Keeble, D. J., Li, Z., & Harmatz, M. (1996). Electron paramagnetic resonance of Cu^{2+} in $PbTiO_3$. *Journal of Physics and Chemistry of Solids, 57*, 1513–1515.
Keeble, D. J., Nielsen, B., Krishnan, A., Lynn, K. G., Madhukar, S., Ramesh, R., et al. (1998). Vacancy defects in (Pb, La)(Zr, Ti)O_3 capacitors observed by positron annihilation. *Applied Physics Letters, 73*, 318–320.
Kingery, W. D., Bowen, H. K., & Uhlmann, D. R. (1976). *Introduction to ceramics*. New York: Wiley.
Kittel, C. (1946). Theory of the structure of ferromagnetic domains in films and small particles. *Physical Review, 70*, 965–971.
Kröger, F. A., & Vink, H. J. (1956). Relations between the concentrations of imperfections in crystalline solids. In E. Seitz, & D. Turnbull (Eds.), *Solid State Physics* (Vol. 3, pp. 307–435). New York: Academic Press.
Landau, L. D. (1937). Theory of phase transformations. II. *Zh. eksp. teor. Fiz.; (Phys. Z. Sowjetunion, 11*, 545 (1937)) *7*, 627, 1937.
Landau, L. D. (1965). *Collected papers of L. D. Landau*. New York: Gordon and Breach.
Lewis, G. V., & Catlow, C. R. A. (1986). Defect studies of doped and undoped barium titanate using computer simulation techniques. *Journal of Physics and Chemistry of Solids, 47*(I), 89–97.

Lines, M. E., & Glass, A. M. (1977). *Principles and applications of ferroelectrics and related materials*. Oxford: Clarendon Press.
Liu, J., Duan, C., Yin, W., Mei, W. N., Smith, R. W., & Hardy, J. R. (2004). Large dielectric constant and Maxwell-Wagner relaxation in $Bi_{23}Cu_3Ti_4O_{12}$. *Physical Review B, 70*(14), 144106.
Lunkenheimer, P., Bobnar, V., Pronin, A. V., Ritus, A. I., Volkov, A. A., & Loidl, A. (2002). Origin of apparent colossal dielectric constants. *Physical Review B, 66*(5), 052105.
Lupascu, D. C., & Rödel, J. (2005). Fatigue in actuator materials: a review. *Advanced Engineering Materials, 7*(10), 882–898.
Lupascu, D. C., Wende, H., Etier, M., Nazrabi, A., Anusca, I., Trivedi, H., et al. (2015). Measuring the magnetoelectric effect across scales. *GAMM-Mitteilungen, 38*, 25–74.
Maksymovych, P., Huijben, M., Pan, M., Jesse, S., Balke, N., Chu, Y.-H., et al. (2012). Ultrathin limit and dead-layer effects in local polarization switching of $BiFeO_3$. *Physical Review B, 85*(1), 014119.
Marti, O. (2016). Universität Ulm, Experimentelle Physik. http://wwwex.physik.uni-ulm.de/lehre/physikalischeelektronik/phys_elektr/phys_elektrse10.html.
Mitsui, T., Tatsuzaki, I., & Nakamura, E. (1976). *An Introduction to the physics of ferroelectrics*. New York: Gordon and Breach Publishers.
Mitsui, R., Fujii, I., Nakashima, K., Kumada, N., Kuroiwa, Y., & Wada, S. (2013). Enhancement in the piezoelectric properties of $BaTiO_3$-$Bi(Mg_{1/2}Ti_{1/2})O_3$-$BiFeO_3$ system ceramics by nanodomain. *Ceramics International, 39*, S695–S699.
Mizuuchi, K., Morikawa, A., Sugita, T., & Yamamoto, K. (2004). Electric-field poling in Mg-doped $LiNbO_3$. *Journal of Applied Physics, 96*(11), 6585–6590.
Morozovska, A. N. (2012). Domain wall conduction in ferroelectrics. *Ferroelectrics, 438*(1), 3–19.
Morozovska, A. N., Vasudevan, R. K., Maksymovych, P., Kalinin, S. V., & Eliseev, Eugene A. (2012). Anisotropic conductivity of uncharged domain walls in $BiFeO_3$. *Physical Review B, 86*, 085315.
Möschwitzer, A. (1992). *Grundlagen der Halbleiter- & Mikroelektronik, Bd.1 (in German)*. München: Hanser Verlag.
Moulson, A. J., & Herbert, J. M. (1989). *Electroceramics: Materials, properties. Applications*. London: Chapman & Hall.
Newnham, R. E. (2005). *Properties of materials: Anisotropy, symmetry structure*. Oxford: Oxford University Press.
Nye, J. F. (1985). *Physical properties of crystals*. Clarendon Press, Oxford: Oxford Science Publications.
Peria, W. T., Bratschun, W. R., & Fenity, R. D. (1961). Possible explanation of positive temperature coefficient in resistivity of semiconducting ferroelectrics. *Journal of the American Ceramic Society, 44*, 249–250.
Prodromakis, T., & Papavassiliou, C. (2009). Engineering the Maxwell Wagner polarization effect. *Applied Surface Science, 255*(15), 6989–6994.
Raevski, I. P., Prosandeev, S. A., Bogatin, A. S., Malitskaya, M. A., & Jastrabik, L. (2003). High dielectric permittivity in $AFe_{1/2}B_{1/2}O_3$ nonferroelectric perovskite ceramics (A=Ba, Sr, Ca; B=Nb, Ta, Sb). *Journal of Applied Physics, 93*(7), 4130–4136.
Reif, F. (1965). *Fundamentals of statistical and thermal physics*. New York: McGraw-Hill.
Rhoderick, E. H., & Williams, R. H. (1988). *Metal-semiconductor contacts* (Vol. 19, 2nd ed.). Clarendon Press and Oxford University Press, Oxford [England] and New York.
Rose, A. (1955). Space-charge limited currents in solids. *Physical Review, 97*, 1538–1544.
Roseman, R. D. (1993). *Domain and grain boundary structural effects on the resistivity behavior in donor doped, PTCR barium titanate*. PhD thesis, University of Illinois at Urbana-Champaign.
Rosen, C. Z., Hiremath, B. V., & Newnham, R. (1992). *Key papers in physics. Piezoelectricity*. New York: American Institute of Physics.
Saburi, O. (1961). Semiconducting bodies in the family of barium titanates. *Journal of the American Ceramic Society, 44*, 54–63.

Schönhals, A., & Kremer, F. (2003). Analysis of dielectric spectra. In F. Kremer, & A. Sch"onhals (Eds.), *Broadband dielectric spectroscopy* (pp. 59–98). Berlin Heidelberg: Springer.
Schröder, J. (2016). Magneto-electro-mechanically couplings: some remarks on transformations, objectivity, and coupling coefficients. *CISM-course*, this volume: 1–1.
Schroeder, M., Haußmann, A., Thiessen, A., Soergel, E., Woike, T., & Eng, L. M. (2012). Conducting domain walls in lithium niobate single crystals. *Advanced Functional Materials*, *22*(18), 3936–3944.
Scott, J. F. (2000). *Ferroelectric memories*. Berlin, Heidelberg: Springer.
Scott, J. F., & Redfern, S. A. T. (2001). Ming Zhang, and M. Dawber. Polarons, oxygen vacancies, and hydrogen in $Ba_xSr_{1-x}TiO_3$. *Journal of the European Ceramic Society*, *21*, 1629–1632.
Seidel, J. (2012). Domain walls as nanoscale functional elements. *Journal of Physical Chemistry Letters*, *3*(19), 2905–2909.
Seidel, J., Martin, L. W., He, Q., Zhan, Q., Chu, Y.-H., Rother, A., et al. (2009). Conduction at domain walls in oxide multiferroics. *Nature Materials*, *8*(3), 229–234.
Seidel, J., Maksymovych, P., Batra, Y., Katan, A., Yang, S.-Y., He, Q., et al. (2010). Domain wall conductivity in La-doped $BiFeO_3$. *Physical Review Letters*, *105*(19), 197603.
Shur, V. Ya., Nikolaeva, E. V., Shishkin, E. I., Kozhevnikov, V. L., & Chernykh, A. P. (2001). Polarization reversal in congruent and stoichiometric lithium tantalate. *Applied Physics Letters*, *79*, 3146–3148.
Shur, V. Ya., Rumyantsev, E. L., Nikolaeva, E. V., Shishkin, E. I., Batchko, R. G., Fejer, M. M., et al. (2002). Domain kinetics in congruent and stoichiometric lithium niobate. *Ferroelectrics*, *269*, 189–194.
Sluka, T., Tagantsev, A. K., Damjanovic, D., Gureev, M., & Setter, Nava. (2012). Enhanced electromechanical response of ferroelectrics due to charged domain walls. *Nature Communications*, *3*, 748.
Sluka, T., Tagantsev, A. K., Bednyakov, P., & Setter, N. (2013). Free-electron gas at charged domain walls in insulating $BaTiO_3$. *Nature Communications*, *4*, 1808.
Smit, J., & Wijn, H. P. J. (1959). *Ferrites*. Philips Technical Library.
Smolenskii, G. A., Bokov, V. A., Isupov, V. A., Krainik, N. N., Pasynkov, R. E., & Sokolov, A. I. (1984). *Ferroelectrics and related materials*. New York: Gordon and Breach Publishers.
Sonin, A. S., & Strukow, B. A. (1974). *Einführung in die Ferroelektrizität*. Berlin: Akademie-Verlag.
Spaldin, N. A., & Fiebig, M. (2005). The renaissance of magnetoelectric multiferroics. *Science*, *309*, 391–392.
Srinivasan, G. (2010). Magnetoelectric composites. *Annual Review of Materials Research*, *40*, 153–178.
Stoneham, A. M. (1975). *Theory of defects in solids*. Oxford: Clarendon Press.
Stowe, K. (2014). *An introduction to thermodynamics and statistical mechanics*. Cambridge UK: Cambridge University Press.
Strukov, B. A., & Levanyuk, A. P. (1998). *Ferroelectric phenomena in crystals*. Heidelberg: Springer.
Suzuki, T., Ueno, M., Nishi, Y., & Fujimoto, M. (2001). Dislocation loop formation in nonstoichiometric (Ba, Ca)TiO_3 and $BaTiO_3$ ceramics. *Journal of the American Ceramic Society*, *84*, 200–206.
Sze, S. M. (1981). *Physics of semiconductor devices*. New York: Wiley.
Tagantsev, A., Cross, L. E., & Fousek, J. (2010). *Domains in ferroic crystals and thin films*. New York: Springer.
Tredgold, R. H. (1966). *Space charge conduction in solids*. Amsterdam: Elsevier.
Trivedi, H., Shvartsman, V. V., Lupascu, D. C., Medeiros, M. S. A., Pullar, R. C., Kholkin, A. L., et al. (2015). Local manifestations of a static magnetoelectric effect in nanostructured $BaTiO_3$ $BaFe_{12}O_9$ composite multiferroics. *Nanoscale*, *7*, 4489–4496.
Tsuda, N. (Ed.) (1991). *Electronic conduction in oxides. Denki-dendōsei-sankabutsu (engl.)*. Springer series in solid state sciences (Vol. 94). Berlin [u.a.]: Springer.
Uchino, K. (1997). *Piezoelectric actuators and ultrasonic motors*. Boston, Dordrecht, London: Kluwer Academic Publ.

Van Run, A. M. J. G., Terrel, D. R., & Scholing, J. H. (1974). An in situ grown eutectic magnetoelectric composite material. *Journal of Materials Science*, *9*, 1710–1714.

Vasudevan, R. K., Morozovska, A. N., Eliseev, E. A., Britson, J., Yang, J.-C., Chu, Y.-H., et al. (2012). Domain wall geometry controls conduction in ferroelectrics. *Nano Letters*, *12*(11), 5524–5531.

Vaz, C. A., Hoffman, J., Ahn, C. H., & Ramesh, R. (2010). Magnetoelectric coupling effects in multiferroic complex oxide composite structures. *Advanced Materials*, *22*, 2900–2918.

Vollman, M., & Waser, R. (1994). Grain boundary defect chemistry of acceptor-doped titanates: Space charge layer width. *Journal of the American Ceramic Society*, *77*(1), 235–243.

Vul, B. M., Guro, G. M., & Ivanchik, I. I. (1973). Encountering domains in ferroelectrics. *Ferroelectrics*, *6*(1), 29–31.

Wadhawan, V. K. (2001). *Introduction to ferroic materials*. Boca Raton, Fl: CRC Press Inc.

Wang, J. W., Narayanan, S., Huang, J. Y., Zhang, Z., Zhu, T., & Mao, S. X. (2013). Atomic-scale dynamic process of deformation-induced stacking fault tetrahedra in gold nanocrystals. *Nature Communications*, *4*, 2340.

Waser, R. (Ed.). (2012). *Nanoelectronics and information technology advanced electronic materials and novel devices*. Wiley-VCH, Weinheim.

Waser, R.. & Smyth, D. M. (1996). Defect chemistry, conduction and breakdown mechanism of perovskite-structure titanates. In C. Paz de Araujo, J.F. Scott, & G.W. Taylor, (Eds.), *Ferroelectric thin films: Synthesis and basic properties, ferroelectricity and related phenomena* (Vol. 10, pp. 47–92). Gordon and Breach, Amsterdam.

Waser, R. (1991). Bulk conductivity and defect chemistry of acceptor-doped strontium titanate in the quenched state. *Journal of the American Ceramic Society*, *74*(8), 1934–1940.

Wechsler, B. A., & Klein, M. B. (1988). Thermodynamic point defect model of barium titanate and application to the photorefractive effect. *Journal of the Optical Society of America B*, *5*(8), 1711–1723.

Weertman, J., & Weertman, J. R. (1992). *Elementary dislocation theory*. New York: Oxford University Press.

Wu, J., Nan, C., Lin, Y., & Deng, Y. (2002). Giant dielectric permittivity observed in Li and Ti doped Nio. *Physical Review Letters*, *89*(21), 217601.

Xu, R., Liu, S., Grinberg, I., Karthik, J., Damodaran, A. R., Rappe, A. M., et al. (2015). Ferroelectric polarization reversal via successive ferroelastic transitions. *Nature Materials*, *14*(1), 79–86.

Ya Shur, V., Ievlev, A. V., Nikolaeva, E. V., Shishkin, E. I., & Neradovskiy, M. M. (2011). Influence of adsorbed surface layer on domain growth in the field produced by conductive tip of scanning probe microscope in lithium niobate. *Journal of Applied Physics*, *110*(5), 052017.

Yang, Y., Infante, I. C., Dkhil, B., & Bellaiche, L. (2015). Strain effects on multiferroic $BiFeO_3$ films. *C. R. Physique, 16*, 193–203.

Yuhuan, X. (1991). *Ferroelectric materials and their applications*. Amsterdam: Elsevier Science Publishers B.V.

Zafar, S., Jones, R. E., Jiang, B., White, B., Chu, P., Taylor, D., et al. (1998a). Oxygen vacancy mobility determined from current measurements in thin $Ba_{0.5}Sr_{0.5}TiO_3$ films. *Applied Physics Letters*, *73*, 175–177.

Zafar, S., Jones, R. E., Jiang, B., White, B., Kaushik, V., & Gillespie, S. (1998b). The electronic conduction mechanism in barium strontium titanate thin films. *Applied Physics Letters*, *73*, 3533–3535.

Zafar, S., Hradsky, B., Gentile, D., Chu, P., Jones, R. E., & Gillespie, S. (1999). Resistance degradation in barium strontium titanate thin films. *Journal of Applied Physics*, *86*, 3890–3894.

Zhan, J., Zhao, P., Guo, F.-F., Liu, W.-C., Ning, H., Chen, Y. B., et al. (2015). Semiconductor/relaxor 0–3 type composites without thermal depolarization in $Bi_{0.5}Na_{0.5}TiO_3$–based lead-free piezoceramics. *Nature Communications, 3*, 857.

Zuo, Y., Genenko, Y. A., Klein, A., Stein, P., & Xu, B. (2014). Domain wall stability in ferroelectrics with space charges. *Journal of Applied Physics*, *115*(8), 084110.

Electromechanical Models of Ferroelectric Materials

J. E. Huber

Abstract Models of the electromechanical behaviour of ferroelectric materials are reviewed. Starting from the constitutive relationships for piezoelectrics and estimates of the response of piezoelectric composites, the development of models is traced from the macro-scale through to the micro-scale. Derivations of models based on extensions of classical plasticity and crystal plasticity theory are given, following the literature, and example applications of these models are shown. The formation of domain patterns is discussed and minimum energy methods based on the concept of compatibility are used to derive typical domain patterns for tetragonal and rhombohedral ferroelectrics. Methods for modelling the evolution of domain patterns are described. Finally the outlook for future directions in modelling of ferroelectrics is discussed.

1 Introduction

Among the ferroic functional materials, the ferroelectrics are distinguished by the presence of a spontaneous dipole moment or *polarization* that can be reversed through the action of electric field. Typical ferroelectrics are non-centrosymmetric crystalline materials, such as the perovskite oxide alloys in the lead zirconate titanate system, barium titanate and others. They exhibit strong coupling between electrical and mechanical effects and this brought them to prominence through diverse applications of the piezoelectric effect in sonar, acoustics and later ultrasound and precision actuators. There is an extensive literature introducing the materials and their physics (Lines and Glass 1977; Xu 1991; Sidorkin 2006; Tagantsev et al. 2010). During the last 30 years interest in the ferroelectrics intensified following the growth of technological opportunities: increasing use of automated control has driven demand for sensors and actuators based on the piezoelectric principle. More recently the ability to produce

J. E. Huber (✉)
Department of Engineering Science, University of Oxford, Parks Rd, Oxford OX1 3PJ, UK
e-mail: john.huber@eng.ox.ac.uk

J. Schröder and Doru C. Lupascu (eds.), *Ferroic Functional Materials*,
CISM International Centre for Mechanical Sciences 581,
https://doi.org/10.1007/978-3-319-68883-1_4

ferroelectrics in a wide variety of forms such as thin films, fibres, micro-machined structures and nano-particles has opened up further opportunities.

While the study of the physics of ferroelectrics has been ongoing for about a century, the advent of practical constitutive models, that can be used to predict the material performance for engineering design purposes, is relatively recent. From the constitutive modelling perspective, the main variables of interest are electrical (electric field/electric displacement), mechanical (stress/strain) and thermal (temperature/entropy). Thermal effects are central to the processing and preparation of ferroelectrics, which commonly have a symmetry breaking phase transition between a high temperature non-ferroelectric phase and a low temperature ferroelectric phase that is key to the appearance of spontaneous polarization during processing. The influence of temperature on polarization through the pyroelectric effect, and on the dielectric permittivity, are also of importance. The well-known phenomonology of Devonshire (1949) captured the dominant effects and this phenomenological approach to modelling remains a key tool for understanding ferroelectrics. However, from the point of view of engineering design, stable isothermal conditions are usually desirable, and can be maintained by suitable thermal controls. In this case, sensitivity of the material behaviour to temperature, and the generation of heat, are still essential considerations, but the key variables for constitutive modelling become the electrical and mechanical variables. In this spirit, the present article reviews constitutive models of ferroelectrics that relate electric field/electric displacement and stress/strain under isothermal conditions.

With the development of devices that have complex structures such as the composite pillar arrangements of sonar transducers and the embedded electrodes used in multi-layer stack actuators, a requirement arises for models that can handle multiaxiality of the field variables. Equally the ability to model non-linearity in the constitutive response due to the intense fields generated near defects, or geometric features has become important. The chapter is structured as an introduction to several modelling approaches, starting with macroscopic models, proceeding to finer scales by considering the effects of ceramic grains and domains, and eventually looking at recent research on the arrangements of domains and domain walls. The focus is on relatively simple models that can be applied to a wide range of materials, but examples are given to illustrate the application to common ferroelectric systems.

2 Origins of Ferroelectricity and Piezoelectricity

The origin of spontaneous polarization and the capability for that polarization to be switched between distinct states can be understood by first considering the crystal structure of a typical ferroelectric. Taking the perovskite oxide $PbTiO_3$ example, the arrangement of a unit cell in the tetragonal phase is shown in Fig. 1. This contains an offset central Ti^{4+} ion, which combined with offsets of other ions and displacement of their charges produces a net polarization aligned with the c-axis of the unit cell. An applied stress produces small displacements of the ions giving rise to the usual elastic

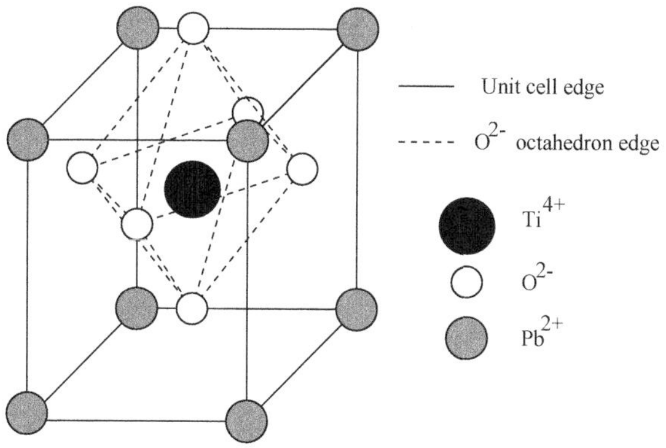

Fig. 1 Unit cell of tetragonal $PbTiO_3$

effects. However, because of the lack of a centre of symmetry, an applied stress can also modify the charge distribution, changing the dipole moment of the cell by the piezoelectric effect. Similarly, an applied electric field will produce linear dielectric effects but can also distort the unit cell, resulting in the converse piezoelectric effect.

The linear, reversible effects can be described by a free energy function of strain ϵ_{ij} and electric displacement D_i written in the form:

$$\Psi(\epsilon, \boldsymbol{D}) = \frac{1}{2}\epsilon_{ij}c^D_{ijkl}\epsilon_{kl} - D_k h_{kij}\epsilon_{ij} + \frac{1}{2}D_i\beta^\epsilon_{ik}D_k \tag{1}$$

Defining stress σ_{ij} as $\partial\Psi/\partial\epsilon_{ij}$ and electric field E_i as $\partial\Psi/\partial D_i$ gives the constitutive relations

$$\sigma_{ij} = c^D_{ijkl}\epsilon_{kl} - h_{kij}D_k \tag{2}$$
$$E_i = -h_{ikl}\epsilon_{kl} + \beta^\epsilon_{ik}D_k \tag{3}$$

where c^D_{ijkl} is the tensor of elastic moduli, h_{kij} are piezoelectric coefficients, and β^ϵ_{ik} is the tensor of dielectric impermeabilities. The superscript D indicates that the elastic moduli are those measured with the electric displacement held constant—that is, in open circuit conditions. Similarly, superscript ϵ indicates that the dielectric impermeabilities are those measured with the strain held constant—that is, with clamped boundary conditions. In (2) and (3) the strain and electric displacement are to be understood as variables measured relative to a reference state that is free of macroscopic stress and electric field. Practical measurements can be achieved by observations of surface displacement and changes in the state of surface charge on a material sample.

For later use, it is worth noting that (2) and (3) can be rearranged and inverted into various forms. Making ϵ_{ij} and D_i the subjects yields equations of the form:

$$\epsilon_{ij} = s^E_{ijkl}\sigma_{kl} + d_{kij}E_k \tag{4}$$
$$D_i = d_{ikl}\sigma_{kl} + \varepsilon^\sigma_{ik}E_k \tag{5}$$

Two further rearrangements are commonly used:

$$\epsilon_{ij} = s^D_{ijkl}\sigma_{kl} + g_{kij}D_k \tag{6}$$
$$E_i = -g_{ikl}\sigma_{kl} + \beta^\sigma_{ik}D_k \tag{7}$$

and

$$\sigma_{ij} = c^E_{ijkl}\epsilon_{kl} - e_{kij}E_k \tag{8}$$
$$D_i = e_{ikl}\epsilon_{kl} + \varepsilon^\epsilon_{ik}E_k \tag{9}$$

Equations (2)–(9) will be used interchangeably, with expressions relating the coefficients given where needed.

When a sufficiently strong electric field is applied to the material, ferroelectric switching can occur, reversing the polarization as shown in Fig. 2. The stable offset of the Ti^{4+} cation from the centre of the unit cell in tetragonal lead titanate is towards one face of the cell. There are six such stable states in polar tetragonal crystals with the corresponding spontaneous polarizations

$$\boldsymbol{P} = \pm P_0 \begin{pmatrix} 1 \\ 0 \\ 0 \end{pmatrix}, \quad \pm P_0 \begin{pmatrix} 0 \\ 1 \\ 0 \end{pmatrix} \text{ and } \pm P_0 \begin{pmatrix} 0 \\ 0 \\ 1 \end{pmatrix} \tag{10}$$

in Cartesian co-ordinates aligned to the unit cell edges. Here P_0 is the magnitude of the spontaneous polarization.

The existence of the set of six polarization states in polar tetragonal crystals allows two distinct kinds of ferroelectric switching: one in which the polarization

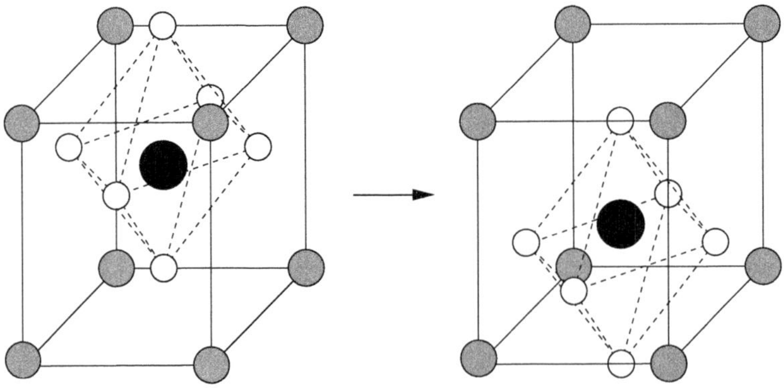

Fig. 2 Ferroelectric switching in the unit cell of tetragonal $PbTiO_3$

vector reverses in sign, or turns through 180° and a second in which the polarization vector turns through 90°. Meanwhile, the corresponding spontaneous strain states of the tetragonal unit cell are given by

$$\epsilon_{ij} = \frac{3}{2}\epsilon_0 \left(p_i p_j - \frac{1}{3}\delta_{ij} \right) \tag{11}$$

where $\boldsymbol{p} = \boldsymbol{P}/P_0$, δ_{ij} is the Kronecker delta, and ϵ_0 is a measure of the spontaneous strain magnitude. It follows from (11) that 180° switching, induces no net change in spontaneous strain. However, 90° switching results in a change in both spontaneous polarization and strain. The term *ferroelastic* is used to describe 90° switching when driven by an applied stress.

A ferroelectric crystal typically contains regions of uniform polarization, or *domains*, separated by thin boundaries (*domain walls*). Each domain has a state of spontaneous strain and polarization that can be changed by an electric field. Hence the overall state of polarization and strain of a ferroelectric crystal can change. Effectively, the macroscopic material stores a state of *remanent* strain $\boldsymbol{\epsilon}^r$ and *remanent* polarization $\boldsymbol{P}^r$. By "remanent" it is meant that the strain and polarization remain after macroscopic electric field and stress are removed. Changing these remanent quantities through the application of stress or electric field can also affect the elastic, piezoelectric and dielectric response. To illustrate this point, take as an example the tetragonal crystal system, where the microscopic piezoelectric response of a single domain is governed by the coefficients

$$\begin{aligned} h_{kij} = h_{33} p_k p_i p_j + h_{31} \left(p_k \delta_{ij} - p_k p_i p_j \right) \\ + h_{15} \left(\delta_{ki} p_j + \delta_{kj} p_i - 2 p_k p_i p_j \right) \end{aligned} \tag{12}$$

h_{33}, h_{31} and h_{15} being material constants. Then 180° switching converts $\boldsymbol{p}$ to $-\boldsymbol{p}$ and since h_{kij} is an odd function of $\boldsymbol{p}$ its sign is also reversed. The h_{kij} coefficients are also changed by 90° switching. By contrast the dielectric and elastic coefficients, β^{ϵ}_{ij} and c^{D}_{ijkl}, are even functions of $\boldsymbol{p}$ and are unchanged by 180° switching. However, they too can be changed by 90° switching.

In order to model the effects of ferroelectric switching, the free energy expression (1) requires modification. This can be achieved by the introduction of higher-order terms to capture non-linearity. The existence of several stable states then leads to a multi-well energy structure that has been used in classical models of ferroelectrics (Devonshire 1949) and can be used to simulate the formation of ferroelectric domains (Li et al. 2001; Wang et al. 2004; Su and Landis 2007). In typical ferroelectrics at equilibrium, it is observed that almost all of the material is in a state at or close to the bottom of one of the energy wells. The energy wells are deep, resulting in thin domain walls (typically only a few lattice parameters across) and high barriers to switching. This motivates modelling approaches analogous to the constrained theory of martensites wherein each material point is assumed to be in a state close to the bottom of an energy well (Ball and James 1987; Shu and Bhattacharya 2001), with a locally convex free energy.

Taking ferroelectric switching into account, (2) and (3) can be rewritten to represent the response of the ferroelectric in the form:

$$\sigma_{ij} = c^D_{ijkl}(\boldsymbol{\epsilon}^r, \boldsymbol{P}^r)\left(\epsilon_{kl} - \epsilon^r_{kl}\right) - h_{kij}(\boldsymbol{\epsilon}^r, \boldsymbol{P}^r)\left(D_k - P^r_k\right) \quad (13)$$

$$E_i = -h_{ikl}(\boldsymbol{\epsilon}^r, \boldsymbol{P}^r)\left(\epsilon_{kl} - \epsilon^r_{kl}\right) + \beta^{\epsilon}_{ik}(\boldsymbol{\epsilon}^r, \boldsymbol{P}^r)\left(D_k - P^r_k\right) \quad (14)$$

where the dependency of elastic, dielectric and piezoelectric coefficients on the state of remanent strain and polarization has been made explicit. In (13) and (14) the strain and electric displacement are referred to a stress and electric field free state with $(\boldsymbol{\epsilon}^r, \boldsymbol{P}^r) = 0$. While (13) and (14) treat $\boldsymbol{\epsilon}^r$ and $\boldsymbol{P}^r$ as state variables, there is no reason to suppose that these variables together contain sufficient information to specify the internal state of the material, or even to specify the coefficients in Eqs. 13 and 14. Treating the two quantities $(\boldsymbol{\epsilon}^r, \boldsymbol{P}^r)$ as state variables is also inconvenient from a practical measurement perspective in that the readily available measurements typically indicate only $(\boldsymbol{\epsilon}, \boldsymbol{D})$, and then only by their changes. Nevertheless, the remanent strain and polarization have been widely used as experimental indicators of the switching process (Lynch 1996; Liu and Huber 2007), and adopted as internal state variables in models of switching (Hwang et al. 1995; Kamlah and Tsakmakis 1999; Landis 2002; Huber and Fleck 2001).

3 Piezoelectric Composites

Taking (2) and (3) as a starting point, it is of interest to consider composites whose components exhibit piezoelectric behaviour. In the first instance, the small signal material response is considered, comprising elastic, dielectric and piezoelectric behaviour, connecting together the stress σ_{ij}, strain ϵ_{ij}, electric field E_i and electric displacement D_i. The strain is taken to be the symmetrized gradient of a continuous displacement field $\boldsymbol{u}$:

$$\epsilon_{ij} = \frac{1}{2}\left(u_{i,j} + u_{j,i}\right) \quad (15)$$

Analogously, electrostatic conditions are assumed such that the electric field can be derived from a scalar potential, ϕ:

$$E_i = -\phi_{,i} \quad (16)$$

Mechanical and electrical equilibrium equations can be written as

$$\sigma_{ij,j} = 0 \quad (17)$$

$$D_{i,i} = 0 \quad (18)$$

where it is assumed that no body forces act and no free charge is present. At internal interfaces, the absence of body forces and charges implies continuity of the normal components of stress and electric displacement:

$$\Delta\sigma_{ij}n_j = 0 \tag{19}$$

$$\Delta D_i n_i = 0 \tag{20}$$

where Δ indicates the jump in each quantity at the interface and $\boldsymbol{n}$ is the interface normal direction. At external free surfaces, the normal component of stress is the surface traction t_i and the jump in normal component of electric displacement is the surface charge density q:

$$\sigma_{ij}n_j = t_i \tag{21}$$

$$\Delta D_i n_i = q \tag{22}$$

Since the permittivity of ferroelectric materials is typically 2–3 orders of magnitude greater than that of air or vacuum, the approximation $D_i n_i = -q$ at free surfaces can often be made with good accuracy. Conditions similar to (19) and (20) apply to the strain and electric field if the displacement and electric potential are continuous at internal interfaces. Let $\boldsymbol{s}$ be a vector tangent to an interface across which the strain and electric field jump. Then

$$\Delta\epsilon_{ij}s_j = 0 \tag{23}$$

$$\Delta E_i s_i = 0 \tag{24}$$

It is convenient to group the constitutive equations together into a matrix form:

$$\begin{pmatrix} \sigma_{ij} \\ E_i \end{pmatrix} = \begin{pmatrix} c^D_{ijkl} & -h_{kij} \\ -h_{ikl} & \beta^\epsilon_{ik} \end{pmatrix} \begin{pmatrix} \epsilon_{kl} \\ D_k \end{pmatrix} \tag{25}$$

and note that, when dealing with a linear piezoelectric material, any spontaneous strain or polarization are taken to provide a constant ground state, relative to which the strain and electric displacement are measured. Likewise, the material state is assumed unchanged by any loading applied, such that the material coefficients or moduli are constants.

The constitutive relation (25) can be thought of as describing the macroscopic behaviour of a bulk material, or (with suitable coefficients) the behaviour of a microscopic region of material. In relating the microscopic to the macroscopic electromechanical response it is convenient to think of the bulk material as a composite comprising various constituent regions, with differing material coefficients. In a polycrystalline ferroelectric, the regions may represent differently oriented ceramic grains, or at a finer scale, individual ferroelectric domains. The composite could also contain non-piezoelectric components ($h_{kij} = 0$) but all constituent regions are assumed to be solid, insulating dielectrics. For a compact notation, rewrite (25) as

$$\boldsymbol{\sigma} = \boldsymbol{C}\boldsymbol{\epsilon} \tag{26}$$

identifying bold symbols with collections of tensor quantities:

$$\boldsymbol{\sigma} = \begin{pmatrix} \sigma_{ij} \\ E_i \end{pmatrix}; \; \boldsymbol{C} = \begin{pmatrix} c^D_{ijkl} & -h_{kij} \\ -h_{ikl} & \beta^{\epsilon}_{ik} \end{pmatrix}; \; \boldsymbol{\epsilon} = \begin{pmatrix} \epsilon_{kl} \\ D_k \end{pmatrix} \tag{27}$$

Taking (26) to represent the microscopic, or local, constitutive relation, a corresponding macroscopic relationship may be written for a bulk region of material, assumed sufficiently large and statistically homogenous that it may be treated as an effective medium:

$$\overline{\boldsymbol{\sigma}} = \boldsymbol{C}^o \overline{\boldsymbol{\epsilon}} \tag{28}$$

Here, the notation $\overline{\bullet} = \frac{1}{V}\int \bullet dV$ has been used to indicate volume averages and the macroscopic variables $\overline{\boldsymbol{\sigma}}$ and $\overline{\boldsymbol{\epsilon}}$ are identified averages of their microscopic counterparts. This step is justified by considering a macroscopic region subject to a charge distribution on its boundary given by $-\tilde{D}_i n_i$ where n_i is the outward surface normal. Then the average of the microscopic electric displacement $D_i(\boldsymbol{x})$ is given by

$$\begin{aligned} \overline{D}_i &= \frac{1}{V}\int_V D_i \, dV = \frac{1}{V}\int_V D_j x_{i,j} \, dV = \frac{1}{V}\int_V (D_j x_i)_{,j} - D_{j,j} x_i \, dV \\ &= \frac{1}{V}\int_S D_j n_j x_i \, dS = \frac{\tilde{D}_j}{V}\int_S x_i n_j \, dS = \frac{\tilde{D}_j}{V}\int_V x_{i,j} \, dV = \tilde{D}_i \end{aligned} \tag{29}$$

Electrical equilibrium (18) has been used along with Gauss' theorem to convert between volume and surface integrals. Also note that $x_{i,j} = \delta_{ij}$, the Kronecker delta. An analogous argument holds for the averaging of the stress field, using (17), and similar arguments apply to the averaging of electric field and strain: consider the macroscopic region having electric potential $\phi = -\tilde{E}_i x_i$ on its boundary. Then

$$\begin{aligned} \overline{E}_i &= \frac{1}{V}\int_V E_i \, dV = \frac{1}{V}\int_V -\phi_{,i} \, dV = \frac{1}{V}\int_S -\phi n_i \, dS \\ &= \frac{1}{V}\int_S \tilde{E}_j x_j n_i \, dS = \frac{\tilde{E}_j}{V}\int_S x_i n_j \, dS = \frac{\tilde{E}_j}{V}\int_V x_{i,j} \, dV = \tilde{E}_i \end{aligned} \tag{30}$$

where (16) has been used.

In general, finding the effective properties $\boldsymbol{C}^o$ from a known spatial distribution of $\boldsymbol{C}$ is a homogenization problem dependent on the detailed arrangement of the components of the composite. Various bounds (Bisegna and Luciano 1996) and estimates (Dunn and Taya 1993) for $\boldsymbol{C}^o$ have been developed for piezoelectric composites. Here, the self-consistent theory of composites, originally derived by by Hill (1965) is described, as this furnishes a practical means of estimating $\boldsymbol{C}^o$ for certain common arrangements of composite that can be approximated as a matrix material containing ellipsoidal inclusions. The self-consistent theory makes the approximation that each component of the composite behaves as if it were an ellipsoidal inclusion embedded in a surrounding matrix with properties $\boldsymbol{C}^o$. Then the difference between the local electromechanical fields, by Eshelby's argument (Eshelby 1957) uniform within the inclusion, and the corresponding mean fields is given by

$$\boldsymbol{\sigma} - \overline{\boldsymbol{\sigma}} = -\boldsymbol{C}^{\star}\left(\boldsymbol{\epsilon} - \overline{\boldsymbol{\epsilon}}\right) \tag{31}$$

Where

$$\boldsymbol{C}^{\star} = \boldsymbol{C}^o(\boldsymbol{\mathcal{S}}^{-1} - \boldsymbol{I}) \tag{32}$$

In (32), $\boldsymbol{I}$ represents the collection of identity tensors given by

$$\boldsymbol{I} = \begin{pmatrix} \frac{1}{2}\left(\delta_{ik}\delta_{jl} + \delta_{jk}\delta_{il}\right) & 0 \\ 0 & \delta_{ik} \end{pmatrix} \tag{33}$$

and $\boldsymbol{\mathcal{S}}$ is the collection of Eshelby tensors relating a notional unconstrained transformation strain and electric displacement, say ϵ^t, of an inclusion in a medium of moduli $\boldsymbol{C}^o$ to the constrained strain and electric displacement of the inclusion. That is, $\epsilon = \boldsymbol{\mathcal{S}}\epsilon^t$.

For particular combinations of the symmetries of $\boldsymbol{C}^o$ and the ellipsoidal inclusion, explicit formulae for the components of $\boldsymbol{\mathcal{S}}$ are known (Dunn and Taya 1993). In the most general case, the piezoelectric Eshelby tensors may be obtained by numerical integration (Huber et al. 1999). The inversion of a collection of tensors such as $\boldsymbol{\mathcal{S}}$ or $\boldsymbol{C}$ is interpreted in the sense that if $\boldsymbol{\sigma} = \boldsymbol{C}\boldsymbol{\epsilon}$ then $\boldsymbol{\epsilon} = \boldsymbol{C}^{-1}\boldsymbol{\sigma}$. This inversion may be accomplished by unpacking the elements of $\boldsymbol{C}$ into a matrix form and using matrix inversion methods.

Making use of (26), (28) and (31) gives

$$\boldsymbol{\epsilon} = \left(\boldsymbol{C} + \boldsymbol{C}^{\star}\right)^{-1}\left(\boldsymbol{C}^o + \boldsymbol{C}^{\star}\right)\overline{\boldsymbol{\epsilon}} = \boldsymbol{A}\overline{\boldsymbol{\epsilon}} \tag{34}$$

which defines $\boldsymbol{A}$ as the set of concentration tensors for strain and electric displacement, specific to a particular ellipsoidal component of the composite. Averaging over the volume of the composite shows that $\overline{\boldsymbol{A}} = \boldsymbol{I}$. Equally, since for each component of the composite $\boldsymbol{\sigma} = \boldsymbol{C}\boldsymbol{\epsilon}$, then

$$\overline{\boldsymbol{\sigma}} = \overline{\boldsymbol{C}\boldsymbol{A}}\overline{\boldsymbol{\epsilon}} \tag{35}$$

and the effective properties $\boldsymbol{C}^o$ can be identified as the volume average $\overline{\boldsymbol{C}\boldsymbol{A}}$.

The estimate of the effective moduli provided by (35) is in implicit form since the concentration tensors $\boldsymbol{A}$ depend on the effective moduli themselves through (32); iterative recalculation of $\boldsymbol{C}^o$ and $\boldsymbol{C}^{\star}$ normally gives rapid convergence. Alternative approaches to the estimation of effective properties include unit cell or representative volume element calculations (Kouznetsova et al. 2001), bounds (Bisegna and Luciano

1996) and asymptotic homogenization (Kanoute et al. 2001). The self-consistent homogenization has the advantage of lying within the well-known Hashin-Shtrikman bounds in cases such as that of elastic spherical inclusions. Also, when there is strong contrast in moduli between the components of the composite the rigorous bounds are typically far apart, while the self-consistent method can still give useful estimates provided that any extremely stiff or compliant phases are dilute. However, in certain cases of strong contrast such as porous composites with significant volume fraction of porosity, the self-consistent estimate becomes inaccurate.

3.1 Example of a Piezoelectric Composite

The self- consistent estimate of $\boldsymbol{C^o}$ can provide insight into the practical design of piezoelectric composites for engineering applications. An example is the piezoelectric hydrophone, used in sonar detection. Of key importance in this application is the piezoelectric response to hydrostatic stress. Rearrangement of (25) gives

$$\begin{pmatrix} \epsilon_{ij} \\ E_i \end{pmatrix} = \begin{pmatrix} s^D_{ijkl} & g_{kij} \\ -g_{ikl} & \beta^\sigma_{ik} \end{pmatrix} \begin{pmatrix} \sigma_{kl} \\ D_k \end{pmatrix} \tag{36}$$

where $s^D_{ijkl} = \left(c^D\right)^{-1}_{ijkl}$, $g_{kij} = \left(c^D\right)^{-1}_{ijpq} h_{kpq}$ and $\beta^\sigma_{ik} = \beta^\epsilon_{ik} - h_{ipq} s^D_{pqrs} h_{krs}$. Taking the axis of piezoelectric poling as the x_3 axis, then the open circuit voltage generated under hydrostatic pressure, which determines the sensitivity of the hydrophone, is governed by the piezoelectric coefficient g_{3kk}, often given the symbol g_h. Typical bulk ferroelectrics have g_{333} coefficients of opposite sign to g_{311} with $g_{311} \approx -g_{333}/2$, resulting in low values of g_h. However, by constructing a composite of ferroelectric fibres or pillars aligned with the x_3 axis and embedded in a compliant matrix, significant enhancements in the macroscopic g_h value can be achieved. To illustrate this an example is calculated using material data taken from Bisegna and Luciano (1996), see Table 1. This allows a direct comparison between self-consistent estimates and the variational bounds computed therein. The composite considered has a matrix of isotropic epoxy resin and contains round fibres of PZT-7A, taken to have transverse isotropy.

Figure 3 shows self-consistent estimates of the dielectric impermeabilities β^ϵ_{11} and β^ϵ_{33} for this composite. To approximate fibres aligned with the x_3 axis, ellipsoidal inclusions with minor radii $a = b$ and major radius $c = 10a$ were used in the calculation, the c axis being aligned with x_3. The volume fraction f of fibres is varied in the range 0–1. Elementary bounds on β^ϵ_{11} and β^ϵ_{33} of the Voigt type ($\boldsymbol{C^V} = \overline{\boldsymbol{C}}$) and Reuss type ($\boldsymbol{C^R} = \left(\overline{\boldsymbol{C^{-1}}}\right)^{-1}$) are also shown, along with tighter variational bounds derived by Bisegna and Luciano at $f = 0.4$. Note that the elementary bounds are far apart in this case; improved variational bounds can be much closer, as in Fig. 3a, but can still show a wide range as in Fig. 3b. The self-consistent estimates give an indication of how the properties of the composite change with volume fraction, and agree well

Table 1 Material data for a piezoelectric composite

	Epoxy	PZT		Epoxy	PZT
c^D_{1111} (GPa)	8.0	157	β^ϵ_{11} (m nF^{-1})	26.9	0.246
c^D_{3333} (GPa)	8.0	175	β^ϵ_{33} (m nF^{-1})	26.9	0.481
c^D_{2323} (GPa)	1.8	47.2	h_{311} (GV m^{-1})	0	−1.02
c^D_{1212} (GPa)	1.8	35.8	h_{333} (GV m^{-1})	0	4.58
c^D_{1122} (GPa)	4.4	85.4	h_{113} (GV m^{-1})	0	2.30
c^D_{1133} (GPa)	4.4	73.0			

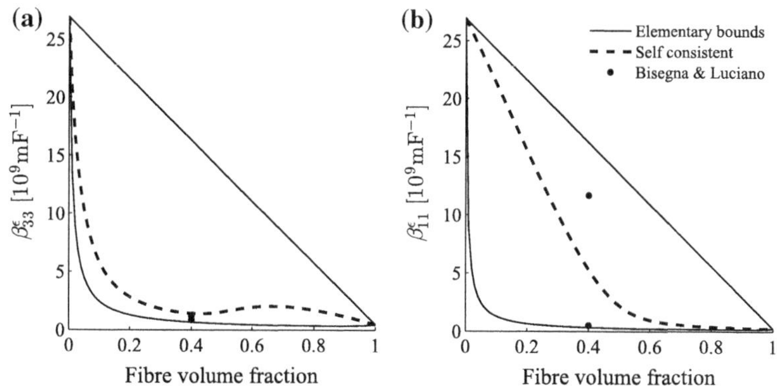

Fig. 3 Self-consistent estimates of impermeabilities **a** β^ϵ_{33} and **b** β^ϵ_{11} for a piezoelectric fibre composite

with the bounds. In particular Fig. 3 shows that β^ϵ_{33} is greatly reduced by adding a small volume of fibres, while β^ϵ_{11} remains fairly high. This may be expected because the fibres provide connectivity in the x_3 direction, such that the electric field component E_3 is similar in the fibres and matrix, while the absence of fibre connectivity in the x_1 direction causes the E_1 component of electric field to be concentrated in the matrix material. Similar anisotropy is seen in the elastic properties of the composite.

A consequence of this strong anisotropy can be seen in the hydrostatic voltage coefficient g_h; self- consistent estimates of g^o_h for the composite, normalized by the g_h coefficient of the piezoelectric component of the composite are shown in Fig. 4. Here, various aspect ratios for the piezoelectric component are compared. In a fibrous composite ($c = 10a$), an order of magnitude enhancement of g_h can be achieved by using a low volume fraction of fibres. The enhancement is mainly due to the elastic anisotropy of the composite. A particulate composite ($c = a$) shows less enhancement of g^o_h and requires a high volume fraction of piezoelectric particles to achieve optimum enhancement, while disc-like piezoelectric particles $c = 0.1a$ are even less effective.

Fig. 4 Self-consistent estimates of enhancement to the hydrostatic voltage coefficient g_h in a piezoelectric fibre composite

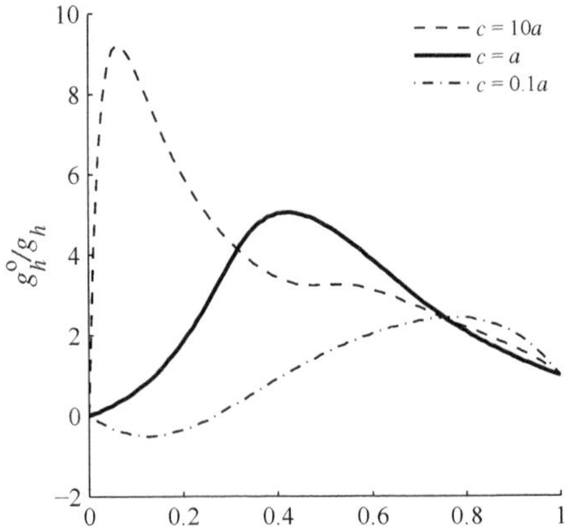

4 Models of Ferroelectric Switching

Now consider an extension of the linear piezoelectric constitutive laws (25) to represent material states with a remanent polarization and strain:

$$\begin{pmatrix} \sigma_{ij} \\ E_i \end{pmatrix} = \begin{pmatrix} c^D_{ijkl} & -h_{kij} \\ -h_{ikl} & \beta^\epsilon_{ik} \end{pmatrix} \begin{pmatrix} \epsilon_{kl} - \epsilon^r_{kl} \\ D_k - P^r_k \end{pmatrix} \tag{37}$$

Or, in the compact notation of Sect. 3,

$$\boldsymbol{\sigma} = \boldsymbol{C}\left(\boldsymbol{\epsilon} - \boldsymbol{\epsilon}^r\right) \tag{38}$$

Changes in $\boldsymbol{\epsilon}^r$ are somewhat analogous to increments of plastic strain in the classical theory of plasticity; there are two significant differences: The first is that here $\boldsymbol{C}$ is dependent on $\boldsymbol{\epsilon}^r$; thus changes in $\boldsymbol{C}$ must be included in the theory. The second is that, unlike classical plasticity, where plastic straining can proceed indefinitely through the nucleation and motion of dislocations, the remanent strain and polarization of ferroelectricity are limited in range. This is because no further ferroelectric switching can happen once all of the material has switched. Thus the P_0 and ϵ_0 of (10) and (11) place limits on $\boldsymbol{\epsilon}^r$. In plasticity theory such a limit on strain could be handled through strain-dependent hardening rules, and similar methods are adopted in the case of ferroelectrics. Finally it should be noted that $\boldsymbol{\epsilon}^r$ itself consists of both a remanent strain and a remanent polarization part, but this added complication has no impact on the theoretical development, except for making the expressions rather cumbersome if written out in full.

A goal of constitutive modelling is to provide tangent moduli $\boldsymbol{C}^t$ to relate increments, or rates of applied loading to the incremental material response such that

$$\dot{\boldsymbol{\sigma}} = \boldsymbol{C}^t \dot{\boldsymbol{\epsilon}} \tag{39}$$

In Sects. 4.1 and 4.2 models of ferroelectric switching are described that are analogous to classical plasticity and crystal plasticity respectively, following closely the corresponding developments in the literature (Bassiouny et al. 1988; Cocks and McMeeking 1999; Huber et al. 1999; Huber and Fleck 2001; Kamlah and Tsakmakis 1999; Landis 2002).

4.1 Classical Plasticity Model

In the spirit of classical plasticity, a constitutive framework for the ferroelectric material can be developed by postulating a free energy expression similar to (1) but including a dissipative part associated with switching (Cocks and McMeeking 1999):

$$\Psi = \frac{1}{2}\left(\boldsymbol{\epsilon} - \boldsymbol{\epsilon}^r\right) \boldsymbol{C} \left(\boldsymbol{\epsilon} - \boldsymbol{\epsilon}^r\right) + \Psi^r(\boldsymbol{\epsilon}^r) \tag{40}$$

As in the linear case, the stress and electric field can be derived from the free energy.

$$\boldsymbol{\sigma} = \frac{\partial \Psi}{\partial \boldsymbol{\epsilon}} \tag{41}$$

Then the second law of thermodynamics requires that the rate of work done by external agencies acting on the material, $\boldsymbol{\sigma}\dot{\boldsymbol{\epsilon}}$, is at least as great as the rate of increase in free energy, or

$$\boldsymbol{\sigma}\dot{\boldsymbol{\epsilon}} - \dot{\Psi} \geq 0 \tag{42}$$

The free energy rate is

$$\dot{\Psi} = \boldsymbol{\sigma}\dot{\boldsymbol{\epsilon}} - \boldsymbol{\sigma}\dot{\boldsymbol{\epsilon}}^r + \frac{\partial \Psi^r}{\partial \boldsymbol{\epsilon}^r}\dot{\boldsymbol{\epsilon}}^r + \frac{\partial \Psi}{\partial \boldsymbol{C}}\frac{\partial \boldsymbol{C}}{\partial \boldsymbol{\epsilon}^r}\dot{\boldsymbol{\epsilon}}^r \tag{43}$$

where the final term is due to the dependence of the elastic, dielectric and piezoelectric moduli on the remanent strain and polarization state. Equally, by defining

$$\boldsymbol{\sigma}^B = \frac{\partial \Psi^r}{\partial \boldsymbol{\epsilon}^r} + \frac{\partial \Psi}{\partial \boldsymbol{C}}\frac{\partial \boldsymbol{C}}{\partial \boldsymbol{\epsilon}^r} \tag{44}$$

and

$$\hat{\boldsymbol{\sigma}} = \boldsymbol{\sigma} - \boldsymbol{\sigma}^B \tag{45}$$

the free energy rate can be written as

$$\dot{\Psi} = \boldsymbol{\sigma}\dot{\boldsymbol{\epsilon}} - \hat{\boldsymbol{\sigma}}\dot{\boldsymbol{\epsilon}}^r \tag{46}$$

and (42) can then be expressed as

$$\hat{\boldsymbol{\sigma}}\dot{\boldsymbol{\epsilon}}^r \geq 0 \tag{47}$$

One way to satisfy (47) is for there to be a convex surface in $\hat{\boldsymbol{\sigma}}$ space enclosing the origin such that switching occurs only at states of $\hat{\boldsymbol{\sigma}}$ that lie on the surface, and such that the associated rates $\dot{\boldsymbol{\epsilon}}^r$ are directed towards the outward normal of the surface at point $\hat{\boldsymbol{\sigma}}$. This is analogous to the convex yield surface and associated flow of classical plasticity. Continuing the development in this way, let switching occur only when

$$G(\hat{\boldsymbol{\sigma}}, \boldsymbol{\epsilon}^r) = 0 \tag{48}$$

and the rates $\dot{\boldsymbol{\epsilon}}^r$ be defined by

$$\dot{\boldsymbol{\epsilon}}^r = \dot{\lambda}\frac{\partial G}{\partial \hat{\boldsymbol{\sigma}}} \tag{49}$$

where $\dot{\lambda}$ is analogous to a plastic multiplier. Then, since $\dot{G} = 0$ during switching,

$$\frac{\partial G}{\partial \hat{\boldsymbol{\sigma}}}\dot{\hat{\boldsymbol{\sigma}}} + \frac{\partial G}{\partial \boldsymbol{\epsilon}^r}\dot{\boldsymbol{\epsilon}}^r = 0 \tag{50}$$

So far, the value of $\dot{\lambda}$ is arbitrary and the determination of the rate of switching corresponding to a given loading rate requires additional information. This can be specified through an analogue of the hardening rules found in classical plasticity, in this case connecting increments in the "back" fields $\boldsymbol{\sigma}^B$ to the increments of the remanent quantities ϵ^r_{ij} and P^r_i. Let

$$\dot{\boldsymbol{\sigma}}^B = \boldsymbol{H}\dot{\boldsymbol{\epsilon}}^r \tag{51}$$

where $\boldsymbol{H}$ is a collection of tensors that depend on the state of the material and can be derived from a specification of Ψ, or defined directly as an alternative to specifying Ψ. Proper specification of the switching surface G and hardening rule, through $\boldsymbol{H}$, will define the non-linear response of the material. The requirement to set G convex motivates the use of relatively simple quadratic expressions for the switching surface (Cocks and McMeeking 1999; Huber and Fleck 2001; Landis 2002), placing the majority of the richness in the material behaviour into $\boldsymbol{H}$. Using (49–51) the rates $\dot{\lambda}$ can be found. First it is convenient to define $\boldsymbol{n} = \partial G/\partial\hat{\boldsymbol{\sigma}}$ and $\boldsymbol{m} = \partial G/\partial\boldsymbol{\epsilon}^r$. Then

$$\boldsymbol{n}\dot{\boldsymbol{\sigma}} = \boldsymbol{n}\dot{\boldsymbol{\sigma}}^B - \boldsymbol{m}\dot{\boldsymbol{\epsilon}}^r = \dot{\lambda}\,(\boldsymbol{n}\boldsymbol{H}\boldsymbol{n} - \boldsymbol{m}\boldsymbol{n}) \tag{52}$$

The rates $\dot{\lambda}$ during switching are given by

$$\dot{\lambda} = \frac{\boldsymbol{n}}{\boldsymbol{n}\boldsymbol{H}\boldsymbol{n} - \boldsymbol{m}\boldsymbol{n}}\dot{\boldsymbol{\sigma}} \tag{53}$$

and thus the rates of the remanent quantities are given by

$$\dot{\epsilon}^r = \mathbf{\Lambda}\dot{\sigma} \tag{54}$$

where

$$\mathbf{\Lambda} = \frac{\mathbf{n} \otimes \mathbf{n}}{\mathbf{n}\mathbf{H}\mathbf{n} - \mathbf{m}\mathbf{n}} \tag{55}$$

Finally, using (38), (39) and (54) the tangent moduli can be derived:

$$\begin{aligned} \dot{\sigma} &= \mathbf{C}^t\dot{\epsilon} = \dot{\mathbf{C}}(\epsilon - \epsilon^r) + \mathbf{C}\dot{\epsilon} - \mathbf{C}\dot{\epsilon}^r \\ &= \left(\frac{\partial \mathbf{C}}{\partial \epsilon^r}\mathbf{C}^{-1}\sigma - \mathbf{C}\right)\mathbf{\Lambda}\dot{\sigma} + \mathbf{C}\dot{\epsilon} \end{aligned} \tag{56}$$

where the inverse of $\mathbf{C}$ represents a set of tensors of the form

$$\mathbf{C}^{-1} = \begin{pmatrix} s^E_{ijkl} & d_{kij} \\ d_{ikl} & \varepsilon^\sigma_{ik} \end{pmatrix} \tag{57}$$

dependent on the current material state. Rearranging (56) gives

$$\dot{\sigma} = \left(\mathbf{I} + \mathbf{C}\mathbf{\Lambda} - \frac{\partial \mathbf{C}}{\partial \epsilon^r}\mathbf{C}^{-1}\sigma\mathbf{\Lambda}\right)^{-1}\mathbf{C}\dot{\epsilon} \tag{58}$$

and the tangent moduli are identified as

$$\mathbf{C}^t = \left(\mathbf{I} + \mathbf{C}\mathbf{\Lambda} - \frac{\partial \mathbf{C}}{\partial \epsilon^r}\mathbf{C}^{-1}\sigma\mathbf{\Lambda}\right)^{-1}\mathbf{C} \tag{59}$$

Care is needed in interpreting the term $(\partial\mathbf{C}/\partial\epsilon^r)\mathbf{C}^{-1}\sigma\mathbf{\Lambda}$—this consists of a collection of tensors of the form

$$\frac{1}{\mathbf{n}\mathbf{H}\mathbf{n} - \mathbf{m}\mathbf{n}}\begin{pmatrix} T''_{ijkl} & T'_{kij} \\ T'_{ikl} & T_{ik} \end{pmatrix} \tag{60}$$

which, if written out in full, produce lengthy expressions such as

$$\begin{aligned} T''_{ijkl} = &\left(\frac{\partial c^D_{ijrs}}{\partial \epsilon^r_{pq}}\frac{\partial G}{\partial \hat{\sigma}_{pq}} + \frac{\partial c^D_{ijrs}}{\partial P^r_q}\frac{\partial G}{\partial \hat{E}_q}\right)\left(s^E_{rsmn}\sigma_{mn} + d_{trs}E_t\right)\frac{\partial G}{\partial \hat{\sigma}_{kl}} \\ &-\left(\frac{\partial h_{rij}}{\partial \epsilon^r_{pq}}\frac{\partial G}{\partial \hat{\sigma}_{pq}} + \frac{\partial h_{rij}}{\partial P^r_q}\frac{\partial G}{\partial \hat{E}_q}\right)\left(d_{rmn}\sigma_{mn} + \varepsilon^\sigma_{rt}E_t\right)\frac{\partial G}{\partial \hat{\sigma}_{kl}} \end{aligned} \tag{61}$$

and so forth. Note that a consequence of the presence of these terms in the tangent moduli is that the tangent moduli depend not only on the material state, represented through the remanent quantities in ϵ^r, but also on the applied stress and electric field loading σ. The physical effect captured by this term is the energy released or absorbed by changes in the material compliance or permittivity due to switching. In most practical ferroelectrics this effect is small because the magnitude of the reversible strain and electric displacement represented by $C^{-1}\sigma$ is much less than the magnitude of the remanent quantities ϵ^r, while the compliances change only fractionally during switching.

Models of the type derived in this section have been illustrated by Huber and Fleck (2001) and Landis (2002). The models have the advantage of simplicity, which enables them to be incorporated readily into finite element codes for engineering design calculations. However, there are disadvantages. Apart from the difficulty of specifying hardening rules that accurately capture both the switching process and the eventual saturation of switching, the main objection to the model leading to (59) is perhaps the assumption that the remanent strain and polarization are sufficient as internal variables to capture the material state. Experimental evidence (Liu and Lynch 2006; Liu and Huber 2007) certainly suggests that the material behaviour is more complex. This motivates models with a greater number of internal variables that build up a macroscopic picture of material behaviour from an understanding of the microscopic process of switching, as explored in the following section.

4.2 *Crystal Plasticity Model*

In the context of crystal plasticity, the starting point is consideration of a single crystal and the microscopic processes that lead to changes in the macroscopic ϵ^r of that crystal. The switching process of a ferroelectric crystal is modelled by replacing the slip systems of classical plasticity with a set of transformation systems that correspond to the motion of domain walls. The macroscopic switching effect arises from a number of these transformation systems acting incrementally and simultaneously. The development here closely follows that in the literature (Huber et al. 1999; Huber and Fleck 2004; Pathak and McMeeking 2008).

First consider a single crystal or grain within a ceramic, in which there are several domains with distinct, symmetry related orientations of the polar axis. In Sect. 2 it was shown that there are six symmetry related variants in the polar tetragonal crystal system. More generally, let there be n variants such that the Ith variant has spontaneous strain ϵ^I_{ij} and polarization P^I_i, the two quantities being represented in compact notation by ϵ^I. Similarly, let each variant have moduli C^I with inverse S^I and occupy volume fraction v^I of the crystal. Define ϵ and σ to represent effective strain/electric displacement and stress/electric field values at the macroscopic level of the entire crystal. Similarly, let the local values in individual domains be ϵ^l, σ^l. Then, by the arguments in (29) and (30), $\epsilon = \overline{\epsilon^l}$ and $\sigma = \overline{\sigma^l}$. But these arguments do not allow ϵ^r to be identified with the volume average $\overline{\epsilon^I} = \sum v^I \epsilon^I$. Instead the effect of averaging can be seen through the constitutive law for the crystal, written as

$$\boldsymbol{\epsilon} - \boldsymbol{\epsilon}^r = \boldsymbol{S}\boldsymbol{\sigma} \tag{62}$$

where $\boldsymbol{S}$ is the inverse of $\boldsymbol{C}$, the set of macroscopic moduli for the crystal. At the microscopic level, each material point has the constitutive relation

$$\boldsymbol{\epsilon}^l - \boldsymbol{\epsilon}^I = \boldsymbol{S}^I \boldsymbol{\sigma}^l \tag{63}$$

Averaging (63) gives

$$\boldsymbol{\epsilon} - \overline{\boldsymbol{\epsilon}^I} = \overline{\boldsymbol{S}^I \boldsymbol{\sigma}^l} \tag{64}$$

Now, if the assumption is made that the stress and electric field are uniform within the crystal, then $\boldsymbol{\sigma}^l = \boldsymbol{\sigma}$ and

$$\boldsymbol{\epsilon} - \overline{\boldsymbol{\epsilon}^I} = \overline{\boldsymbol{S}^I}\, \boldsymbol{\sigma} \tag{65}$$

so that it is then possible to identify $\boldsymbol{S} = \overline{\boldsymbol{S}^I}$ and $\boldsymbol{\epsilon}^r = \overline{\boldsymbol{\epsilon}^I}$. This is the Reuss approximation at the level of the single crystal or grain. Proceeding on this basis, define a set of transformations $\alpha := I \rightarrow J$ that correspond to the actions of moving domain walls in transforming material of crystal variant I into material of crystal variant J. Let $\dot{f}^\alpha$ be the rate of volume fraction transformation by α. Then

$$\dot{\boldsymbol{\epsilon}}^r = \dot{\overline{\boldsymbol{\epsilon}^I}} = \sum_\alpha \dot{f}^\alpha \boldsymbol{\Delta\epsilon}^\alpha \tag{66}$$

where $\boldsymbol{\Delta\epsilon}^\alpha = \boldsymbol{\epsilon}^J - \boldsymbol{\epsilon}^I$. Similarly, defining $\boldsymbol{\Delta S}^\alpha = \boldsymbol{S}^J - \boldsymbol{S}^I$ gives

$$\dot{\boldsymbol{S}} = \dot{\overline{\boldsymbol{S}^I}} = \sum_\alpha \dot{f}^\alpha \boldsymbol{\Delta S}^\alpha \tag{67}$$

Considering energy, the electromechanical work stored in the crystal by virtue of its elastic, dielectric and piezoelectric moduli is

$$w = \frac{1}{2}\boldsymbol{\sigma}\boldsymbol{S}\boldsymbol{\sigma} \tag{68}$$

Now suppose that each transformation system dissipates energy of at least $G^{\alpha c}$ per unit volume in converting a volume of material from variant I to variant J. The second law of thermodynamics requires that

$$\boldsymbol{\sigma}\dot{\boldsymbol{\epsilon}} - \dot{w} - \sum_\alpha \dot{f}^\alpha G^{\alpha c} \geq 0 \tag{69}$$

Noting, from (62), that

$$\boldsymbol{\sigma}\dot{\boldsymbol{\epsilon}} = \boldsymbol{\sigma}\boldsymbol{S}\dot{\boldsymbol{\sigma}} + \boldsymbol{\sigma}\dot{\boldsymbol{S}}\boldsymbol{\sigma} + \boldsymbol{\sigma}\dot{\boldsymbol{\epsilon}}^r \tag{70}$$

and, from (68),

$$\dot{w} = \boldsymbol{\sigma}\boldsymbol{S}\dot{\boldsymbol{\sigma}} + \frac{1}{2}\boldsymbol{\sigma}\dot{\boldsymbol{S}}\boldsymbol{\sigma} \tag{71}$$

allows (69) to be written as

$$\sum_{\alpha} \dot{f}^{\alpha} G^{\alpha} - \sum_{\alpha} \dot{f}^{\alpha} G^{\alpha c} \geq 0 \tag{72}$$

where G^{α} is the thermodynamic driving force for switching, given by

$$G^{\alpha} = \boldsymbol{\sigma}\boldsymbol{\Delta}\boldsymbol{\epsilon}^{\alpha} + \frac{1}{2}\boldsymbol{\sigma}\boldsymbol{\Delta}\boldsymbol{S}^{\alpha}\boldsymbol{\sigma} \tag{73}$$

A criterion for ferroelectric switching that satisfies (72) is that switching occurs only when

$$G^{\alpha} = G^{\alpha c} \text{ and } \dot{f}^{\alpha} > 0 \tag{74}$$

Following classical crystal plasticity, the set of tangent moduli for the crystal can be found by connecting increments in G^{α} to the increments $\dot{f}^{\alpha}$ via a hardening rule. For simplicity, consider independent hardening, with

$$\dot{G}^{\alpha c} = h^{\alpha} \dot{f}^{\alpha} \tag{75}$$

Then, from (73)

$$\dot{G}^{\alpha} = \dot{\boldsymbol{\sigma}}\hat{\boldsymbol{\epsilon}}^{\alpha} \tag{76}$$

where

$$\hat{\boldsymbol{\epsilon}}^{\alpha} = \boldsymbol{\Delta}\boldsymbol{\epsilon}^{\alpha} + \boldsymbol{\Delta}\boldsymbol{S}^{\alpha}\boldsymbol{\sigma} \tag{77}$$

and since during switching $\dot{G}^{\alpha} = \dot{G}^{\alpha c}$

$$\dot{f}^{\alpha} = \frac{1}{h^{\alpha}}\hat{\boldsymbol{\epsilon}}^{\alpha}\dot{\boldsymbol{\sigma}} \tag{78}$$

The incremental constitutive relation derived from (62) is

$$\dot{\boldsymbol{\epsilon}} = \boldsymbol{S}\dot{\boldsymbol{\sigma}} + \dot{\boldsymbol{S}}\boldsymbol{\sigma} + \dot{\boldsymbol{\epsilon}}^{r} \tag{79}$$

or, making use of (77),

$$\dot{\boldsymbol{\epsilon}} = \boldsymbol{S}\dot{\boldsymbol{\sigma}} + \sum_{\alpha} \dot{f}^{\alpha}\hat{\boldsymbol{\epsilon}}^{\alpha} \tag{80}$$

Substituting from (78) gives

$$\dot{\epsilon} = \left(S + \sum_{\alpha} \frac{1}{h^{\alpha}} \hat{\epsilon}^{\alpha} \otimes \hat{\epsilon}^{\alpha} \right) \dot{\sigma} \tag{81}$$

and the tangent moduli are thus given by

$$S^{t} = S + \sum_{\alpha} \frac{1}{h^{\alpha}} \hat{\epsilon}^{\alpha} \otimes \hat{\epsilon}^{\alpha} \tag{82}$$

The summation in (82) is taken only over those transformation systems that are active by satisfaction of (74). So far, no attention has been given to the requirement that the switching process saturates when any of the v^{I} falls to zero. This requirement can be met by making the h^{α} functions of v^{I} that grow large as $v^{I} \to 0$. Alternatively, incremental calculations can be carried out by testing for $v^{I} > 0$ at each step and suppressing any transformation that violates this condition. Care must also be taken to arrange that, during cyclical loading, the accumulated hardening of each transformation system is reversed during the cycle to allow for the stable cyclic hysteresis observed in most ferroelectrics. Since the transformation systems that convert $I \to J$ and $J \to I$ are separate, this requires softening of inactive transformation systems when the corresponding v^{I} move away from zero. Approximate incremental calculations have been achieved by setting the h^{α} values to be small, and neglecting changes in $G^{\alpha c}$ (Huber and Fleck 2004).

Having established the tangent moduli for a single crystal or grain, the response of a polycrystal can be computed by treating the polycrystal as a composite of a number of grains, using the methods discussed in Sect. 3. If the self-consistent method is used, iteration is required to find the stress and electric field in each grain as the tangent moduli are dependent on $\boldsymbol{\sigma}$ through (77); this method has been shown to be in good agreement with more sophisticated, but computationally expensive, finite element calculations by Haug et al. (2007). Also, Huber and Fleck (2004) have demonstrated that the model can be fitted to various ferroelectric materials, though the fitting process is not straightforward, since neither the microscopic moduli $\boldsymbol{S}^{I}$ nor the $G^{\alpha c}$ are readily available for most materials. The rate-independent formulation developed here can be replaced with a rate-dependent formulation by making the transformation rates $\dot{f}^{\alpha}$ depend directly on G^{α} without requiring a switching criterion to be satisfied. This method has the advantage of avoiding the need for hardening relationships, and providing an explicit, forward calculation; the rate-dependent formulation has been used by several researchers (Huber and Fleck 2001; Kamlah et al. 2005; Haug et al. 2007; Pathak and McMeeking 2008) for practical calculations.

4.3 Example of Crystal Plasticity Model

To illustrate the crystal plasticity approach to modelling ferroelectrics, the response of a single crystal of polar tetragonal material is simulated here. The six crystal variants have spontaneous polarization states as given by (10) and the corresponding spontaneous strain states given by (11). In the initial state the volume fractions of the crystal variants are taken to be $v^I = 1/6$, $I = 1\ldots6$. For simplicity the variants are here modelled as having isotropic elasticity and dielectric permittivity given by

$$s^E_{ijkl} = \frac{1+\nu}{2Y}\left(\delta_{ik}\delta_{jl} + \delta_{jk}\delta_{il}\right) - \frac{\nu}{Y}\delta_{ij}\delta_{kl} \tag{83}$$

$$\varepsilon^\sigma_{ij} = \varepsilon\delta_{ij} \tag{84}$$

where Y is Young's modulus, ν is Poisson's ratio and ε is the dielectric permittivity at constant stress. The piezoelectric tensor is expected to be strongly anisotropic and is given by

$$\begin{aligned} d^I_{kij} = {} & d_{33} p^I_k p^I_i p^I_j + d_{31}\left(p^I_k \delta_{ij} - p^I_k p^I_i p^I_j\right) \\ & + d_{15}\left(\delta_{ki} p^I_j + \delta_{kj} p^I_i - 2 p^I_k p^I_i p^I_j\right) \end{aligned} \tag{85}$$

where $\boldsymbol{p}^I = \boldsymbol{P}^I / P_0$ is the unit vector in the direction of the spontaneous polarization for each crystal variant; d_{33}, d_{31} and d_{15} are piezoelectric constants. With these definitions, $\Delta\boldsymbol{S}^\alpha$ becomes

$$\Delta\boldsymbol{S}^\alpha = \boldsymbol{S}^J - \boldsymbol{S}^I = \begin{pmatrix} 0 & d^J_{kij} - d^I_{kij} \\ d^J_{ikl} - d^I_{ikl} & 0 \end{pmatrix} \tag{86}$$

At each step in the simulation, (73) is used to compute the driving force for switching by each transformation system, but instead of finding the tangent moduli of (82), a rate-dependent formulation is used by setting (Huber and Fleck 2001; Pathak and McMeeking 2008)

$$\dot{f}^\alpha = H^\alpha \left|\frac{G^\alpha}{G^{\alpha c}}\right|^{n-1} \frac{G^\alpha}{G^{\alpha c}} \tag{87}$$

which, for values of $n \gg 1$ has the effect of suppressing switching when $G^\alpha < G^{\alpha c}$ and allowing switching to proceed rapidly whenever $G^\alpha > G^{\alpha c}$. To provide saturation of switching as $v^I \to 0$, Pathak and McMeeking (2008) introduced the hardening rule

$$H^\alpha = 1 - \mathrm{e}^{-v^I/v^0} \tag{88}$$

with $v^0 = 0.01$. Material parameters similar to those of Pathak and McMeeking (2008) have been used (see Table 2), except for the use of isotropic elastic and dielectric moduli. Note that distinct values of $G^{\alpha c}$ are used for 90° switching and 180° switching. This has the effect that the critical value of electric field at which rapid ferroelectric switching occurs is about the same for the two distinct types of switching. In the model, the crystal is subjected to cyclic electric field loading of the form $E_1 = 2E_0 \sin(\omega t)$ with $\omega = 2\pi \mathrm{rad\,s^{-1}}$ and $E_0 = G^{\alpha c}(90°)/P_0$. The stress is held at zero throughout. The $\dot{f}^\alpha$ are computed using (87) and integrated over small

Table 2 Material data and model parameters for the tetragonal crystal plasticity model

Y (GPa)	60	P_0 (C m^{-2})	0.25
ν	0.3	ϵ_0	0.2%
ε (nF m^{-1})	22.5	$G^{\alpha c}(180°)$ (MJ m^{-3})	0.5
d_{33} (pm V^{-1})	315	$G^{\alpha c}(90°)$ (MJ m^{-3})	0.25
d_{31} (pm V^{-1})	−128	n	5
d_{15} (pm V^{-1})	482		

time steps to update the volume fractions. At each time step, the remanent quantities ϵ^r_{ij} and P^r_i are updated using (66) and the material moduli are updated using (67), with only the piezoelectric moduli changing, as in (86).

Starting from the initial condition with all $v^I = 1/6$ the electric field E_1 increases, causing switching to increase the volume fraction of the crystal variant with $P^J_1 = P_0$ at the expense of the other five variants. Eventually this process saturates with $\boldsymbol{P}^r = P_0\,(1\;0\;0)^T$. Later in the cycle, when $E_1 < 0$ the switching process is reversed. The volume fraction of the variant with $P^J_1 = -P_0$ increases, and similarly the volume fractions of variants with polarization states orthogonal to the applied field increase. However, as this happens, further switching transfers volume fraction from these orthogonal variants into the $P^J_1 = -P_0$ variant, which eventually reaches a volume fraction of unity, so that $\boldsymbol{P}^r = -P_0\,(1\;0\;0)^T$. As the electric field is cycled, the polarization state cycles between $\pm P_0\,(1\;0\;0)^T$. Figure 5 shows the resulting hysteresis in electric displacement versus electric field, calculated using (62). Also shown is the variation of the strain component ϵ_{11} which follows the "butterfly" hysteresis pattern characteristic of ferroelectrics. This arises due to a combination of effects: the piezoelectric moduli of the crystal reverse during each half cycle of loading, giving rise to regions with positive and regions with negative slope $\partial\epsilon_{11}/\partial E_1$. The changes in volume fractions also affect the remanent strain, producing a large increase in strain during the first quarter cycle of loading and affecting the shape of the subsequent hysteresis.

Figure 5 also shows, in dashed lines, the effect on the hysteresis curves of a moderate compressive stress $\sigma_{11} = -0.01\,P_0/d_{33}$, which is equivalent to about 8MPa compression. This stress is held constant throughout, while the electric field is again cycled. The compression has only a slight effect on the dielectric hysteresis curve, but the strain hysteresis is substantially changed. Of interest, the presence of compressive stress greatly increases the amplitude of strain variation during the stable cyclic hysteresis. This happens because the compressive stress stabilizes the four crystal variants with polarization orthogonal to the electrical loading axis. These crystal variants have negative ϵ^I_{11} components which causes the strain response to curve downwards during switching, and then turn sharply upwards as switching proceeds favouring the crystal variant whose spontaneous polarization is aligned with the electric field. A practical consequence is that the strain of piezoelectric actuators can be enhanced by holding them under moderate compression. This encourages an

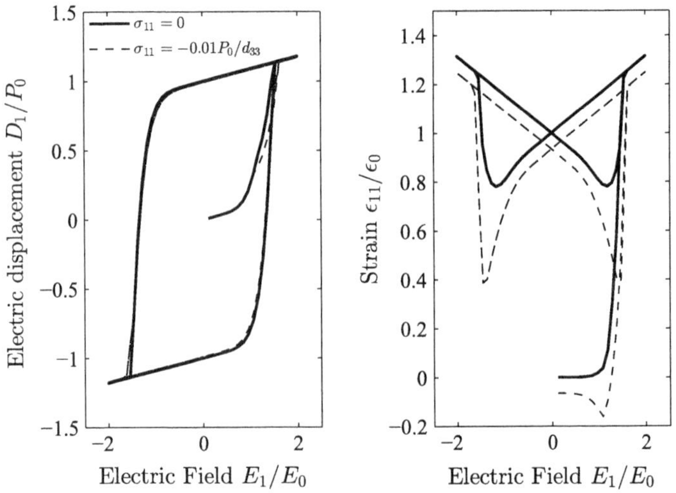

Fig. 5 Electric displacement and strain hysteresis loops for a tetragonal single crystal, modelled by the crystal plasticity method, and showing the effect of a compressive stress

"extrinsic" contribution to the piezoelectric response, caused by the movement of domain walls. Piezoelectric actuators are commonly held under such a bias stress to improve their performance. Similar effects were observed in barium single titanate crystals by Burcsu et al. (2004).

The model described in this section can be incorporated into finite element calculations (Haug et al. 2007; Pathak and McMeeking 2008) enabling simulation of the response of polycrystalline materials, and allowing complex boundary conditions to be handled. The crystal plasticity approach is also attractive in that the model contains an ample quantity of internal variables in the form of the volume fractions v^I, and these relate naturally to both the switching process and the saturation of switching. Furthermore, the v^I and crystallographic orientations can be inferred from X-ray or neutron diffraction studies (Hall et al. 2005; Jones et al. 2005) enabling comparison of models with experimental data at both the macroscopic and microscopic scales. However, as noted by Pathak and McMeeking (2008), there are deficiencies. The Reuss-type assumption of uniform stress and electric field at the grain level is unlikely to be accurate. Also, the model allows that any crystal variant can transform into any other. In practice, specific structures or patterns of domains form within ferroelectric crystals and these limit the available transformation systems. This factor motivates modelling at a finer scale to capture the effects of domain pattern.

5 Models of Ferroelectric Domain Patterns

In Sect. 4.3 the observation was made that specific structures or patterns of domains may influence the behaviour of a ferroelectric crystal. Hence there arises the possibility of arranging the domains within a crystal in an advantageous way. For example, in a crystal with *engineered domain configuration* (Park and Shrout 2007; Liu and Lynch 2006; Wu et al. 2011), enhanced piezoelectric properties are achieved by arranging a pattern of favourably oriented domains, or aligning a crystal such that favourable transformations happen when electric field is applied. There is also the opportunity to design new ferroelectrics with favourable properties by understanding the process by which domain patterns form and their consequences for the mechanism of ferroelectric switching—similar methods have been applied in the context of martensites (James and Hane 2000; Zhang and James 2009). Finally, an improved representation of domain patterns could be incorporated into models of the type developed in Sect. 4.2 for better accuracy.

Experimental evidence (Arlt 1990; Tsou et al. 2011) makes clear that laminated patterns of domains are common and this observation can be interpreted as a consequence of energy minimization, similar to the formation of laminates in the theory of martensite (Ball and James 1987; Bhattacharya 2003). The energy minimization approach has been used to predict domain patterns in ferroelectric crystals (Shu and Bhattacharya 2001) and hence obtain the piezoelectric properties of poled crystals with engineered domain configurations (Li and Liu 2004). The key step in the energy minimization approach is to recognize (Shu and Bhattacharya 2001; Ball and James 1987) that the overall energy is minimized when the strain and polarization state at almost every material point belongs to one of the set of minima of a multi-well free energy function. Consider a ferroelectric occupying volume V which is part of the extended space (volume $\tilde{V}$) that contains free space outside V. Let the material be subjected to tractions t_i and surface charges of density q over a portion of its surface. Then write the energy to be minimized as:

$$\int_V \Psi_{\text{bulk}}(\epsilon) + \Psi_{\text{grad}}\, dV - \int_S t_i u_i + \phi q\, dS + \int_{\tilde{V}} \frac{1}{2}\varepsilon_0 \phi_{,i}\phi_{,i}\, d\tilde{V} \tag{89}$$

The physical meaning of the terms in (89) is as follows: Ψ_{bulk} is the bulk energy, depending on the strain and electric displacement at each point in the region V occupied by the ferroelectric. This term has the multi-well structure discussed in Sect. 2. In the ferroelectric phase, Ψ_{bulk} can without loss of generality be defined such that $\Psi_{\text{bulk}} \geq 0$ with $\Psi_{\text{bulk}}(\epsilon) = 0$ for any $\epsilon = \epsilon^I$, $I \in 1 \dots N$, corresponding to the set of spontaneous strain and polarization states of the crystal variants. The term Ψ_{grad} is the additional energy due to local gradients in polarization. Typically, this term is assumed to depend only on the gradient $P_{i,j}$ of the conventional polarization, which is that part of the electric displacement due to the presence of the material in free space; that is,

$$D_i = P_i + \varepsilon_0 E_i \tag{90}$$

where ε_0 is the permittivity of free space, $8.854 \times 10^{-12}\mathrm{F\,m}^{-1}$. The meaning of polarization in this context is different from the *remanent* polarization discussed in Sects. 2–4: P_i includes both the spontaneous polarization P_i^I of the domain and that part of the reversible electric displacement resulting from the dielectric properties of the material. For example in a linear piezoelectric, using (9)

$$P_i = e_{ikl}\epsilon_{kl} + \left(\varepsilon_{ik}^{\epsilon} - \varepsilon_0\delta_{ik}\right) E_k + P_i^I \tag{91}$$

Including the gradient energy is physically reasonable on the grounds that domain walls, where $P_{i,j}$ is non-zero, contribute to the overall energy of the system. But the domain walls have finite thickness, introducing a length scale that is associated with Ψ_{grad}. Continuing with the terms in (89), the surface integral contains the potential energy of external loads in the form of tractions t_i displaced by u_i and surface free charge density q at voltage ϕ, applied to the material. The final term includes the stored electrostatic energy of free space. This term should be modified if there is any material other than free space surrounding the ferroelectric in region V. However, changes in this final term due to loading on V or rearrangement of the domains in V are typically much smaller than the resulting changes in the bulk energy term and so the energy of free space is often (but not always—see Dayal and Bhattacharya (2007)) neglected.

Given a specific form for Ψ_{bulk} and Ψ_{grad}, (89) can be minimized to yield patterns of polarization throughout a ferroelectric body, consistent with applied boundary conditions. The use of gradient methods to relax the energy towards the minimum results in Ginzburg-Landau type phase-field models which have been widely used to study ferroelectric domains. For periodic problems, Fourier spectral methods are expedient (Wang et al. 2013), while real space solutions can be found using finite difference, finite element and boundary element methods (Wang et al. 2009; Su and Landis 2007; Dayal and Bhattacharya 2007). Phase-field models are not described in further detail in this chapter. However, it should be noted that, in numerical models, it is necessary to discretize space at a scale finer than the domain wall width. Typically, a minimum of about 3 elements across a domain wall is needed (Völker and Kamlah 2012), resulting in a mesh spacing of $<10^{-9}$ m. The resulting models are computationally expensive, but have the advantage of allowing a great diversity of problems to be studied. A further point to note is that the phase-field models will typically identify a unique minimum, or in time-dependent formulations, a unique path of pattern evolution. This is of course desirable on physical grounds. However, due to the multi-well structure of the bulk energy, multiple local minima may exist and repeat runs of these models with slightly different starting conditions can produce qualitatively different results.

5.1 Theory of Compatibility

The remainder of this section is devoted to the constrained theory of ferroelectric domains. The constrained theory arises from neglecting the term Ψ_{grad} in (89), recognizing that the domain walls in ferroelectrics are of thickness only a few times the lattice parameter and the domain wall energy makes no significant contribution when the size of region V is much greater than the domain wall thickness. Then, considering the case without external loads, the minimisation problem reduces to finding an arrangement of ϵ throughout V that minimizes $\int_V \Psi_{\text{bulk}}\, dV$. This can be achieved by giving each point (except those within domain walls, which are taken to constitute a negligible volume) a strain and polarization belonging to one of the set of ϵ^I of the symmetry related variants. Setting

$$\epsilon = \epsilon^I, \quad I \in 1, 2, \ldots, N \tag{92}$$

gives immediately $\int_V \Psi_{\text{bulk}}\, dV = 0$. But this also produces particular requirements on the arrangement of domains and of any domain walls present. Consider a domain wall with normal $\boldsymbol{n}$ separating domains in states ϵ^1 and ϵ^2. Then, from (20):

$$\left(P_i^2 - P_i^1\right) n_i = 0 \tag{93}$$

Similarly, considering the jump in strain and requiring continuity of displacement, (23) results in the Hadamard jump condition (Shu and Bhattacharya 2001) which can be written, in the context of small strains, as

$$\epsilon_{ij}^2 - \epsilon_{ij}^1 = n_i a_j + n_j a_i \tag{94}$$

for some vector $\boldsymbol{a}$. Now, since ϵ^1 and ϵ^2 must be drawn from the set of spontaneous states of the domains, it is possible that, for a given pair of domains, the compatibility equations (93) and (94) have no solutions. There can be at most two non-trivial solutions for $\boldsymbol{n}$ (Ball and James 1987). However, there are also degenerate cases. For example it is common in ferroelectrics to have two variants with identical spontaneous strain states but opposite polarization. Then (94) is satisfied by any $\boldsymbol{n}$ with $\boldsymbol{a} = 0$ and (93) only restricts $\boldsymbol{n}$ to lie in the plane orthogonal to $\boldsymbol{P}^2 - \boldsymbol{P}^1$, producing a continuous set of domain wall normals. This results in 180° walls without a unique orientation, and gives rise to wavy patterns commonly seen in ferroelectric microstructure. More generally, the left hand side of (94) is symmetric so that orthogonal eigenvectors $\boldsymbol{e}_1$, $\boldsymbol{e}_2$ and $\boldsymbol{e}_3$ can be identified, with corresponding eigenvalues λ_1, λ_2 and λ_3. For solutions of (94) to exist, there is at least one eigenvalue that is zero: let $\lambda_3 = 0$. Additionally, because the spontaneous strains are those of symmetry related variants any one of which can be derived from any other by a pure rotation, the eigenvalues sum to zero, so $\lambda_1 = -\lambda_2$. Then solutions of (94) give the interface normal as

$$\boldsymbol{n} = \frac{\boldsymbol{e}_1 \pm \boldsymbol{e}_2}{\sqrt{2}} \tag{95}$$

and the resulting $\boldsymbol{n}$ can be tested for electrical compatibility using (93). Ball and James(1987) and Shu and Bhattacharya (2001) give a fuller discussion of the compatibility equations in the context of finite deformation gradients. Note that if a unique $\boldsymbol{n}$ results as the solution to (93) and (94), then the pair of domains can form only one orientation of domain wall consistent with energy minimization. This typically results in laminated microstructure with the layers alternating between two crystal variants.

What states of macroscopic remanent strain and polarization can a crystal achieve while satisfying the requirements of compatibility? Suppose that a region of a ferroelectric crystal in volume V is subjected to surface displacements and charges u_i, q, relative to a reference state with $\epsilon = 0$, such that $u_i = \tilde{\epsilon}_{ij} x_j$ and $-q = \tilde{D}_i n_i$ for some $\tilde{\epsilon} = (\tilde{\epsilon}_{ij}, \tilde{D}_i)$, where n_i is the surface normal. Let the resulting pattern of domains form on a scale much finer than that of the region V, and homogeneous across the region. Then a compatible arrangement of domains with average strain and polarization given by $\overline{\epsilon} = \tilde{\epsilon}$ that can be periodically repeated across the region achieves a minimum energy state in V with stress and electric field everywhere equal to zero. Care must be taken in applying the small strain formulation of the compatibility equations (93) and (94) to this problem. Specifically, (94) neglects the rotations in crystal lattice orientation that may occur at domain walls. This point will briefly be discussed later on. First, however, consider specific arrangements of domains in the form of a hierarchical laminate that satisfies (93) and (94) in an average sense at every internal interface.

5.2 Average Compatibility

A hierarchical arrangement of laminations can guarantee satisfying compatibility conditions and matching a prescribed macroscopic state of remanent strain and polarization ϵ^r by construction. First note that within the constrained theory, each material point has strain and electric displacement matching one of the N states ϵ^I, $I = 1 \ldots N$. Implicitly, each domain has zero stress and electric field, so that, in the absence of external loading, $\epsilon^r = \overline{\epsilon} = \tilde{\epsilon} = \overline{\epsilon^I}$. That is, the strain and electric displacement is entirely due to spontaneous states of the crystal variants.

To form a hierarchical laminate, a pair of crystal variants are first laminated with domain walls normal to $\boldsymbol{n}$, derived by solving (93) and (94). The resulting laminate is formed with sufficiently fine layers to be treated as a homogenized medium with effective ϵ that can be computed as a volume average over the constituent layers. The result is a rank-1 laminate. Higher rank laminates can be formed by laminating together layers of two materials, each of which is formed as a lower rank lamination—a rank R lamination being formed from fine layers comprising two rank $R-1$ laminations. The resulting hierarchy of laminates can be visualized using a binary tree diagram (see Fig. 6) in which the root node (node number 1) represents the macroscopic material, comprising laminations of materials represented by nodes 2

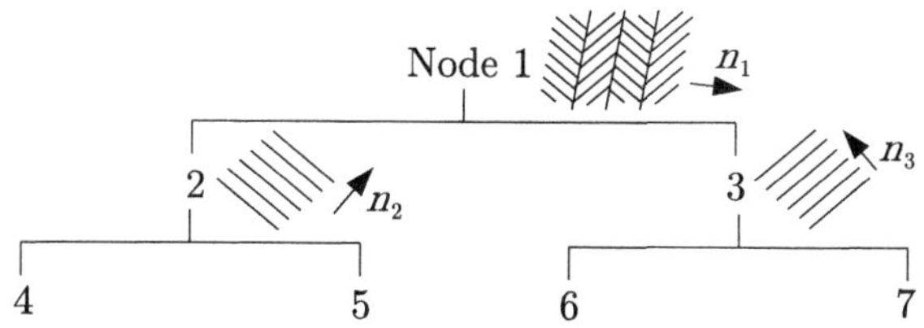

Fig. 6 Tree diagram corresponding to a rank-2 laminate, showing interface normals and pattern of laminations

and 3 and so forth. A rank R lamination has a corresponding tree diagram with $R+1$ layers and $2^{R+1}-1$ nodes.

The nodes in layer $R+1$ each represent material comprising a single variant drawn from the set of N crystal variants. Numbering the nodes in the tree diagram from the top down and from left to right as shown in Fig. 6, each node i in layers $1 \dots R$ has an associated interface normal, $\boldsymbol{n}_i$ defining the orientation of its constituent layers. Each node also has an associated volume fraction f_i such that $f_1 = 1 = f_2 + f_3$ and so forth:

$$f_i = f_{2i} + f_{2i+1}, \quad i = 1 \dots 2^R - 1 \tag{96}$$

and since each lamination satisfies the compatibility conditions,

$$f_i \epsilon_i^r = f_{2i} \epsilon_{2i}^r + f_{2i+1} \epsilon_{2i+1}^r, \quad i = 1 \dots 2^R - 1 \tag{97}$$

As usual, ϵ^r is used as compact notation representing both remanent strain and polarization. Each of the N variants has some volume fraction $f^I \geq 0$, $I = 1 \dots N$ where the superscript I distinguishes volume fractions of the crystal variants from the volume fractions f_i associated with the nodes. Again, the compatibility conditions enforce

$$\epsilon^r = \sum_1^N f^I \epsilon^I \tag{98}$$

with

$$\sum_1^N f^I = 1 \tag{99}$$

If the macroscopic remanent strain and polarization are specified, (98) and (99) give rise to a system of linear equations for the N volume fractions f^I. The strain component of (98) gives 6 linear equations; the polarization part gives 3 equations and (99) provides one further equation. Hence there are 10 equations for the N unknowns, with coefficients dependent on the spontaneous strain and polarization states of the crystal. It is immediately evident that the crystal parameters can result in (98) and (99) being invertible, singular or over-determined. Particular cases will be examined in detail in Sect. 5.4.

Suppose now that the f^I are known. Following Bhattacharya (1993) and Li and Liu (2004) it is possible to construct a compatible lamination with these overall variant volume fractions as follows. Rewrite (98) as

$$\epsilon^r = \mu_1 \epsilon^1 + \mu_2(1-\mu_1)\epsilon^2 + \cdots + \mu_{N-1} \prod_{m=1}^{N-2} (1-\mu_m)\epsilon^{N-1} + \prod_{m=1}^{N-1} (1-\mu_m)\epsilon^N \tag{100}$$

where the coefficients μ_m are given by

$$\mu_m = 0 \quad \text{if} \sum_{I=1}^{m-1} f^I = 1,$$
$$\mu_m = \frac{f^m}{1 - \sum_{I=1}^{m-1} f^I} \quad \text{otherwise} \tag{101}$$

The volume fractions f_i of the material represented by each node in the lowest level of the rank-R binary tree diagram can then be reconstructed using (Tsou and Huber 2010):

$$f_i = \prod_{k=1}^{R} (1-2b)\mu_k + b, \quad i = 2^R \ldots 2^{R+1} - 1 \tag{102}$$

with

$$b = \left\lfloor \frac{i - 2^R}{2^{k-1}} \right\rfloor - 2\left\lfloor \frac{i - 2^R}{2^k} \right\rfloor \tag{103}$$

where $\lfloor x \rfloor$ represents the floor function that gives the greatest integer less than or equal to x. Volume fractions associated with node numbers $i < 2^R$ are recovered by successive applications of (96). The binary digit b defined in (103) takes on the values zero or unity to indicate the branch of the tree diagram in which node i resides. If node i is reached via the left hand branch at level $k-1$ then $b = 0$; otherwise $b = 1$. The resulting pattern of volume fractions in a rank-2 tree diagram is shown in Fig. 7, along with labels (A,B,C) to indicate the crystal variants represented at each node. Provided that the individual crystal variants are pairwise compatible, that is, there exists a domain wall orientation satisfying (93) and (94) for any pairing of crystal variants, then the arrangement of volume fractions specified by (100–103) guarantees that a fully compatible hierarchical laminate can form with average strain and polarization state ϵ^r. The way this arrangement produces compatibility is evident from Fig. 7. Variant A is repeated in laminations A/B and A/C with identical volume fractions. Since A can form a notional interface with itself in any orientation, the only requirement to form a lamination between A/B and A/C is that B can form an interface with C; this requirement is met by pairwise compatibility. In terms of the average strains and polarizations,

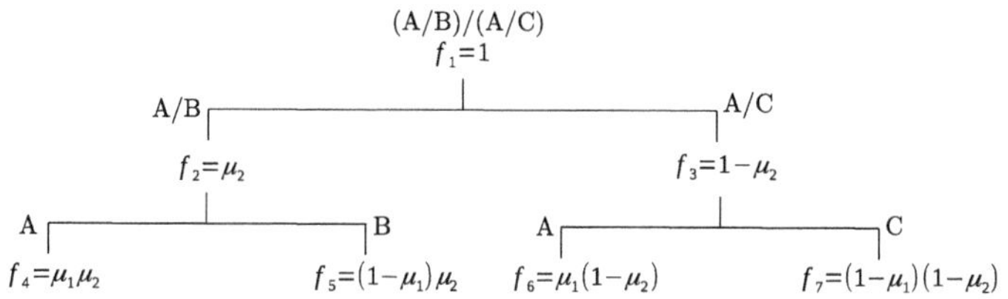

Fig. 7 A rank-2 tree diagram showing the construction of volume fractions f_i from μ_k

$$\epsilon_2 = \mu_1\epsilon_4 + (1-\mu_1)\epsilon_5 = \mu_1\epsilon^A + (1-\mu_1)\epsilon^B \tag{104}$$
$$\epsilon_3 = \mu_1\epsilon_6 + (1-\mu_1)\epsilon_7 = \mu_1\epsilon^A + (1-\mu_1)\epsilon^C \tag{105}$$

where superscripts on ϵ identify particular crystal variants, but subscripts identify node numbers in the binary tree. Now, from pairwise compatibility

$$\epsilon_{ij}^B - \epsilon_{ij}^C = n_i^{BC}a_j + n_j^{BC}a_i \tag{106}$$

and

$$\left(P_i^B - P_i^C\right)n_i^{BC} = 0 \tag{107}$$

From which it follows that

$$(\epsilon_2 - \epsilon_3)_{ij} = n_i^{BC}a_j' + n_j^{BC}a_i' \tag{108}$$

with $a_i' = (1-\mu_1)a_i$, and

$$(\boldsymbol{P}_2 - \boldsymbol{P}_3)_i\, n_i^{BC} = 0 \tag{109}$$

Similar compatible interfaces are guaranteed throughout the laminate. Since the μ_m in (100) can be chosen arbitrarily in the range $0 \le \mu \le 1$ and a compatible laminate can still be found, equivalently the volume fractions f^I can be arbitrarily chosen. It follows that the states of remanent strain and polarization ϵ^r accessible by a zero energy compatible laminate include all states within the convex hull of the set of ϵ^I, the spontaneous states of the individual crystal variants, given by

$$\epsilon^r = \sum_{i=1}^{N} f^I\epsilon^I,\ \ f^I \ge 0,\ \ \sum_{i=1}^{N} f^I = 1 \tag{110}$$

Compatibility is achieved in an *average* sense: The homogenized materials represented by the nodes can form an interface satisfying electrical and mechanical compatibility conditions, but the fine scale domain walls formed at the interface between two laminates do not necessarily satisfy compatibility conditions. The resulting pattern is an energy minimizer provided that there is a separation of length scales between successive laminations, allowing each lamination to be treated as a homogenized medium. This separation of length scales is consistent with observed microstructure (Liu and Lynch 2006; Arlt and Sasko 1980) where layers of very fine domains are seen to form.

Note, however, that the resulting pattern of crystal variants requires a lamination of rank $N - 1$ to accommodate N distinct variants. It is rare in practice to see such multiple levels of length scale separation in ferroelectric crystals, and this motivates consideration of other ways that energy minimizing domain patterns can form, such as the formation of laminates without separation of length scales, but with compatibility satisfied exactly at every domain wall.

5.3 Exact Compatibility

Next consider conditions for the formation of a laminate in which the compatibility conditions are satisfied exactly at every domain wall. Tsou and Huber (2010) identified a set of sufficient conditions which guarantee such exact compatibility: First, wherever two laminations meet to form a higher-rank lamination, there should be a matching of volume fractions across the higher-rank interface. This avoids the type of mismatch indicated in Fig. 8a which, though it does not prevent the formation of exactly compatible interfaces, places extremely restrictive conditions upon their formation. With matched volume fractions, fewer pairings of crystal variants are formed. By reference to the binary tree diagram of rank-R, this first condition can be written as (Tsou and Huber 2010):

$$\frac{f_{2^{R-r}}}{f_{2^{R-r}+1}} = \frac{f_{2^{R-r}+2n-2}}{f_{2^{R-r}+2n-1}}, \quad n = 2, \ldots, 2^{R-r-1}, \quad r = 0 \ldots R-2 \tag{111}$$

This imposes a requirement upon all pairs of materials that are laminated together in a given level r of the binary tree to share a common volume fraction ratio, allowing matched patterns such as Fig. 8b.

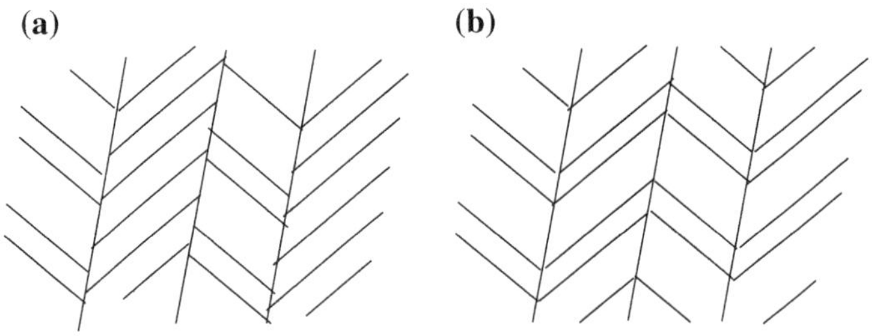

Fig. 8 Rank-2 laminations with **a** unmatched volume fractions and **b** matched volume fractions according to Eq. (111)

A second condition identified by Tsou and Huber (2010) is that pairs of domain walls in the two laminations that are joined to form a higher rank lamination intersect the common interface on the same line. This restricts the domain walls to have a common orientation when projected into the plane of that interface across which the higher rank lamination is formed. The condition forces the domain patterns on either side of an interface to match not only in spacing, as specified by (111) but also in orientation. The requirement can be written as a coplanarity condition on the interface normals associated with particular nodes in the binary tree:

$$\boldsymbol{n}_i \cdot \left(\boldsymbol{n}_p \times \boldsymbol{n}_q\right) = 0 \tag{112}$$

where i, p and q are node numbers selected as follows:

$$p = 2^{j+1} i + k, \quad q = p + 2^j, \tag{113}$$

with i, j and k in the ranges

$$i = 1, \ldots, 2^{R-1} - 1, \quad j = 0, \ldots, r-2, \quad k = 0, \ldots, 2^j - 1 \tag{114}$$

and tree level $r = R - \lfloor \log_2 i \rfloor$. The pattern of nodes picked out by (112–114) can be illustrated by application to a rank-2 laminate. With $R = 2$, the range of values specified in (114) becomes simply $i = 1$, $j = 0$, $k = 0$. Then the node numbers i, p, q for use in (112) are just $i = 1$, $p = 2$, $q = 3$, requiring that $\boldsymbol{n}_1$, $\boldsymbol{n}_2$, and $\boldsymbol{n}_3$ are coplanar. When $R = 3$, (112–114) produce five coplanarity equations, with node numbers i, p, q equal to

1, 2, 3;
1, 4, 6;
1, 5, 7;
2, 4, 5;
3, 6, 7.

Higher rank laminations lead to greater numbers of coplanarity equations and hence reduce the chance of finding an exactly compatible arrangement.

The final exact-compatibility condition for a rank-R lamination is that each domain walls satisfies (93) and (94), which brings in the spontaneous strain and polarization states of the individual domains. Once again, the set of nodes over which the compatibility equations apply must be specified. Write the compatibility equations as:

$$\left(\boldsymbol{P}_p - \boldsymbol{P}_q\right) \cdot \boldsymbol{n}_i = 0 \tag{115}$$

and

$$\boldsymbol{\epsilon}_p - \boldsymbol{\epsilon}_q = \boldsymbol{n}_i \boldsymbol{a}_i + \boldsymbol{a}_i \boldsymbol{n}_i \tag{116}$$

for nodes i, p and q. The nodes p, q are always drawn from the pure crystal variants in the lowest level of the tree diagram—nodes $2^R, \ldots, 2^{R+1} - 1$. Meanwhile, the domain wall normal $\boldsymbol{n}_i$ is always associated with a higher level node $i < 2^R$. The set of domain walls formed in the rank-R lamination is specified by:

$$i = 1, \ldots, 2^R - 1, \quad p = 2^r i + k, \quad q = p + 2^{r-1} \tag{117}$$

with

$$k = 0, \ldots, 2^{r-1} - 1, \quad r = R - \lfloor \log_2 i \rfloor \tag{118}$$

In the case of a rank-2 laminate, (117) and (118) produce the node numbers i, p, q equal to

1, 4, 6;
1, 5, 7;
2, 4, 5;
3, 6, 7;

while in a rank-R laminate, $R \cdot 2^{R-1}$ sets of node numbers are produced. These conditions become highly restrictive in laminates of rank $R > 2$ and are unlikely to be satisfied by sets of crystal variants with arbitrary spontaneous states ϵ^I. However, the crystal variants of typical ferroelectrics are both symmetric and symmetry related, with the result that exactly compatible laminates can often form.

It should be emphasized that the conditions specified by (111–118) are sufficient conditions for an exactly compatible laminate pattern, but are not necessary conditions for exactly compatible patterns of domains: other patterns such as crossing domains (Shu and Bhattacharya 2001) can form and the microstructure is of course not restricted to form only binary laminates. However, Tsou et al. (2011) have shown that this formulation generates many of the commonly observed patterns of domains, and can give insight into observed domain patterns.

Returning to the problem of accessible states of macroscopic remanent strain and polarization, it was noted in Sect. 5.2 that any state ϵ^r that can be made up as the volume average over a set of pairwise compatible variants with states ϵ^I admits an averagely compatible laminate to form. However, the same is not true for exactly compatible laminates and the set of accessible states is in this case not known.

The theory of compatibility described in Sects. 5.1–5.3 uses the linear approximation in which deformation gradients are infinitesimal and deformation is represented by a strain tensor that is the symmetrized deformation gradient: the skew-symmetric part of the deformation gradient, representing rotation, is neglected. In typical ferroelectrics such rotations are small and the predictions of the linear theory are similar to those of the nonlinear theory. However, as noted by Bhattacharya (1993), significant errors can occur in the linear theory where there are multiple rotations. This can arise at the junction of domains. To see this, consider a path crossing a domain wall at which there is a net rotation of crystal lattice planes. Now a closed path around the junction of domains can cross several such walls. If the total lattice rotation is non-zero, then the formation of the junction from a continuous region of crystal results in a disclination and the loss of stress/electric field-free perfect compatibility.

To account for the finite rotations, the use of the non-linear theory (Shu and Bhattacharya 2001) is expedient. Alternatively, using the linear theory, Tsou and Huber (2010) computed the net rotation of lattice planes at each domain wall and used this to check for the existence of disclinations.

5.4 Examples of Compatible Laminates

To illustrate the use of the theory of compatibility, consider ferroelectric crystals in the polar tetragonal system, which includes $BaTiO_3$ and many other ferroelectrics. Here the crystal has six variants with spontaneous polarization states expressed in co-ordinates aligned with the crystallographic axes as follows:

$$
\begin{aligned}
\boldsymbol{P}^1 &= P_0 \begin{pmatrix} 1 \\ 0 \\ 0 \end{pmatrix}; \quad \boldsymbol{P}^2 = P_0 \begin{pmatrix} -1 \\ 0 \\ 0 \end{pmatrix}; \\
\boldsymbol{P}^3 &= P_0 \begin{pmatrix} 0 \\ 1 \\ 0 \end{pmatrix}; \quad \boldsymbol{P}^4 = P_0 \begin{pmatrix} 0 \\ -1 \\ 0 \end{pmatrix}; \\
\boldsymbol{P}^5 &= P_0 \begin{pmatrix} 0 \\ 0 \\ 1 \end{pmatrix}; \quad \boldsymbol{P}^6 = P_0 \begin{pmatrix} 0 \\ 0 \\ -1 \end{pmatrix}
\end{aligned} \tag{119}
$$

The corresponding spontaneous strain states are

$$
\begin{aligned}
\epsilon^1 = \epsilon^2 &= \epsilon_0 \begin{pmatrix} 1 & 0 & 0 \\ 0 & -\frac{1}{2} & 0 \\ 0 & 0 & -\frac{1}{2} \end{pmatrix}; \\
\epsilon^3 = \epsilon^4 &= \epsilon_0 \begin{pmatrix} -\frac{1}{2} & 0 & 0 \\ 0 & 1 & 0 \\ 0 & 0 & -\frac{1}{2} \end{pmatrix}; \\
\epsilon^5 = \epsilon^6 &= \epsilon_0 \begin{pmatrix} -\frac{1}{2} & 0 & 0 \\ 0 & -\frac{1}{2} & 0 \\ 0 & 0 & 1 \end{pmatrix}
\end{aligned} \tag{120}
$$

Application of (93) and (94) to the polar tetragonal system shows that 180° domain walls can form between variants 1 and 2, variants 3 and 4 or variants 5 and 6 while 90° domain walls form between any other pairs of variants. Application of (98) shows immediately that all ϵ^r_{ij}, $i \neq j$ must be zero: the crystal cannot achieve remanent strain states that include shear components in co-ordinates aligned with the crystallographic axes. The remaining, non-trivial equations derived from (98) and (99) can be written as:

$$\begin{bmatrix} \epsilon^r_{11}/\epsilon_0 \\ \epsilon^r_{22}/\epsilon_0 \\ \epsilon^r_{33}/\epsilon_0 \\ P^r_1/P_0 \\ P^r_2/P_0 \\ P^r_3/P_0 \\ 1 \end{bmatrix} = \begin{bmatrix} 1 & 1 & -\frac{1}{2} & -\frac{1}{2} & -\frac{1}{2} & -\frac{1}{2} \\ -\frac{1}{2} & -\frac{1}{2} & 1 & 1 & -\frac{1}{2} & -\frac{1}{2} \\ -\frac{1}{2} & -\frac{1}{2} & -\frac{1}{2} & -\frac{1}{2} & 1 & 1 \\ 1 & -1 & 0 & 0 & 0 & 0 \\ 0 & 0 & 1 & -1 & 0 & 0 \\ 0 & 0 & 0 & 0 & 1 & -1 \\ 1 & 1 & 1 & 1 & 1 & 1 \end{bmatrix} \begin{bmatrix} f^1 \\ f^2 \\ f^3 \\ f^4 \\ f^5 \\ f^6 \end{bmatrix} \tag{121}$$

This has the form $\boldsymbol{e} = \boldsymbol{M} \boldsymbol{f}$ where $\boldsymbol{M}$ is a 7×6 matrix of rank 6. Given $\boldsymbol{e}$, this yields a unique solution for the six volume fractions in $\boldsymbol{f}$ provided that the remanent strain and polarization state specified in $\boldsymbol{e}$ is within the convex hull of the set of spontaneous states. This result is special to the polar tetragonal system: the macroscopic ϵ^r prescribes the volume fractions f^I, but the arrangement of domains is non-unique. Consider, for example, a fully "unpoled" state, in which

$$\epsilon^r = 0 \tag{122}$$

so that

$$\boldsymbol{e} = [0\,0\,0\,0\,0\,0\,1]^T \tag{123}$$

For this state, (121) gives the volume fractions

$$\boldsymbol{f} = \left[\frac{1}{6}\ \frac{1}{6}\ \frac{1}{6}\ \frac{1}{6}\ \frac{1}{6}\ \frac{1}{6}\right]^T \tag{124}$$

Now the construction of Li and Liu (2004) and Bhattacharya (1993) gives the μ values for an averagely compatible laminate with this state of macroscopic ϵ^r. Using (101)

$$\mu_1 = \frac{1}{6};\ \ \mu_2 = \frac{1}{5};\ \ \mu_3 = \frac{1}{4};\ \ \mu_4 = \frac{1}{3};\ \ \mu_5 = \frac{1}{2}; \tag{125}$$

An averagely compatible laminate of rank-5 can match the "unpoled" state. The resulting laminate is, however, highly complicated, with the requirement for separation of length scales over multiple levels of hierarchical lamination. Considering exactly compatible laminates that may match the same ϵ^r, Tsou and Huber (2010) showed that a rank-3 exactly compatible laminate can form, see Fig. 9. The rank-3 solution is not unique and many alternate arrangements are possible. The arrangement shown in Fig. 9 can be represented by a binary tree diagram with the crystal variants in the lowest level of the tree having numbers 1, 2, 6, 2, 4, 3, 5, 3 reading across the tree diagram from left to right. Use the notation "IJ" to represent a rank-1 lamination of variants I and J, "$IJKL$" to represent a rank-2 lamination comprising alternating layers of "IJ" and "KL", and so forth. Then, Fig. 9 shows the laminate "12624353". Note, however, that this numbering specifies only the arrangement of the crystal variants, not the orientation of domain walls or the volume fractions. In

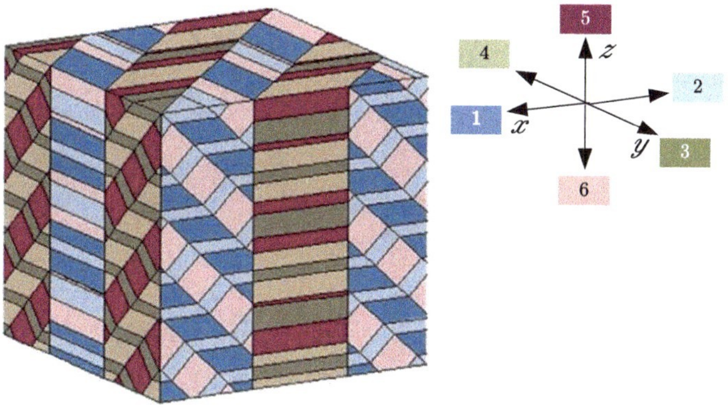

Fig. 9 A rank-3 laminate in the polar tetragonal crystal system that achieves $\epsilon^r = 0$. Spontaneous polarization directions of the individual domains are indicated using the colour coding

Fig. 9, $f^I = 1/6$ for each variant. The corresponding nodal volume fractions in the lowest level of the tree diagram are $f_8 = 1/6$, $f_9 = 1/12$, $f_{10} = 1/6$, $f_{11} = 1/12$, $f_{12} = 1/6$, $f_{13} = 1/12$, $f_{14} = 1/6$, $f_{15} = 1/12$.

It is of interest to study the variety of exactly compatible laminates that can form within a given crystal system. From consideration of the binary tree diagram, each node in the lowest level of the tree could represent any of the individual crystal variants. A rank-R binary tree has 2^R nodes in its lowest level, so in a crystal system with N variants the lowest level of the tree diagram has $N^{(2^R)}$ possible arrangements. However, many of these may represent identical laminates, differing only by rotations or reflections. Furthermore, many of the arrangements may not satisfy the conditions for compatibility given in Sect. 5.3.

Using the 6 variants of the polar tetragonal system defined by (119) and (120) as an example, the rank-1 arrangements can easily be enumerated. The 36 possible arrangements of the rank-1 tree include 6 of the form "II" that pair a crystal variant with itself and thus have no domain wall, representing rank-0 laminates, or single crystal variants. Another 6 contain pairings such as "12" that represent laminations of a crystal variant with another variant having the same spontaneous strain but opposite spontaneous polarization. In each case, a 180° domain wall can form satisfying the exact compatibility conditions. These laminates are congruent, differing only by rotations. The remaining 24 laminates contain pairings such as "13" that represent laminations of one crystal variant with another having different spontaneous strain and polarization. Once again, the exact compatibility conditions can be met, this time by forming 90° domain walls. These 24 laminates are again congruent by rotation. Effectively, then, there are just two families of rank-1 laminate in the polar tetragonal system, those with 90° and those with 180° domain walls. The notation {12} and {13} can be used to indicate these families.

Tsou et al. (2011) have similarly enumerated the rank-2 laminates in the polar tetragonal system. At first sight there are $6^4 = 1296$ arrangements. Removing those

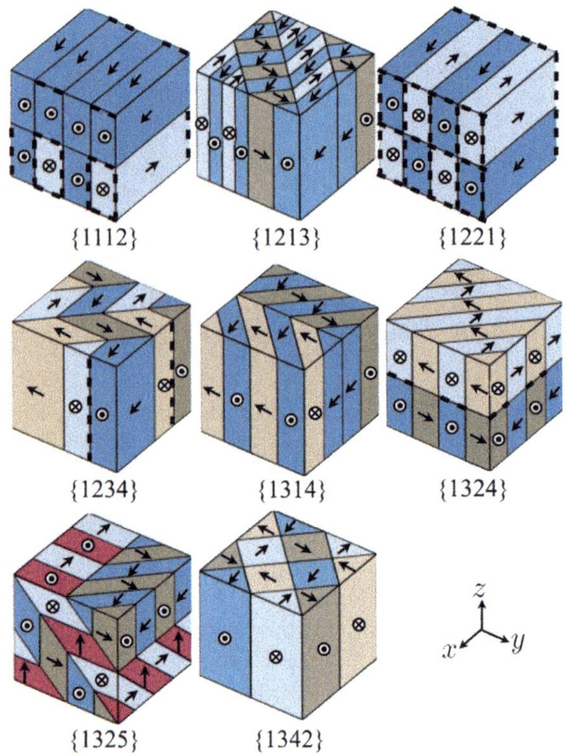

Fig. 10 The periodic repeating cells of the set of distinct rank-2 laminations in the tetragonal system (Tsou et al. 2011)

that either (i) represent lower rank laminations such as "1111", or (ii) cannot satisfy exact compatibility conditions, or (iii) are rotations or reflections of some lower numbered pattern, leaves just eight distinct families. These are illustrated in Fig. 10.

The domain walls shown using dashed lines in Fig. (10) are 180° walls without a unique orientation. Consequently these walls can form wavy patterns, producing the "watermarks" commonly observed in ferroelectric microstructure. Figure (10) also shows notional domain walls where a crystal variant meets itself across an interface introduced by the laminate. In practice, no domain wall would form in these circumstances and the crystal would be continuous, without an interface. Upon checking for the presence of disclinations, it is found that only patterns {1112}, {1221}, {1234} and {1324} are disclination free. These are zero-energy laminations within the constrained theory and are among the commonest patterns of domains observed in tetragonal ferroelectric crystals such as $BaTiO_3$ (Tsou et al. 2011; McGilly et al. 2010; Hooton and Merz 1955; Tagantsev et al. 2010).

The polar tetragonal crystal structure is among the most important of the ferroelectric crystal systems, being found in $BaTiO_3$ and in the PZT alloys $PbZr_xTiO_{(1-x)}O_3$ as well as many others. The PZT alloy series also includes rhombohedral crystal structures, which have trigonal symmetry. The polar rhombohedral crystal system has eight variants as follows:

$$\boldsymbol{P}^1 = \frac{P_0}{\sqrt{3}}\begin{pmatrix}1\\1\\1\end{pmatrix};\quad \boldsymbol{P}^2 = \frac{P_0}{\sqrt{3}}\begin{pmatrix}-1\\-1\\-1\end{pmatrix};$$
$$\boldsymbol{P}^3 = \frac{P_0}{\sqrt{3}}\begin{pmatrix}-1\\1\\1\end{pmatrix};\quad \boldsymbol{P}^4 = \frac{P_0}{\sqrt{3}}\begin{pmatrix}1\\-1\\-1\end{pmatrix};$$
$$\boldsymbol{P}^5 = \frac{P_0}{\sqrt{3}}\begin{pmatrix}-1\\-1\\1\end{pmatrix};\quad \boldsymbol{P}^6 = \frac{P_0}{\sqrt{3}}\begin{pmatrix}1\\1\\-1\end{pmatrix};$$
$$\boldsymbol{P}^7 = \frac{P_0}{\sqrt{3}}\begin{pmatrix}1\\-1\\1\end{pmatrix};\quad \boldsymbol{P}^8 = \frac{P_0}{\sqrt{3}}\begin{pmatrix}-1\\1\\-1\end{pmatrix}; \tag{126}$$

with corresponding spontaneous strain states

$$\boldsymbol{\epsilon}^1 = \boldsymbol{\epsilon}^2 = \epsilon_0\begin{pmatrix}0&1&1\\1&0&1\\1&1&0\end{pmatrix}$$
$$\boldsymbol{\epsilon}^3 = \boldsymbol{\epsilon}^4 = \epsilon_0\begin{pmatrix}0&-1&-1\\-1&0&1\\-1&1&0\end{pmatrix}$$
$$\boldsymbol{\epsilon}^5 = \boldsymbol{\epsilon}^6 = \epsilon_0\begin{pmatrix}0&1&-1\\1&0&-1\\-1&-1&0\end{pmatrix}$$
$$\boldsymbol{\epsilon}^7 = \boldsymbol{\epsilon}^8 = \epsilon_0\begin{pmatrix}0&-1&1\\-1&0&-1\\1&-1&0\end{pmatrix} \tag{127}$$

The crystal variants are pairwise compatible and from (93) and (94) it can be found that there are three distinct types of domain wall that can form. The corresponding rank-1 laminates are {12} with 180° domain walls, {13} with 70.5° domain walls, and {14} with 109.5° domain walls. In this crystal system, equation (98) indicates that $\epsilon^r_{ij} = 0$ whenever $i = j$ and so the macroscopic remanent strain includes only shear components in co-ordinates aligned with the crystallographic axes. Then (98)and (99) give:

$$\begin{bmatrix}\epsilon^r_{23}/\epsilon_0\\ \epsilon^r_{31}/\epsilon_0\\ \epsilon^r_{12}/\epsilon_0\\ \sqrt{3}P^r_1/P_0\\ \sqrt{3}P^r_2/P_0\\ \sqrt{3}P^r_3/P_0\\ 1\end{bmatrix} = \begin{bmatrix}1&1&1&1&-1&-1&-1&-1\\1&1&-1&-1&-1&-1&1&1\\1&1&-1&-1&1&1&-1&-1\\1&-1&-1&1&-1&1&1&-1\\1&-1&1&-1&-1&1&-1&1\\1&-1&1&-1&1&-1&1&-1\\1&1&1&1&1&1&1&1\end{bmatrix}\begin{bmatrix}f^1\\f^2\\f^3\\f^4\\f^5\\f^6\\f^7\\f^8\end{bmatrix} \tag{128}$$

Noting once again that this has the form $\boldsymbol{e} = \boldsymbol{M}\boldsymbol{f}$, $\boldsymbol{M}$ is in this case a 7×8 matrix of rank 7. Thus there is a degree of freedom in the solution to (128), the general solution being

$$\boldsymbol{f} = \boldsymbol{f}_0 + \gamma\,[1\ \ -1\ \ -1\ 1\ 1\ \ -1\ \ -1\ 1]^T \tag{129}$$

where $\boldsymbol{f}_0$ is any particular solution for the set of volume fractions, and γ is a scalar parameter that is limited in range by the requirement for all $f^I \geq 0$. Hence for any feasible macroscopic state of ϵ^r it is possible to find a set of volume fractions f^I with at least one of the f^I equal to zero. Averagely compatible laminates can be found for any ϵ^r in the convex hull of the set of ϵ^I, but these will typically have high rank and so are unlikely to form naturally. The macroscopic strain and polarization states that can be achieved with exactly compatible laminates have been explored by Tsou and Huber (2010). In considering the rank-2 exactly compatible laminates that can form, Tsou (2011) found the 14 families of laminates shown in Fig. 11.

These rank-2 laminates have some remarkable properties. For example, the laminate designated "1458" can, through the choice of nodal volume fractions $f_4 = f_5 = f_6 = f_7 = 1/4$, produce the "unpoled" macroscopic state $\epsilon^r = 0$. But by varying the volume fractions, a continuous range of laminates can be produced including both the "unpoled" state and a single variant state, representing a fully "poled" condition. Possibilities of this kind indicate the potential utility of the constrained theory in identifying low energy pathways that the domain patterns in ferroelectric crystals could follow, potentially enabling ferroelectric switching with very low coercive field. A discussion of the design of martensitic materials to achieve such low energy transformation paths has been given by Zhang and James (2009). In the case of ferroelectrics, the possibility of domain walls migrating through a laminate, transforming it continuously from one state of ϵ^r to another, provides a means to model the effect of applied macroscopic loads in the form of electric field and stress. The next section discusses the evolution of laminate domain patterns under applied loads.

5.5 *Evolution of Laminate Domain Patterns*

A number of constitutive models for ferroelectric crystals have been developed around the concepts of compatible laminates (Yen et al. 2008; Weng and Wong 2009; Tsou et al. 2013). The advantage of this approach, relative to the crystal plasticity methods described in Sect. 4.1, is that some account is taken of the pattern of domains within a crystal, or within individual grains of a polycrystal. This is significant because the presence of particular arrangements of domains can affect the switching process. To illustrate models of this kind, a general model of ferroelectric switching based on the incremental motion of a known pattern of domain walls is first developed here. Following Huber and Cocks (2008), rate potentials are used to characterize the mobility of the domain walls.

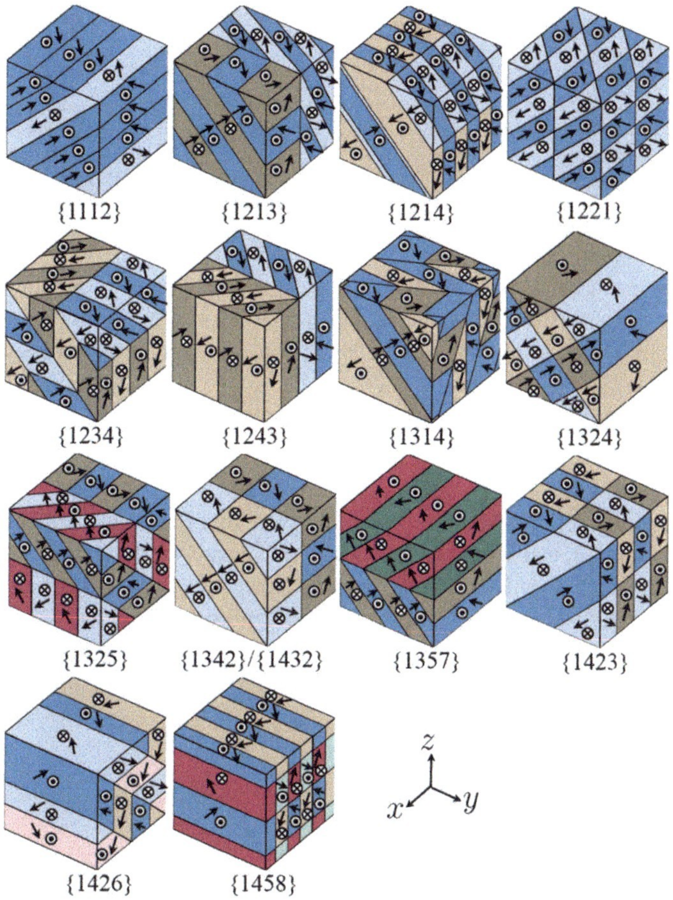

Fig. 11 The periodic repeating cells of the set of distinct rank-2 laminations in the rhombohedral system, after Tsou (2011). Arrows show the projections of each polarization vector onto the surface of the cell

Consider a region of a ferroelectric crystal occupying volume V, bounded by surface S, containing domains and domain walls such that the positions of the domain walls are defined by a set of m configurational variables a^i, with $i = 1 \dots m$. As in the constrained theory, the domain walls are taken to be thin such that each material point lies within one of the domains, with spontaneous strain state ϵ^I_{ij} and spontaneous polarization P^I_i, corresponding to one of the N crystal variants ($I \in 1 \dots N$). The domain walls themselves occupy area A. Let each material point experience local stress σ_{ij} and electric field E_i. The Gibbs free energy of the system G is the sum of an internal energy $U(\boldsymbol{\sigma}, \boldsymbol{E}, a^i)$ and potential energy, Ω, due to boundary tractions t_i and surface charge density q. The stored dielectric energy outside region V is neglected here, but should be included whenever changes in external stored energy significantly influence the material behaviour. The Gibbs free energy in the current configuration may be written as

$$G\left(\boldsymbol{\sigma}, \boldsymbol{E}, a^i\right) = \frac{1}{2}\int_V \sigma_{ij}\left(\epsilon_{ij} - \epsilon_{ij}^I\right) + E_i\left(D_i - P_i^I\right) dV - \int_S \phi q + t_i u_i \, dS \quad (130)$$

where I varies from point to point, indicating which crystal variant is present at each point.

A change in the configurational variables δa^i causes points on the domain walls to displace by δx resulting in a change δG in the free energy. Define an equivalent thermodynamic pressure $F(v)$ on the domain walls, moving at speed v such that

$$\int_A F(v)\delta x \, dA = -\delta G \quad (131)$$

dA being an element of the domain wall area. Note that $F(v)$ can vary from point to point on the domain walls; also G depends on the current configuration and loading, but is independent of v. Hence if the dependence of $F(v)$ on v is specified, (131) ensures that the displacements δx are energetically consistent with δG. Thus the motion of domain walls is driven by changes in G, but their increment of position is governed by $F(v)$. If the increment δx happens over time increment δt, then in the limit $\delta t \to 0$

$$\int_A F(v) v \, dA = -\dot{G} \quad (132)$$

Now define a rate potential $\Psi(a^i, \dot{a}^i)$ as

$$\Psi = \int_A \int_0^v F(\zeta) \, d\zeta \, dA \quad (133)$$

A virtual variation $\delta\dot{a}^i$ in the rates of the configurational variables corresponds to variation δv of the domain wall velocity at each point on the walls, and a variation $\delta\dot{G}$ in the rate of change of Gibbs free energy. Using (132) and (133)

$$\delta\Psi = \int_A F(v)\delta v \, dA = -\delta\dot{G} \quad (134)$$

Hence, defining a functional Π by

$$\Pi = \Psi + \dot{G} \quad (135)$$

the rates $\dot{a}^i$ are such that Π is stationary to variations $\delta\dot{a}^i$; that is

$$\delta\Pi = 0 \tag{136}$$

This variational principle provides a general framework for the thermodynamic evolution of configurational variables and similar methods have been widely applied in the study of microstructure evolution (Fischer et al. 2014). In the present context, since G is independent of the $\dot{a}^i$ then

$$\dot{G} = \sum_i \frac{\partial G}{\partial a^i}\dot{a}^i \tag{137}$$

or writing the a^i as a row matrix $\boldsymbol{a}$,

$$\dot{G} = \frac{\partial G}{\partial \boldsymbol{a}}\dot{\boldsymbol{a}}^T \tag{138}$$

Then the variation in $\dot{G}$ is

$$\delta\dot{G} = \frac{\partial G}{\partial \boldsymbol{a}}\delta\dot{\boldsymbol{a}}^T \tag{139}$$

Application of (136) provides a set of equations for the $\dot{a}^i$, which can be coupled and non-linear. However, a particularly simple expression results if the system has linear kinetics, of the form

$$\Psi = \frac{1}{2}\dot{\boldsymbol{a}}\boldsymbol{M}\dot{\boldsymbol{a}}^T \tag{140}$$

where $\boldsymbol{M}$ is a symmetric matrix of constant coefficients that defines the mobility of the domain walls. Then

$$\delta\Psi = \dot{\boldsymbol{a}}\boldsymbol{M}\delta\dot{\boldsymbol{a}}^T \tag{141}$$

and so

$$-\frac{\partial G}{\partial \boldsymbol{a}}\delta\dot{\boldsymbol{a}}^T = \dot{\boldsymbol{a}}\boldsymbol{M}\delta\dot{\boldsymbol{a}}^T \tag{142}$$

As this holds for any variation $\delta\dot{a}^i$ about the true rates $\dot{a}^i$, then the $\dot{a}^i$ can be found directly by

$$\dot{\boldsymbol{a}}^T = -\boldsymbol{M}^{-1}\left(\frac{\partial G}{\partial \boldsymbol{a}}\right)^T \tag{143}$$

5.6 Example of Domain Pattern Evolution

The model of evolution of domain patterns outlined in Sect. 5.5 can be illustrated by considering the evolution of a rank-2 laminate under simple loading conditions. Take as an example a laminate in the family {1458} of the polar rhombohedral crystal

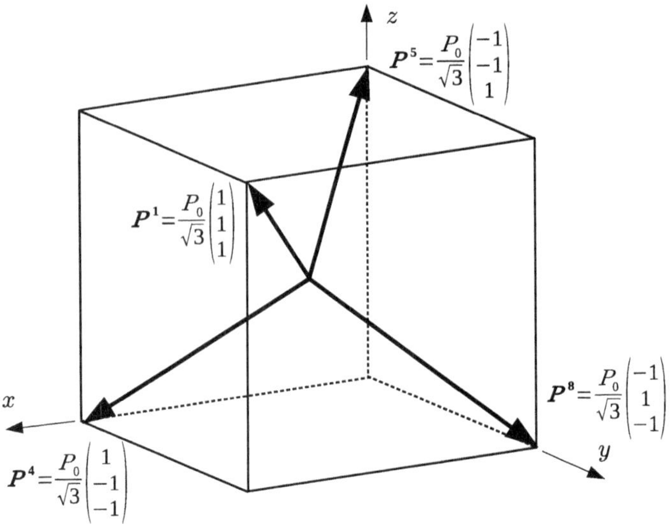

Fig. 12 The polarization states of the four rhombohedral crystal variants in the laminate "1458"

system. This family was shown in Sect. 5.4 to allow a state of zero average remanent strain and polarization when the four crystal variants have equal volume fractions. These laminates also allow a fully polarized state when the volume fraction of any variant tends to unity. It can easily be shown that the {1458} family is unique among rank-2 rhombohedral laminates in allowing $\epsilon^r = 0$. To see this, consider the general solution for the volume fractions of the crystal variants given in (129). When $\epsilon^r = 0$ the volume fractions take the form

$$\boldsymbol{f} = \frac{1}{8}[1\ 1\ 1\ 1\ 1\ 1\ 1\ 1]^T + \gamma[1\ -1\ -1\ 1\ 1\ -1\ -1\ 1]^T \tag{144}$$

with γ limited such that $0 \leq f^I \leq 1$. Setting γ such that only four crystal variants are present gives

$$\boldsymbol{f} = \frac{1}{4}[1\ 0\ 0\ 1\ 1\ 0\ 0\ 1]^T \text{ or } \boldsymbol{f} = \frac{1}{4}[0\ 1\ 1\ 0\ 0\ 1\ 1\ 0]^T \tag{145}$$

The first of these possibilities has only variant numbers 1, 4, 5, and 8 present and so includes the {1458} family. Since these four crystal variants produce a set of polarizations $\boldsymbol{P}^I$ with tetrahedral symmetry (see Fig. 12), it is evident that all permutations such as "1458", "1548", and so forth, are related by pure rotations or reflections. Similarly it can be seen that the second arrangement of volume fractions in (145) produces laminates such as "2376" which is a rotation of "1458", and permutations thereof.

Now consider the evolution of the specific domain pattern "1458". A binary tree diagram for this laminate is shown in Fig. 13a. Application of the compatibility equations from Sect. 5.3 shows that the domain wall orientations are given by

$$\boldsymbol{n}^1 = \begin{pmatrix} 0 \\ 0 \\ 1 \end{pmatrix}; \quad \boldsymbol{n}^2 = \boldsymbol{n}^3 = \begin{pmatrix} 1 \\ 0 \\ 0 \end{pmatrix} \tag{146}$$

The resulting pattern of domains is a rectangular grid or checkerboard as shown in cross-section in the x-z plane in Fig. 13b. All of the internal interfaces are 109.5° domain walls. For simplicity, let the material experience uniform stress and electric field with all components equal to zero except σ_{11} and E_1. Also, let each crystal variant have identical material properties s^E_{ijkl} and ε^σ_{ij} in x, y, z co-ordinates aligned with the crystallographic axes. Define the positions of the domain walls within a unit volume of the material by two configurational variables (a^1, a^2) as shown in Fig. 13b, such that the volume fractions of the individual crystal variants are:

$$f^1 = a^1 a^2; \quad f^4 = a^1(1 - a^2); \quad f^5 = a^2(1 - a^1); \quad f^8 = (1 - a^1)(1 - a^2) \tag{147}$$

In order to find $\partial G/\partial \boldsymbol{a}$, recall that $G = U + \Omega$, and note that for the specified loading with only σ_{11} and E_1 the internal energy U is given by

$$U = \int_V \frac{1}{2}\sigma_{11} s^E_{1111} \sigma_{11} + \frac{1}{2} E_1 \varepsilon^\sigma_{11} E_1 + E_1 d_{111} \sigma_{11} \, dV \tag{148}$$

while the potential energy of boundary tractions and charges is

$$\begin{aligned} \Omega &= -\int_S \phi q + t_i u_i \, dS \\ &= -\int_V \sigma_{11} s^E_{1111} \sigma_{11} + E_1 \varepsilon^\sigma_{11} E_1 + 2E_1 d_{111} \sigma_{11} \\ &\quad + E_1 P^I_1 + \sigma_{11} \epsilon^I_{11} \, dV \end{aligned} \tag{149}$$

where use has been made of Gauss' theorem and electromechanical equilibrium to convert to a volume integral. The simplifying assumptions of uniform s^E_{ijkl} and ε^σ_{ij} imply that the corresponding terms in (148) and (149) will not enter into the expression for $\partial G/\partial \boldsymbol{a}$. However, it is expected that the piezoelectric coefficient d_{111} will differ among the variants because of their differing polar axes. From symmetry, it can be seen that the d_{111} coefficient will be identical in variants 1 and 4, and have an opposite (negative) value in variants 5 and 8. Let variants 1 and 4 have $d_{111} = d_0$ and variants 5 and 8 have $d_{111} = -d_0$. Also, $\epsilon^I_{11} = 0$ for all four variants. Then, for a unit volume of the material, using (147)–(149):

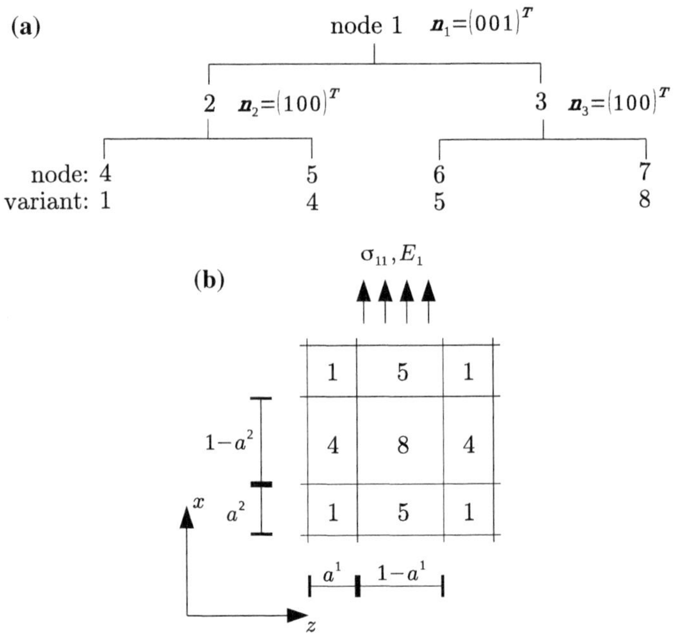

Fig. 13 **a** Binary tree diagram for the laminate "1458", and **b** arrangement of crystal variants in the $x - z$ plane

$$\frac{\partial G}{\partial a^1} = -2E_1 \left(\sigma_{11} d_0 + \frac{P_0}{\sqrt{3}} \right) \tag{150}$$

and

$$\frac{\partial G}{\partial a^2} = 0 \tag{151}$$

The uniaxial loading case considered here does not drive changes in the second degree of freedom, as indicated by (151). This is as expected from the symmetry of the loading and the crystal variants. The rate $\dot{a}^1$ is then specified by rate potential Ψ. Taking the simple case of linear kinetics, given by

$$\Psi = \frac{1}{2} \left(\dot{a}^1 \; \dot{a}^2 \right) \begin{pmatrix} m & 0 \\ 0 & m \end{pmatrix} \begin{pmatrix} \dot{a}^1 \\ \dot{a}^2 \end{pmatrix} \tag{152}$$

per unit area, the evolution of domains is governed by

$$\dot{a}^1 = \frac{2E_1}{m} \left(\sigma_{11} d_0 + \frac{P_0}{\sqrt{3}} \right) \tag{153}$$

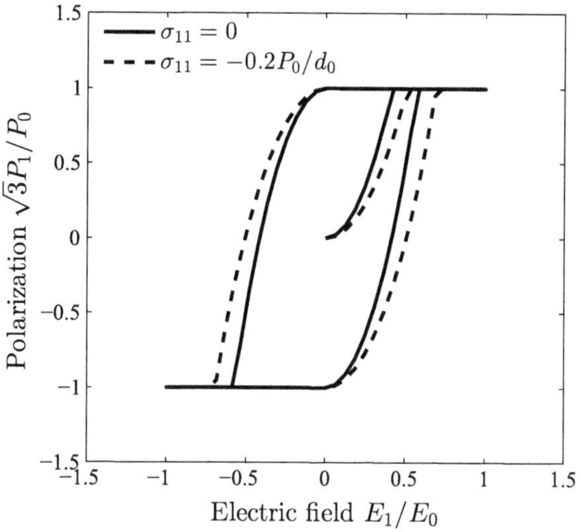

Fig. 14 Polarization response of the "1458" laminate

The evolution of the laminate can be simulated by integrating (153), using (147) and (98) to reconstruct the remanent strain and polarization state. Additionally, the constraint $0 \leq a^i \leq 1$ must be enforced. Figure 14 shows the polarization response of the laminate when subjected to sinusoidal loading $E_1 = E_0 \sin(\omega t)$ with $\omega = 2\pi \text{rad s}^{-1}$ and $E_0 P_0/m = 30$, starting from a state with $a^1 = a^2 = 0.5$. The response is shown with no stress and also with a compressive stress $\sigma_{11} = -0.2P_0/d_0$, which reduces the rate of ferroelectric switching through piezoelectric coupling.

Other examples have been given by Huber and Cocks (2008) and Tsou et al. (2013).

6 Summary and Outlook

This chapter has reviewed a variety of methods for modelling the electromechanical response of ferroelectrics. The methods are derived from classical theories of plasticity, energy minimization, and the use of thermodynamic principles that have wide application in mechanics. It is expected that extensions of these methods can also provide insight into other ferroic materials and multiferroics. While the chapter makes a general progression from macroscopic towards microscopic models, it does not address finer scale modelling methods such as Ginzburg-Landau phase field models, molecular dynamics or atomistics, each of which has made a substantial contribution to the understanding of ferroic materials. At present the fine-scale models are restricted by the limits of practical computation to regions of material of size in the order of nanometres. However, coupling of fine-scale to coarse-scale is an ongoing endeavor that has already made substantial progress. With continued growth in the applications of and the demand for ferroic materials, it can be expected that models of ferroelectrics and related materials will continue to provide a rich source of insight for engineering design.

Acknowledgements Thanks are due to Nien-Ti Tsou for preparing and allowing the use of figures showing domain patterns, as well as for many helpful and informative discussions.

References

Arlt, G. (1990). Twinning in ferroelectric and ferroelastic ceramics: Stress relief. *Journal of Materials Science*, *25*, 2655–2666.

Arlt, G., & Sasko, P. (1980). Domain configuration and equilibrium size of domains in $BaTiO_3$ ceramics. *Journal of Applied Physics*, *51*, 4956–4960.

Ball, J. M., & James, R. D. (1987). Fine phase mixtures as minimizers of energy. *Archive for Rational Mechanics and Analysis*, *11*, 13–52.

Bassiouny, E., Ghaleb, A. F., & Maugin, G. A. (1988). Thermodynamical formulation for coupled electromechanical hysteresis effects 1. Basic equations. *International Journal of Engineering Science*, *26*, 1279–1295.

Bhattacharya, K. (1993). Comparison of the geometrically nonlinear and linear theories of martensitic transformtion. *Continuum Mechanics and Thermodynamics*, *5*, 205–242.

Bhattacharya, K. (2003). *Microstructure of Martensite, Why it Forms and How It Gives Rise to the Shape-memory Effect*. New York: Oxford University Press.

Bisegna, P., & Luciano, R. (1996). Variational bounds for the overall properties of piezoelectric composites. *Journal of the Mechanics and Physics of Solids*, *44*, 583–602.

Burcsu, E., Ravichandran, G., & Bhattacharya, K. (2004). Large electrostrictive actuation of barium titanate single crystals. *Journal of the Mechanics and Physics of Solids*, *52*, 823–846.

Cocks, A. C. F., & McMeeking, R. M. (1999). A phenomenological constitutive law for the behaviour of ferroelectric ceramics. *Ferroelectrics*, *228*, 219–228.

Dayal, K., & Bhattacharya, K. (2007). A real-space non-local phase-field model of ferroelectric domain patterns in complex geometries. *Acta Materialia*, *55*, 1907–1917.

Devonshire, A. F. (1949). Theory of barium titanate 1. *Philosophical Magazine*, *40*, 1040–1063.

Dunn, M. L., & Taya, M. (1993). Micromechanics predictions of the effective electroelastic mouli of piezoelectric composites. *International Journal of Solids and Structures*, *30*, 161–175.

Eshelby, J. D. (1957). The detrmination of the elastic field of an ellipsoidal inclusion, and related problems. *Proceedings of the Royal Society of London A*, *241*, 376–396.

Fischer, F. D., Svoboda, J., & Petryk, H. (2014). Thermodynamic extremal principles for irreversible processes in materials science. *Acta Materialia*, *67*, 1–20.

Hall, D. A., Steuwer, A., Cherdhirunkorn, B., Withers, P., & Mori, T. (2005). Micromechanics of residual stress and texture development due to poling in polycrystalline ferroelectric ceramics. *Journal of the Mechanics and Physics of Solids*, *53*, 249–260.

Haug, A., Huber, J. E., Onck, P. R., & Van der Giessen, E. (2007). Multi-grain analysis versus self-consistent estimates of ferroelectric polycrystals. *Journal of the Mechanics and Physics of Solids*, *55*, 648–665.

Hill, R. (1965). A self-consistent mechanics of composite materials. *Journal of the Mechanics and Physics of Solids*, *13*, 213–222.

Hooton, J. A., & Merz, W. J. (1955). Etch patterns and ferroelectric domains in $BaTiO_3$ single crystals. *Physical Review*, *98*, 409–413.

Huber, J.E., & Cocks, A.C.F. (2008). A variational model of ferroelectric microstructure. In *Proceedings of ASME SMASIS08*, 28–30 October 2008. Ellicott City, MD, USA.

Huber, J. E., & Fleck, N. A. (2001). Multi-axial electrical switching of a ferroelectric: Theory versus experiment. *Journal of the Mechanics and Physics of Solids*, *49*, 785–811.

Huber, J. E., & Fleck, N. A. (2004). Ferroelectric switching: A micromechanics model versus measured behaviour. *European Journal of Mechanics A-Solids*, *23*, 203–217.

Huber, J. E., Fleck, N. A., Landis, C. M., & McMeeking, R. M. (1999). A constitutive model for ferroelectric polycrystals. *Journal of the Mechanics and Physics of Solids*, *47*, 1663–1697.

Hwang, S. C., Lynch, C. S., & McMeeking, R. M. (1995). Ferroelectric/ferroelastic interactons and a polarization switching model. *Acta Metallurgica et Materialia*, *43*, 2073–2084.

James, R. D., & Hane, K. F. (2000). Martensitic transformations and shape-memory materials. *Acta Materialia*, *48*, 197–222.

Jones, J. L., Slamovich, E. B., & Bowman, K. J. (2005). Domain texture distributions in tetragonal lead zirconate titanate by X-ray and neutron diffraction. *Journal of Applied Physics*, *97*, 034113.

Kamlah, M., Liskowsky, A. C., McMeeking, R. M., & Balke, H. (2005). Finite element simulation of a polycrystalline ferroelectric based on a multidomain single crystal switching model. *International Journal of Solids and Structures*, *42*, 2949–2964.

Kamlah, M., & Tsakmakis, C. (1999). Phenomenological modelling of the non-linear electromechanical coupling in ferroelectrics. *International Journal of Solids and Structures*, *36*, 669–695.

Kanoute, P., Boso, D. P., Chaboche, J. L., & Schrefler, B. A. (2001). Multiscale methods for composites: A review. *Archives of Computational Methods in Engineering*, *16*, 31–75.

Kouznetsova, V., Brekelmans, W. A. M., & Baaijens, F. P. T. (2001). An approach to micro-macro modeling of heterogeneous materials. *Computational Mechanics*, *27*, 37–48.

Landis, C. M. (2002). Fully coupled, multi-axial, symmetric constitutive laws for polycrystalline ferroelectric ceramics. *Journal of the Mechanics and Physics of Solids*, *50*, 127–152.

Li, J. Y., & Liu, D. (2004). On ferroelectric crystals with engineered domain configurations. *Journal of the Mechanics and Physics of Solids*, *52*, 1719–1742.

Li, Y. L., Hu, S. Y., Liu, Z. K., & Chen, L. Q. (2001). Phase-field model of domain structures in ferroelectric thin films. *Applied Physics Letters*, *78*, 3878–3880.

Lines, M.E., & Glass, A.M. (1977). *Principles and Applications of Ferroelectrics and Related Materials*. New York: Oxford University Press.

Liu, Q. D., & Huber, J. E. (2007). State dependent linear moduli in ferroelectrics. *International Journal of Solids and Structures*, *44*, 5635–5650.

Liu, T., & Lynch, C. S. (2006). Domain engineered relaxor ferroelectric single crystals. *Continuum Mechanics and Thermodynamics*, *18*, 119–135.

Loge, R. E., & Suo, Z. (1996). Nonequilibrium thermodynamics of ferroelectric domain evolution. *Acta Materialia*, *44*, 3429–3438.

Lynch, C. S. (1996). The effect of uniaxial stress on the electro-mechanical response of 8/65/35 PLZT. *Acta Materialia*, *44*, 4138–4148.

McGilly, L. J., Schilling, A., & Gregg, J. M. (2010). Domain bundle boundaries in single crystal $BaTiO_3$ lamellae: searching for naturally forming dipole flux-closure/quadrupole chains. *Nano Letters*, *10*, 4200–4205.

Park, S. E., & Shrout, T. R. (2007). Ultrahigh strain and piezoelectric behaviour in relaxor based ferroelectric single crystals. *Journal of Applied Physics*, *82*, 1804–1811.

Pathak, A., & McMeeking, R. M. (2008). Three-dimensional finite element simulations of ferroelectric poycrystals under electrical and mechanical loading. *Journal of the Mechanics and Physics of Solids*, *56*, 663–683.

Shu, Y. C., & Bhattacharya, K. (2001). Domain patterns and macroscopic behaviour of ferroelectric materials. *Philosophical Magazine B*, *81*, 2021–2054.

Sidorkin, A. S. (2006). *Domain Structure in Ferroelectric and Related Materials*. Cambridge International Science Publishing.

Su, Y., & Landis, C. M. (2007). Continuum thermodynamics of ferroelectric domain evolution: Theory, finite element implementation, and application to domain wall pinning. *Journal of the Mechanics and Physics of Solids*, *55*, 280–305.

Tangantsev, A. K., Cross, L. E., & Fousek, J. (2010). *Domains in Ferroic Crystals and Thin Films*. New York: Springer.

Tsou, N.T. (2011) *Compatible domain structures in ferroelectric single crystals*, Ph.D. thesis, University of Oxford.

Tsou, N. T., & Huber, J. E. (2010). Compatible domain structures and the poling of single crystal ferroelectrics. *Mechanics of Materials*, *42*, 740–753.

Tsou, N. T., Huber, J. E., & Cocks, A. C. F. (2008). Evolution of compatible laminate domain structures in ferroelectric single crystals. *Acta Materialia*, *56*, 2117–2135.

Tsou, N. T., Potnis, P. R., & Huber, J. E. (2011). Classification of laminate domain patterns in ferroelectrics. *Physical Review B*, *83*, 184120.

Völker, B., & Kamlah, M. (2012). Large-signal analysis of typical ferroelectric domain structures using phase-field modeling. *Smart Materials and Structures*, *21*, 055013.

Wang, J. J., Ma, X. Q., Li, Q., Britson, J., & Chen, L. Q. (2013). Phase-transitions and domain structures of ferroelectric nanoparticales: Phase-field model incorporating strong elastic and dielectric inhomogeneity. *Acta Materialia*, *61*, 7591–7603.

Wang, J., Kamlah, M., & Zhang, T.-Y. (2009). Phase field simulations of ferroelectric nanoparticales with different long-range-electrostatic and -elastic interactions. *Journal of Applied Physics*, *105*, 014104.

Wang, J., Shi, S. Q., Chen, L. Q., Li, Y. L., & Zhang, T. Y. (2004). Phase-field simulations of ferroelectric/ferroelastic polarization switching. *Acta Materialia*, *52*, 749–764.

Weng, G. J., & Wong, D. T. (2009). Thermodynamic driving force in ferroelectric crystals with a rank-2 laminated domain pattern, and a study of enhanced electrostriction. *Journal of the Mechanics and Physics of Solids*, *57*, 571–597.

Wu, T., Zhao, P., Bao, M. Q., Bur, A., Hockel, J. L., Wong, K., et al. (2011). Domain engineered switchable strain states in ferroelectric(011) $[Pb(Mg_{1/3}Nb_{2/3})O_3]_{(1-x)}$-$[PbTiO_3]_{(x)}$(PMN-PT, $x \approx 0.32$) single crystals. *Journal of Applied Physics*, *109*, 124101.

Xu, Y. (1991). *Ferroelectric Materials and their Applications*, North Holland Press.

Yen, J. H., Shu, Y. C., Shieh, J., & Yeh, J. H. (2008). A study of electromechanical switching in ferroelectric single crystals. *Journal of the Mechanics and Physics of Solids*, *56*, 2117–2135.

Zhang, Z. Y., James, R. D., & Muller, S. (2009). Energy barriers and hysteresis in martensitic phase transformations. *Acta Materialia*, *57*, 4332–4352.

An FE^2-Scheme for Magneto-Electro-Mechanically Coupled Boundary Value Problems

Matthias Labusch, Jörg Schröder and Marc-André Keip

Abstract The magneto-electric coupling in materials can find several applications in future technologies and could furthermore improve devices, for instance in data storage media. Since all natural single-phase magneto-electric (ME) multiferroics and most of the synthetic single-phase ME materials show a magneto-electric coupling far below room temperature, composite materials are manufactured which consist of a ferroelectric and a magnetostrictive phase. Both phases interact with each other in consequence of transferred deformations, such that these composites produce strain-induced ME properties at room temperature. In order to predict a realistic material behavior, it is necessary to implement appropriate numerical models to reflect the natural behavior of the single phases. This chapter will focus on the numerical implementation of different numerical models for the description of ferroic materials into the two-scale finite element homogenization approach (FE^2-method) for the simulation of magneto-electro-mechanically coupled boundary value problems. In detail, we investigate the magneto-electric response in consideration of piezoelectric, electrostrictive and ferroelectric models in combination with a piezomagnetic phase. Reliable results for the ME coefficient are obtained by using a ferroelectric model which takes into account the microscopic properties over single tetragonal unit cells, which are allocated in the three dimensional space based on an orientation distribution function. As a result of switching-events of microscopic polarization vectors, we

M. Labusch, J. Schröder and M.-A. Keip—The authors greatly appreciates the "Deutsche Forschungsgemeinschaft" (DFG) for the financial support under the research grant SCHR 570/12-2 and KE 1849/2-2 within the research group FOR 1509 on "Ferroische Funktionsmaterialien–Mehrskalige Modellierung und experimentelle Charakterisierung".
M.-A. Keip—This research has been facilitated through financial funding of the Cluster of Excellence EXC 310 in *Simulation Technology*) as well as the "Ministerium für Wissenschaft, Forschung und Kunst des Landes Baden-Württemberg". This funding is gratefully acknowledged.

M. Labusch (✉) · J. Schröder
Institute of Mechanics, University of Duisburg-Essen, Essen, Germany
e-mail: matthias.labusch@uni-due.de

M.-A. Keip
Institute of Applied Mechanics (CE), Chair I, University of Stuttgart, Stuttgart, Germany

J. Schröder and Doru C. Lupascu (eds.), *Ferroic Functional Materials*,
CISM International Centre for Mechanical Sciences 581,
https://doi.org/10.1007/978-3-319-68883-1_5

obtain the macroscopic dielectric and butterfly hysteresis curves of the ferroelectric phase. Thereby, the model for the computations of the ME coefficient is physically motivated.

1 Introduction

Ferroic materials are characterized by particular coupling behaviors, such as electro-mechanical interactions in piezoelectric materials or magneto-mechanical interactions in magnetostrictive materials. They provide many applications in modern technical devices, which are used for sensors, actuators or data storage devices. A further special phenomenon, which could improve these technical devices and extend the range of application, is the magneto-electric (ME) coupling. It is denoted by an interaction between magnetic fields and electric polarization or electric fields and magnetization. Such ME multiferroics allow for new applications in various fields, such as electrical magnetic-field sensors or electric-write/magnetic-read-memories (Magneto-Electric Random Access Memory; MERAM) (Eerenstein et al. 2007; Spaldin and Fiebig 2005; Bibes and Barthélémy 2008). Therefore, these multiferroics have been investigated intensively, in e.g. Schmid (1994), Hill (2000), Cheong and Mostovoy (2007), Ramesh and Spaldin (2007), Khomskii (2009), Martin et al. (2010), Zohdi (2010, 2012). The interest in magneto-electric coupling phenomenon goes back in the end of the 19th century as Röntgen discovered that a moving dielectric became magnetized when placed in an electric field, see Röntgen (1888). Further experiments showed the existence of a magneto-electric coupling behavior, e.g. Wilson (1905), and theoretical advisements supposed an emerging ME behavior in single-phase materials, see for instance Debye (1926), Dzyaloshinskii (1959). Landau and Lifshitz (1960) have shown that there may occur a magneto-electric effect in some antiferromagnetic crystals and in Dzyaloshinskii (1959) the author explicitly mentioned the material chromium oxide (Cr_2O_3) with possible ME properties. These assumptions were proven a few years later with the first successful evidences of an ME coupling in Cr_2O_3, see Astrov (1960, 1961), Rado and Folen (1961), Folen et al. (1961). In the following decades various investigations on Cr_2O_3 could be reported and further materials with magneto-electric properties have been observed, see for instance Al'shin and Astrov (1963), Shtrikman and Treves (1963), Rado (1964), Ascher et al. (1966), O'Dell (1967), Howes et al. (1984), Rado et al. (1984), Watanabe and Kohn (1989), Rivera (1994a, b), Ye et al. (1994). Table 1 gives an overview of measured ME-coefficients. A comprehensive historical overview can be found in Fiebig (2005).

However, due to physical reasons, only very few natural materials with magneto-electric properties exist. To explain this reason, the origin of ferroelectric and ferromagnetic properties have to be considered in more detail. Ferroelectric materials obtain electro-mechanical properties due to their crystal structure, where the center of the positive and negative charged ions do not coincide, they exhibit a distance (off-center displacement). The resulting spontaneous polarization can be influenced

Table 1 Measured ME coefficients of single-phase materials

Single-phase magneto-electric materials			
Material	ME coef. s/m	Temp. K	Reference
Cr_2O_3	$4 \cdot 10^{-14}$	273.2	Astrov (1960)
Cr_2O_3	$1.43 \cdot 10^{-12}$	293.2	Astrov (1961)
$TbPO_4$	$3.67 \cdot 10^{-11}$	1.9	Rado et al. (1984)
$LiCoPO_4$	$18.40 \cdot 10^{-12}$	4.2	Rivera (1994b)
$Cr_3B_7O_{13}Cl$	$5 \cdot 10^{-14}$	4.2	Ye et al. (1994)
$LiCoPO_4$	$30.60 \cdot 10^{-12}$	4.2	Rivera (1994a)
Cr_2O_3	$4.13 \cdot 10^{-12}$	263.0	Rivera (1994a)
$BiFeO_3$	$0.52 \cdot 10^{-12}$	18.0	Rivera and Schmid (1997)
$Mn_3B_7O_{13}I$	$9.50 \cdot 10^{-14}$	4.2	Crottaz et al. (1997)
Cr_2O_3	$4.17 \cdot 10^{-12}$	270.0	Vaz et al. (2010)

by applied stresses or electric fields going along with mechanical deformations. On the other hand, an arising magneto-mechanical coupling has its origin in the atomic orbitals which are partially filled with electrons. These electrons, which generate a magnetic moment due to their spin and movements, interact with each other such that the magnetic moments of the electrons in the partially filled orbitals are aligned in one direction. These interactions proceed between different atoms resulting in a macroscopic magnetization. Applied magnetic fields change the orientation of the microscopic magnetic moments involving mechanical deformations. Although, partially filled orbitals are necessary for a remaining magnetic moment they eliminate in most cases the distortion, the off-center displacement, which removes the center of symmetry. However, filled orbitals yield a higher driving force for off-center displacements and are beneficial for ferroelectric properties. A combination of both properties in single-phase materials are rare, because the associated mechanisms are mutually orthogonal to each other, thus only a very small number of natural ME multiferroics exists. For a more detailed explanation of the ferroelectric and ferromagnetic origins we refer to Hill (2000). Furthermore, magneto-electric properties arise far below room temperature in most of the single-phase ME multiferroics, see Table 1. Chromium(III)-oxide (Cr_2O_3) also shows an ME coefficient at room temperature (293.15 K), but it quickly decreases for slightly higher temperatures and finally vanishes at a Néel temperature of about $T_n = 307$ K, see Astrov (1961). The disadvantage of the very low coefficients has been circumvented successfully by manufacturing *magneto-electric composites* which consist of magneto-mechanically and electro-mechanically coupled phases, see Fig. 1. The idea behind the development of such composites is to generate the desired magneto-electric effect as a *strain-induced product property* (Nan 1994; Ryu et al. 2002; Eerenstein et al. 2006; Priya et al. 2007; Nan et al. 2008; Srinivasan 2010; Wang et al. 2010). Generally, a product property of a composite is defined as an effective property which is not present in each of its individual phases, it only appears effectively through their interaction, see van Suchtelen (1972) for a general treatment on possible product properties. In the case of ME composites, we distinguish between the direct and converse ME effect. The direct effect characterizes magnetically induced electric polarization: an applied magnetic field yields a deformation of the magneto-mechanically coupled phase which is transferred to the electro-mechanically coupled phase. As a result, a *strain-induced polarization* in the electric phase is observed. On the other hand, the converse effect characterizes electrically activated magnetization: an applied electric field yields a deformation of the electro-mechanical phase which is then transferred to the magneto-mechanical phase. The result is thus a *strain-induced magnetization*. Several experiments on ME composites showed remarkable ME coefficients, which are orders of magnitudes higher than those of single-phase materials (Fiebig 2005). In this work we consider two-phase ME composites consisting of a barium titanate ($BaTiO_3$) matrix with cobalt ferrite inclusions ($CoFe_2O_4$), see for instance Shvartsman et al. (2011), Etier et al. (2013). Due to the fact that the effective magneto-electric coupling in composite materials is a strain-induced property, it is obvious that it strongly depends on the characteristics of the individual phases and the morphology of the microstructure. In order to predict the effective properties of composites, a variety of analytical methods has been developed, for example by Harshé et al. (1993a, b), Huang and Kuo (1997), Bichurin et al. (2003), Nan et al. (2005), Srinivas et al. (2006), Shi and Gao (2014), Jayachandran et al. (2013), Jin and Aboudi (2015) and Bichurin et al. (2010).

In Kuo and Wang (2012) an optimization of the magneto-electric coefficient of fiber-reinforced composites with respect to the volume fractions and the crystallographic orientations of the phases is discussed. Most of these methods, however, obtain their full predictability only when idealized microstructures are considered, such as perfect laminates or particulate composites with prefect ellipsoidal inclusions. Thus, when more complex microstructures have to be analyzed, analytical methods experience certain limitations. In that case, more flexible computational tools become attractive since they are not restricted to specific microstructural morphologies. Innovative contributions on the computational prediction of ME composites are given by Lee et al. (2005) and Miehe et al. (2011). In these works, the effective properties of piezomagnetic/piezoelectric ME composites are determined through simulations based on the Finite Element Method (FEM).

The goal of this paper is to derive a computational method which allows for the *direct incorporation of ME composite microstructures in macroscopic simulations*. This idea traces back to the fundamental works on the so-called FE^2-method (also denoted as multi-level FEM) developed in the late 90 s of the last century. In this method a representative volume element ($\mathcal{RVE}$) is attached at each material point of the macrostructure (which means at each Gauss point of the discrete version obtained from the FE^2-method). At this point, we would like to mention explicitly the early works on nonlinear heterogeneous problems by Smit et al. (1998), on a general framework for geometrically and physically linear/nonlinear homogenization by Miehe et al. (1999a, b) and on elasto-viscoplastic material behavior of fiber-reinforced composites by Feyel and Chaboche (2000). The latter work also presents

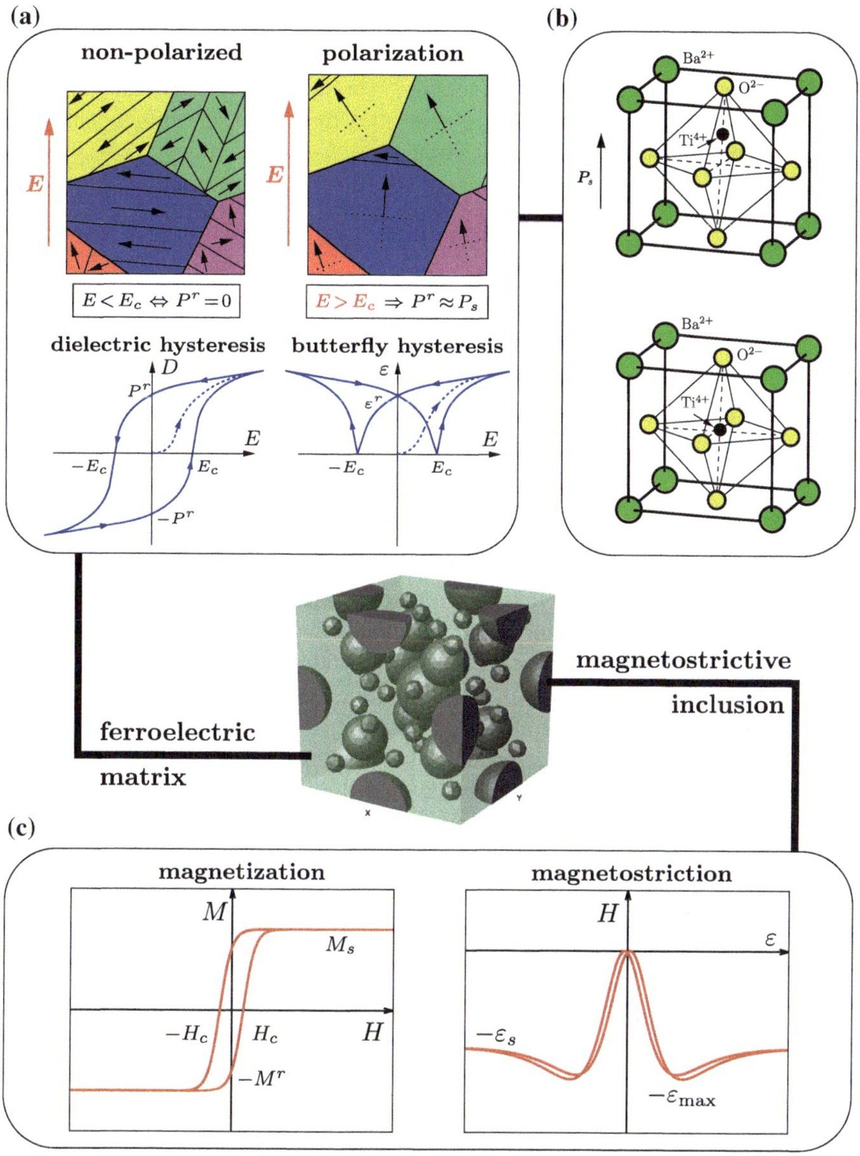

Fig. 1 Design of two-phase magneto-electric composites: **a** Typical macroscopic dielectric and butterfly hysteresis of $BaTiO_3$, resulting from microscopic domain movements, **b** Unit cell structure on nanoscopic level and **c** typical macroscopic magnetization and magnetostriction of $CoFe_2O_4$

a closed-form representation of the macroscopic (effective) tangent moduli, which is of particular relevance for the present paper. Further important works on FE^2-methods are e.g. Miehe and Koch (2002), van der Sluis et al. (2000), Terada et al. (2000, 2013),

Kouznetsova et al. (2001, 2002), Miehe and Bayreuther (2007), Perić et al. (2011), Miehe (2003) and Temizer and Wriggers (2008). For a general introduction to this method, including a comprehensive literature overview, see Schröder (2014).

A recent route of developments is concerned with the two-scale homogenization of coupled problems. Here, we want to mention the recent extensions to thermo-mechanical coupling by Özdemir et al. (2008), to electro-mechanically coupling by Schröder (2009), Schröder and Keip (2012), Keip et al. (2014) and to magneto-mechanical coupling by Javili et al. (2013), Keip and Rambausek (2016, 2017).

In the present contribution we will apply the FE^2-method to *magneto-electro-mechanically coupled solids*, based on the contribution of Schröder et al. (2016). Here, we will assume classical continua on both scales and suppose the existence of scale separation. To be precise and to embed our method into the above strategies, we propose a *magneto-electro-mechanically coupled first-order homogenization scheme with directly coupled scales*.

The outline of the paper is as follows. In Sect. 2, we will discuss in detail the two-scale method. This includes the micro-macro transition conditions, which are based on an extended form of the classical Hill-Mandel-condition. Furthermore, we derive an expression for the effective tangent moduli, which include the algorithmically consistent magneto-electric coupling coefficients. Section 3 gives a brief overview of the used material models which are implemented into the FE^2-method. This contains a piezoelectric-/magnetic, an electrostrictive and a ferroelectric material model. In Sect. 4, we apply the method to the two-scale simulation of magneto-electro-mechanically coupled boundary value problems. In order to show the influence of the used material model on the resulting strain-induced ME coefficient, we considered the aforementioned material models for the electro-active phase. Thus, we could make a statement of the pros and cons of the different models and the prediction of a realistic material response. All simulations are performed on the basis of experimental material parameters. Since the corresponding literature is vast and since some of the parameters which are often employed in numerical simulations in fact violate fundamental thermodynamic principles, we select a set of parameters which is physically reasonable. A more detailed comment on this can be found in Schröder et al. (2016). The Sect. 5 will give a short summary and a conclusion.

2 Theory of the Two-Scale Homogenization Scheme

In order to simulate the magneto-electro-mechanically coupled properties, we make use of a finite element two-scale transition, where we connect the macroscopic material behavior with an underlying microstructure. In that sense, we do not have to derive a macroscopic thermodynamically consistent energy function describing the effective material behavior. Instead the macroscopic properties are obtained due to a homogenization procedure over the microscale, which also includes the determination of the magneto-electric characteristics. Therefore, we consider on the microscopic level a two-phase composite consisting of a ferroelectric matrix with piezomagneticbreak

Table 2 Magneto-electro-mechanical quantities and their Units

Macro- and micro-symbol		Continuum mechanical description	Unit
$\overline{\boldsymbol{u}}$	$\boldsymbol{u}$	Displacement vector	m
$\overline{\boldsymbol{\varepsilon}}$	$\boldsymbol{\varepsilon}$	Linear strain tensor	1
$\overline{\boldsymbol{\sigma}}$	$\boldsymbol{\sigma}$	Cauchy stress tensor	N/m^2
$\overline{\boldsymbol{t}}$	$\boldsymbol{t}$	Traction vector	N/m^2
$\overline{\boldsymbol{f}}$	$\boldsymbol{f}$	Mechanical body forces	N/m^3
$\overline{\phi}$	ϕ	Electric potential	V
$\overline{\boldsymbol{E}}$	$\boldsymbol{E}$	Electric field vector	V/m
$\overline{\boldsymbol{D}}$	$\boldsymbol{D}$	Electric displacement vector	C/m^2
$\overline{Q}$	Q	Electric surface flux density	C/m^2
$\overline{q}$	q	Density of free charge carriers	C/m^3
$\overline{\varphi}$	φ	Magnetic potential	A
$\overline{\boldsymbol{H}}$	$\boldsymbol{H}$	Magnetic field	A/m
$\overline{\boldsymbol{B}}$	$\boldsymbol{B}$	Magnetic flux density	T
$\overline{\zeta}$	ζ	Magnetic surface flux density	Vs/m^2

inclusions. These phases are determined by specific material models. In order to ease the readability of the following sections, we summarize the basic magneto-electro-mechanical quantities in Table 2.

The FE^2 approach is based on consecutive localization and homogenization steps, which are listed below and visualized in Fig. 2.

(i) At each macroscopic material point, suitable macroscopic quantities such as the strains, the electric and the magnetic field have to be localized to the microscale. In other words, energetically consistent boundary conditions must be applied to an $\mathcal{RVE}$.

(ii) In order to obtain the dual microscopic quantities such as the stresses, the electric displacements and the magnetic induction, the weak forms of the balance equations have to be solved on the microscale.

(iii) A homogenization step is performed, in which average values of the dual quantities on the microscale are determined. Then, the homogenized variables have to be transferred to the associated points of the macroscale.

(iv) Finally, the coupled boundary value problem on the macroscale has to be solved. Once the solution is obtained, steps (i) to (iv) have to be carried out until an equilibrium state on both scales is reached.

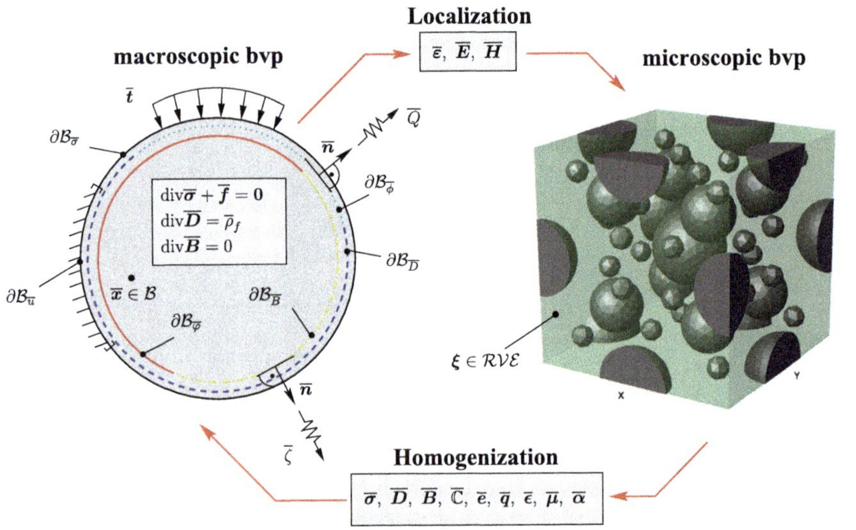

Fig. 2 Visualized consecutive steps of the two-scale homogenization procedure including the macroscopic body (left) and attached $\mathcal{RVE}$ (right). The macroscopic boundary conditions can be decomposed into a mechanical, electrical and magnetical part with $\partial\mathcal{B}_{\overline{u}}\cup\partial\mathcal{B}_{\overline{\sigma}}=\partial\mathcal{B}$, $\partial\mathcal{B}_{\overline{\phi}}\cup\partial\mathcal{B}_{\overline{D}}=\partial\mathcal{B}$, $\partial\mathcal{B}_{\overline{\varphi}}\cup\partial\mathcal{B}_{\overline{B}}=\partial\mathcal{B}$ and $\partial\mathcal{B}_{\overline{u}}\cap\partial\mathcal{B}_{\overline{\sigma}}=\emptyset$, $\partial\mathcal{B}_{\overline{\phi}}\cap\partial\mathcal{B}_{\overline{D}}=\emptyset$, $\partial\mathcal{B}_{\overline{\varphi}}\cap\partial\mathcal{B}_{\overline{B}}=\emptyset$

2.1 Boundary Value Problems and Scale Transition

The macroscopic body $\mathcal{B}\subset\mathbb{R}^3$ is parameterized in the Cartesian coordinates $\overline{\boldsymbol{x}}$. The macroscopic fundamental balance laws are given by the balance of linear momentum as well as Gauß's laws of electro- and magneto-statics

$$\mathrm{div}_{\overline{x}}[\overline{\boldsymbol{\sigma}}]+\overline{\boldsymbol{f}}=\boldsymbol{0},\quad \mathrm{div}_{\overline{x}}[\overline{\boldsymbol{D}}]=0,\quad \text{and}\quad \mathrm{div}_{\overline{x}}[\overline{\boldsymbol{B}}]=0\quad \text{in}\quad \mathcal{B}\,. \tag{1}$$

The macroscopic gradient fields, which are the strain as well as the electric and magnetic field are defined as

$$\overline{\varepsilon}(\overline{\boldsymbol{x}}):=\mathrm{sym}[\overline{\nabla}\overline{\boldsymbol{u}}(\overline{\boldsymbol{x}})]\,,\quad \overline{\boldsymbol{E}}(\overline{\boldsymbol{x}}):=-\overline{\nabla}\,\overline{\phi}(\overline{\boldsymbol{x}})\,,\quad \overline{\boldsymbol{H}}(\overline{\boldsymbol{x}}):=-\overline{\nabla}\,\overline{\varphi}(\overline{\boldsymbol{x}})\,, \tag{2}$$

with the macroscopic gradient operator $\overline{\nabla}$ defined with respect to $\overline{\boldsymbol{x}}$. The macroscopic boundary conditions are prescribed through the mechanical displacement and the surface traction

$$\overline{\boldsymbol{u}}=\overline{\boldsymbol{u}}_0\quad \text{on}\ \partial\mathcal{B}_{\overline{u}}\quad \text{and}\quad \overline{\boldsymbol{t}}_0=\overline{\boldsymbol{\sigma}}\cdot\overline{\boldsymbol{n}}\quad \text{on}\ \partial\mathcal{B}_{\overline{\sigma}} \tag{3}$$

with the relations $\partial\mathcal{B}_{\overline{u}}\cup\partial\mathcal{B}_{\overline{\sigma}}=\partial\mathcal{B}$ and $\partial\mathcal{B}_{\overline{u}}\cap\partial\mathcal{B}_{\overline{\sigma}}=\emptyset$, through the electric potential and the electric surface flux density

$$\overline{\phi} = \overline{\phi}_0 \quad \text{on } \partial\mathcal{B}_{\overline{\phi}} \quad \text{and} \quad -\overline{Q}_0 = \overline{\boldsymbol{D}} \cdot \overline{\boldsymbol{n}} \quad \text{on } \partial\mathcal{B}_{\overline{D}} \tag{4}$$

with the relations $\partial\mathcal{B}_{\overline{\phi}} \cup \partial\mathcal{B}_{\overline{D}} = \partial\mathcal{B}$ and $\partial\mathcal{B}_{\overline{\phi}} \cap \partial\mathcal{B}_{\overline{D}} = \emptyset$, as well as through the magnetic potential and the magnetic surface flux density

$$\overline{\varphi} = \overline{\varphi}_0 \quad \text{on } \partial\mathcal{B}_{\overline{\varphi}} \quad \text{and} \quad -\overline{\zeta}_0 = \overline{\boldsymbol{B}} \cdot \overline{\boldsymbol{n}} \quad \text{on } \partial\mathcal{B}_{\overline{B}} \tag{5}$$

with the relations $\partial\mathcal{B}_{\overline{\varphi}} \cup \partial\mathcal{B}_{\overline{B}} = \partial\mathcal{B}$ and $\partial\mathcal{B}_{\overline{\varphi}} \cap \partial\mathcal{B}_{\overline{B}} = \emptyset$, where $\overline{\boldsymbol{n}}$ denotes the outward unit normal vector of the surfaces $\partial\mathcal{B}_{\overline{B}}$, see Fig. 2.

To solve the macroscopic boundary value problem the overall material tangent modulus is necessary. But, in the sense of the FE^2-method no macroscopic energy function, which describes the material behavior, is defined. Instead, a representative volume element $\mathcal{RVE} \subset \mathbb{R}^3$ at each macroscopic point is attached, such that the macroscopic response depends on the underlying microstructure. The simulation across both scales requires the solution of the micro- and macroscopic boundary value problem. The representative volume element on the microscale $\mathcal{RVE} \subset \mathbb{R}^3$ is parameterized in $\boldsymbol{x}$. The balance of momentum as well as the Gauß's law of electro- and magneto-statics on microscopic level are given by

$$\text{div}_{\boldsymbol{x}}[\boldsymbol{\sigma}] = \boldsymbol{0}\,, \quad \text{div}_{\boldsymbol{x}}[\boldsymbol{D}] = 0 \quad \text{and} \quad \text{div}_{\boldsymbol{x}}[\boldsymbol{B}] = 0 \quad \text{in} \quad \mathcal{RVE}. \tag{6}$$

The linear strains as well as the microscopic electric and magnetic fields are defined by

$$\boldsymbol{\varepsilon}(\boldsymbol{x}) := \text{sym}[\nabla \boldsymbol{u}(\boldsymbol{x})]\,, \quad \boldsymbol{E}(\boldsymbol{x}) := -\nabla\,\phi(\boldsymbol{x})\,, \quad \boldsymbol{H}(\boldsymbol{x}) := -\nabla\,\varphi(\boldsymbol{x})\,. \tag{7}$$

To solve the microscopic boundary value problem, suitable energy functions for the different phases are defined, which describe the corresponding material moduli.

The considered $\mathcal{RVE}$ on the microscale should represent the overall material behavior and especially, for the purpose of ME composites, the magneto-electric coupling coefficient. As a consequence, all the above quantities are defined through an averaging process over the volume of the $\mathcal{RVE}$. Assuming continuity of the displacements as well as the electric and magnetic potential, we can express the macroscopic variables in terms of simple volume integrals as

$$\overline{\boldsymbol{\lambda}} = \langle \boldsymbol{\lambda} \rangle_V := \frac{1}{V_{\mathcal{RVE}}} \int_{\mathcal{RVE}} \boldsymbol{\lambda}\, dv \quad \text{with} \quad \boldsymbol{\lambda} := \{\boldsymbol{\varepsilon}, \boldsymbol{\sigma}, \boldsymbol{E}, \boldsymbol{D}, \boldsymbol{H}, \boldsymbol{B}\}\,, \tag{8}$$

see for example Schröder et al. (2016). In order to determine the microscopic quantities by solving the boundary value problem of the $\mathcal{RVE}$, we have to define boundary conditions on microscopic level. Starting from the fundamental works of Hill (1963) and Mandel and Dantu (1963) we assume that the individual parts of a generalized magneto-electro-mechanical Hill-Mandel condition of the form

$$\overline{\boldsymbol{\sigma}} : \dot{\overline{\boldsymbol{\varepsilon}}} = \langle \boldsymbol{\sigma} : \dot{\boldsymbol{\varepsilon}} \rangle_V, \quad \overline{\boldsymbol{D}} \cdot \dot{\overline{\boldsymbol{E}}} = \langle \boldsymbol{D} \cdot \dot{\boldsymbol{E}} \rangle_V, \quad \text{and} \quad \overline{\boldsymbol{B}} \cdot \dot{\overline{\boldsymbol{H}}} = \langle \boldsymbol{B} \cdot \dot{\boldsymbol{H}} \rangle_V \tag{9}$$

have to be fulfilled independently, see Schröder (2009) for the electro-mechanical case and Schröder et al. (2016) for the magneto-electro-mechanical case. A reformulation of the latter condition for the mechanical part yields

$$\mathcal{P}_{mech} = \overline{\boldsymbol{\sigma}} : \dot{\overline{\varepsilon}} - \langle \boldsymbol{\sigma} : \dot{\varepsilon} \rangle_V \tag{10}$$

with $\mathcal{P}_{mech} = 0$. Using the local balance law of linear momentum as well as the Cauchy theorem $\boldsymbol{t} = \boldsymbol{\sigma} \cdot \boldsymbol{n}$ we reformulate the above equation to

$$\mathcal{P}_{mech} = \langle (\boldsymbol{t} - \overline{\boldsymbol{\sigma}} \cdot \boldsymbol{n})(\dot{\boldsymbol{u}} - \dot{\overline{\varepsilon}} \cdot \boldsymbol{x}) \rangle_\Gamma \tag{11}$$

where $\langle \bullet \rangle_\Gamma$ denotes the surface average over the representative volume element, i.e. $\langle \bullet \rangle_\Gamma := \frac{1}{V_{\mathcal{RVE}}} \int_{\partial \mathcal{RVE}} (\bullet)\, da$, and $\boldsymbol{n}$ the outward unit normal on $\partial \mathcal{RVE}$. Here we refer to (ref. Keip et al. 2014) for a step-by-step derivation in case of electro-mechanical coupling. An evaluation of $\mathcal{P}_{mech} = 0$ leads to the Reuss- and Voigt bounds (constraint conditions)

$$\boldsymbol{\sigma} = \overline{\boldsymbol{\sigma}} = \text{const.} \quad \text{or} \quad \dot{\varepsilon} = \dot{\overline{\varepsilon}} = \text{const.} \quad \forall \boldsymbol{x} \in \mathcal{RVE}\,. \tag{12}$$

Possible periodic boundary conditions, which fulfill $\mathcal{P}_{mech} = 0$, can be derived effectively by assuming a decomposition of the microscopic strains in an affine part $\overline{\varepsilon} \cdot \boldsymbol{x}$ and a fluctuation field $\widetilde{\varepsilon}$, satisfying $\langle \varepsilon \rangle_V = \overline{\varepsilon}$ and $\langle \widetilde{\varepsilon} \rangle_V = 0$, holding the relation

$$\dot{\varepsilon} = \dot{\overline{\varepsilon}} + \dot{\widetilde{\varepsilon}}\,. \tag{13}$$

The associated periodic boundary conditions for the mechanical part are then given by

$$\widetilde{\boldsymbol{u}}(\boldsymbol{x}^+) = \widetilde{\boldsymbol{u}}(\boldsymbol{x}^-) \quad \text{and} \quad \boldsymbol{t}(\boldsymbol{x}^+) = -\boldsymbol{t}(\boldsymbol{x}^-) \quad \text{on} \quad \boldsymbol{x}^\pm \in \Gamma^\pm \tag{14}$$

respectively, see Fig. 3 for a visualization of the mechanical part. Here, $\boldsymbol{x}^+$ and $\boldsymbol{x}^-$ define associated points at Γ^+ and Γ^-, respectively, satisfying $\boldsymbol{n}^+ = -\boldsymbol{n}^-$. Analogous to the mechanical part, we can derive the boundary conditions for the electric and magnetic fields by reformulating Eq. $(9)_2$ and $(9)_3$

$$\mathcal{P}_{elec} = \overline{\boldsymbol{D}} \cdot \dot{\overline{\boldsymbol{E}}} - \langle \boldsymbol{D} \cdot \dot{\boldsymbol{E}} \rangle_V \quad \text{and} \quad \mathcal{P}_{magn} = \overline{\boldsymbol{B}} \cdot \dot{\overline{\boldsymbol{H}}} - \langle \boldsymbol{B} \cdot \dot{\boldsymbol{H}} \rangle_V \tag{15}$$

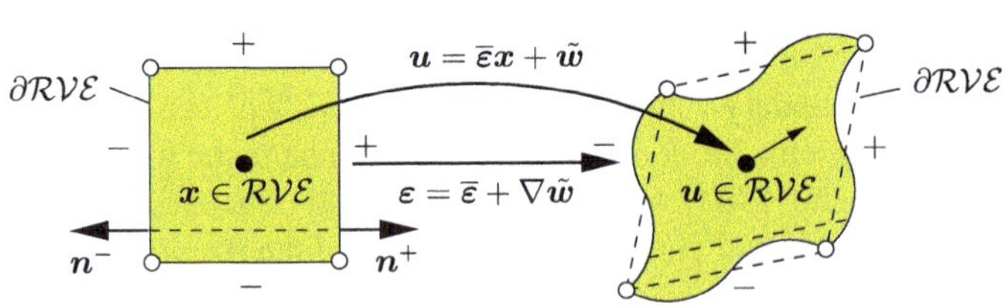

Fig. 3 Possible mechanical periodic boundary conditions on the $\mathcal{RVE}$

which have to fulfill $\mathcal{P}_{elec} = 0$ and $\mathcal{P}_{magn} = 0$, respectively. Using the Maxwell equations of electro- and magnetostatics and the definitions of the surface tractions Q and ζ yield

$$\mathcal{P}_{elec} = \langle (Q + \overline{\boldsymbol{D}} \cdot \boldsymbol{n})(\dot{\phi} + \dot{\overline{\boldsymbol{E}}} \cdot \boldsymbol{x}) \rangle_\Gamma \quad \& \quad \mathcal{P}_{magn} = \langle (\zeta + \overline{\boldsymbol{B}} \cdot \boldsymbol{n})(\dot{\varphi} + \dot{\overline{\boldsymbol{H}}} \cdot \boldsymbol{x}) \rangle_\Gamma \,. \tag{16}$$

The evaluation of $\mathcal{P}_{elec} = 0$ and $\mathcal{P}_{magn} = 0$ leads to the constraint conditions for the electric and magnetic fields as

$$\begin{aligned} \boldsymbol{D} = \overline{\boldsymbol{D}} = \text{const. or } \dot{\boldsymbol{E}} = \dot{\overline{\boldsymbol{E}}} = \text{const. } \forall \boldsymbol{x} \in \mathcal{RVE} \,, \\ \boldsymbol{B} = \overline{\boldsymbol{B}} = \text{const. or } \dot{\boldsymbol{H}} = \dot{\overline{\boldsymbol{H}}} = \text{const. } \forall \boldsymbol{x} \in \mathcal{RVE} \,. \end{aligned} \tag{17}$$

For the periodic boundary conditions of the electric and magnetic fields we assume analogous to the mechanic fields the following decompositions

$$\dot{\boldsymbol{E}} = \dot{\overline{\boldsymbol{E}}} + \dot{\widetilde{\boldsymbol{E}}} \quad \text{and} \quad \dot{\boldsymbol{H}} = \dot{\overline{\boldsymbol{H}}} + \dot{\widetilde{\boldsymbol{H}}} \,, \tag{18}$$

and the associated periodic conditions given by

$$\begin{aligned} \widetilde{\phi}(\boldsymbol{x}^+) = \widetilde{\phi}(\boldsymbol{x}^-) \text{ and } Q(\boldsymbol{x}^+) = -Q(\boldsymbol{x}^-) \text{ on } \boldsymbol{x}^\pm \in \Gamma^\pm \,, \\ \widetilde{\varphi}(\boldsymbol{x}^+) = \widetilde{\varphi}(\boldsymbol{x}^-) \text{ and } \zeta(\boldsymbol{x}^+) = -\zeta(\boldsymbol{x}^-) \text{ on } \boldsymbol{x}^\pm \in \Gamma^\pm \,, \end{aligned} \tag{19}$$

respectively. In the following we introduce the discretizations of the microscopic boundary value problems. Furthermore, the determination of the macroscopic material tangent, obtained by a homogenization over the $\mathcal{RVE}$ involving a consistent linearization of the macroscopic constitutive quantities, is shown.

2.2 *Discretizations of the Boundary Value Problems*

For the discretizations of the boundary value problem on the microscale, we apply for actual, virtual and incremental deformations, electric as well as magnetic potential the following discretizations

$$\begin{aligned} \{\widetilde{\underline{u}}, \delta\widetilde{\underline{u}}, \Delta\widetilde{\underline{u}}\} &= \sum_{I=1}^{\mathrm{n}_{node}} \mathbb{N}_u^I \{\widetilde{\underline{d}}_u^I, \delta\widetilde{\underline{d}}_u^I, \Delta\widetilde{\underline{d}}_u^I\} \\ \{\widetilde{\phi}, \delta\widetilde{\phi}, \Delta\widetilde{\phi}\} &= \sum_{I=1}^{\mathrm{n}_{node}} \mathbb{N}_\phi^I \{\widetilde{d}_\phi^I, \delta\widetilde{d}_\phi^I, \Delta\widetilde{d}_\phi^I\} \\ \{\widetilde{\varphi}, \delta\widetilde{\varphi}, \Delta\widetilde{\varphi}\} &= \sum_{I=1}^{\mathrm{n}_{node}} \mathbb{N}_\varphi^I \{\widetilde{d}_\varphi^I, \delta\widetilde{d}_\varphi^I, \Delta\widetilde{d}_\varphi^I\} \,, \end{aligned} \tag{20}$$

where $\mathbb{N}^I$ contains the classical shape-functions associated with node I and the expressions $\{\underline{\widetilde{\boldsymbol{d}}}_u, \underline{\widetilde{d}}_\phi, \underline{\widetilde{d}}_\varphi\}$ the nodal displacement, the electric and magnetic potential. We approximate the actual, virtual and incremental fluctuation fields of the deformation, electric as well as magnetic field with

$$\begin{aligned}
\underline{\widetilde{\varepsilon}} &= \mathbb{B}^e_u \underline{\widetilde{\boldsymbol{d}}}^e_u , \quad \delta\underline{\widetilde{\varepsilon}} = \mathbb{B}^e_u \delta\underline{\widetilde{\boldsymbol{d}}}^e_u , \quad \Delta\underline{\widetilde{\varepsilon}} = \mathbb{B}^e_u \Delta\underline{\widetilde{\boldsymbol{d}}}^e_u , \\
\underline{\widetilde{\boldsymbol{E}}} &= \mathbb{B}^e_\phi \underline{\widetilde{d}}^e_\phi , \quad \delta\underline{\widetilde{\boldsymbol{E}}} = \mathbb{B}^e_\phi \delta\underline{\widetilde{d}}^e_\phi , \quad \Delta\underline{\widetilde{\boldsymbol{E}}} = \mathbb{B}^e_\phi \Delta\underline{\widetilde{d}}^e_\phi , \\
\underline{\widetilde{\boldsymbol{H}}} &= \mathbb{B}^e_\varphi \underline{\widetilde{d}}^e_\varphi , \quad \delta\underline{\widetilde{\boldsymbol{H}}} = \mathbb{B}^e_\varphi \delta\underline{\widetilde{d}}^e_\varphi , \quad \Delta\underline{\widetilde{\boldsymbol{H}}} = \mathbb{B}^e_\varphi \Delta\underline{\widetilde{d}}^e_\varphi ,
\end{aligned} \tag{21}$$

with the $\mathbb{B}^e$-matrices containing the partial derivatives of the shape-functions with respect to the reference coordinates. The following expressions

$$\underline{\widetilde{\mathbb{X}}} = \mathbb{B}^e_\xi \underline{\widetilde{\boldsymbol{d}}}^e_\xi , \quad \delta\underline{\widetilde{\mathbb{X}}} = \mathbb{B}^e_\xi \delta\underline{\widetilde{\boldsymbol{d}}}^e_\xi , \quad \Delta\underline{\widetilde{\mathbb{X}}} = \mathbb{B}^e_\xi \Delta\underline{\widetilde{\boldsymbol{d}}}^e_\xi , \tag{22}$$

give a general form of the discretizations in order to simplify the derivation of the following derivation of the algorithmic consistent tangent moduli. Here, we introduce the abbreviations $\underline{\mathbb{X}} = \{\varepsilon, \ \boldsymbol{E}, \ \boldsymbol{H}\}$ and $\boldsymbol{\xi} = \{\boldsymbol{u}, \ \phi, \ \varphi\}$. In analogy to the discretizations on the microscale we discretize the macroscopic boundary value problem.

2.3 Consistent Linearization of Macroscopic Field Equations

The boundary value problems on the macroscale as well as on the microscale are solved using the Finite Element Method and involve the derivation of the weak forms on both scales. To obtain the solution of the problems the Newton-Raphson iteration scheme is used, where we want to achieve quadratic convergence on both scales. Therefore, a consistent linearization of the macroscopic constitutive quantities is required. The incremental constitutive equations on the macroscopic level can be expressed as

$$\underbrace{\begin{bmatrix} \Delta\overline{\boldsymbol{\sigma}} \\ -\Delta\overline{\boldsymbol{D}} \\ -\Delta\overline{\boldsymbol{B}} \end{bmatrix}}_{\Delta\underline{\overline{\boldsymbol{\Sigma}}}} = \underbrace{\begin{bmatrix} \overline{\mathbb{C}} & -\overline{\boldsymbol{e}}^T & -\overline{\boldsymbol{q}}^T \\ -\overline{\boldsymbol{e}} & -\overline{\epsilon} & -\overline{\alpha}^T \\ -\overline{\boldsymbol{q}} & -\overline{\alpha} & -\overline{\mu} \end{bmatrix}}_{\underline{\overline{\mathbb{Z}}}} \underbrace{\begin{bmatrix} \Delta\overline{\varepsilon} \\ \Delta\overline{\boldsymbol{E}} \\ \Delta\overline{\boldsymbol{H}} \end{bmatrix}}_{\Delta\underline{\overline{\mathbb{X}}}} , \tag{23}$$

with the macroscopic and microscopic elasticity tangent moduli $\overline{\mathbb{C}}$ and $\mathbb{C}$, the piezoelectric tangent moduli $\overline{\boldsymbol{e}}$ and $\boldsymbol{e}$, the piezomagnetic tangent moduli $\overline{\boldsymbol{q}}$ and $\boldsymbol{q}$, the electric permittivities $\overline{\epsilon}$ and ϵ, the magnetic permeabilities $\overline{\mu}$ and μ, as well as the magnetoelectric tangent moduli $\overline{\alpha}$. In the following we use a matrix-notation denoted by $(\underline{\bullet})$, where we neglect the notations for the single or double contraction. For the determination of the macroscopic constitutive tangent moduli the increments of the macroscopic stresses, electric displacement and magnetic induction are expressed by the volume averages of the corresponding microscopic variables

$$\overline{\underline{\mathbb{Z}}} = \frac{\partial \langle \underline{\boldsymbol{\Sigma}} \rangle_V}{\partial \overline{\underline{\mathbb{X}}}} \tag{24}$$

The computation of the macroscopic moduli is obviously not straight-forward. The macroscopic stresses, electric displacements and magnetic inductions are defined through volume integrals over the microscopic counterparts. With the application of the chain rule and the additive decomposition of the microscopic quantities into a macroscopic and a fluctuation part, we obtain the overall material tangent as

$$\begin{aligned} \overline{\underline{\mathbb{Z}}} &= \frac{\partial \langle \underline{\boldsymbol{\Sigma}}(\underline{\mathbb{X}}) \rangle_V}{\partial \overline{\underline{\mathbb{X}}}} = \left\langle \frac{\partial \underline{\boldsymbol{\Sigma}}(\underline{\mathbb{X}})}{\partial \underline{\mathbb{X}}} \, \frac{\partial (\overline{\underline{\mathbb{X}}} + \widetilde{\underline{\mathbb{X}}})}{\partial \overline{\underline{\mathbb{X}}}} \right\rangle_V \\ &= \underbrace{\langle \underline{\mathbb{Z}} \rangle_V}_{\underline{\mathbb{Z}}^{\text{Voigt}}} + \underbrace{\left\langle \underline{\mathbb{Z}} \, \frac{\partial \widetilde{\underline{\mathbb{X}}}}{\partial \overline{\underline{\mathbb{X}}}} \right\rangle_V}_{\underline{\mathbb{Z}}^{\text{Soft}}} . \end{aligned} \tag{25}$$

The first part of the latter equation $\underline{\mathbb{Z}}^{\text{Voigt}}$ is denoted as the Voigt upper bound of the material tangent and is defined as a simple volume average. The second part $\underline{\mathbb{Z}}^{\text{Soft}}$ is called the softening term of the overall tangent and reduces the material stiffness. In order to derive the softening part, we linearize the microscopic weak forms at an equilibrium state. In the following we define a general weak form for the balance of momentum, Gauß's law of electrostatics and Gauß's law of magnetostatics as

$$\mathbb{G}_\xi = -\int_{\mathcal{RVE}} \delta \widetilde{\underline{\boldsymbol{\xi}}} \, \text{div}\, [\underline{\boldsymbol{\Sigma}}] \, dv = \int_{\mathcal{RVE}} \delta \widetilde{\underline{\mathbb{X}}} \, \underline{\boldsymbol{\Sigma}} \, dv - \int_{\partial \mathcal{RVE}} \delta \widetilde{\underline{\boldsymbol{\xi}}} \, (\underline{\boldsymbol{\Sigma}} \, \boldsymbol{n}) \, da \, . \tag{26}$$

The linearization of the above general weak form with respect to the microscopic quantities yields

$$\Delta \mathbb{G}_\xi = \int_{\mathcal{RVE}} \delta \widetilde{\underline{\mathbb{X}}} \, \underline{\mathbb{Z}} \, \Delta \underline{\mathbb{X}} \, dv = \int_{\mathcal{RVE}} \delta \widetilde{\underline{\mathbb{X}}} \, \underline{\mathbb{Z}} \, (\Delta \overline{\underline{\mathbb{X}}} + \Delta \widetilde{\underline{\mathbb{X}}}) \, dv = 0 \, , \tag{27}$$

where we decomposed the microscopic quantities into a macroscopic and a fluctuation part. After a reformulation of the latter equation and the insertion of the discretizations we obtain

$$\begin{aligned} \Delta \mathbb{G}_\xi &= \int_{\mathcal{RVE}} \delta \widetilde{\underline{\mathbb{X}}} \, \underline{\mathbb{Z}} \, dv \, \Delta \overline{\underline{\mathbb{X}}} + \int_{\mathcal{RVE}} \delta \widetilde{\underline{\mathbb{X}}} \, \underline{\mathbb{Z}} \, \Delta \widetilde{\underline{\mathbb{X}}} \, dv \\ &= \sum_{e=1}^{num_{ele}} \delta \widetilde{\underline{\boldsymbol{d}}}_\xi^{eT} \Big\{ \underbrace{\int_{\mathcal{RVE}} \mathbb{B}_\xi^{eT} \, \underline{\mathbb{Z}} \, dv}_{\underline{\boldsymbol{l}}^e} \; \Delta \overline{\underline{\mathbb{X}}} + \underbrace{\int_{\mathcal{RVE}} \mathbb{B}_\xi^{eT} \, \underline{\mathbb{Z}} \, \mathbb{B}_\xi^e \, dv}_{\underline{\boldsymbol{k}}^e} \; \Delta \widetilde{\underline{\boldsymbol{d}}}_\xi^e \Big\} = 0 \end{aligned} \tag{28}$$

with the number of microscopic finite elements num_{ele}, the element stiffness matrices $\underline{\boldsymbol{k}}^e$ and the sensitivity of the moduli of the finite elements $\underline{\boldsymbol{l}}^e$. The contracted matrix notation of the latter expression can be reformulated with the application of a standard assembling procedure as

$$\sum_{e=1}^{num_{ele}} \delta \underline{\widetilde{\boldsymbol{d}}}_{\xi}^{eT} \{ \underline{\boldsymbol{l}}^e \, \Delta \overline{\underline{\mathbb{X}}} + \underline{\boldsymbol{k}}^e \, \Delta \underline{\widetilde{\boldsymbol{d}}}_{\xi}^e \} = \delta \underline{\widetilde{\boldsymbol{D}}}_{\xi}^T (\underline{\boldsymbol{L}} \, \Delta \overline{\underline{\mathbb{X}}} + \underline{\boldsymbol{K}} \, \Delta \underline{\widetilde{\boldsymbol{D}}}_{\xi}) = 0 \, . \tag{29}$$

From this we achieve the microscopic fluctuations depending on the incremental macroscopic process variables as

$$\Delta \underline{\widetilde{\boldsymbol{D}}}_{\xi} = -\underline{\boldsymbol{K}}^{-1} \underline{\boldsymbol{L}} \, \Delta \overline{\underline{\mathbb{X}}} \quad \Leftrightarrow \quad \frac{\partial \Delta \underline{\widetilde{\boldsymbol{D}}}_{\xi}}{\partial \overline{\underline{\mathbb{X}}}} = -\underline{\boldsymbol{K}}^{-1} \underline{\boldsymbol{L}} \, . \tag{30}$$

Inserting Eq. $(30)_2$ and the discretizations (22) in (25) we obtain the final expression of the effective magneto-electro-mechanical moduli as

$$\begin{aligned} \overline{\underline{\mathbb{Z}}} &= \langle \underline{\mathbb{Z}} \rangle_V + \left\langle \sum_{e=1}^{num_{ele}} \underline{\mathbb{Z}} \, \frac{\partial \underline{\mathbb{B}}_{\xi}^e \Delta \underline{\widetilde{\boldsymbol{d}}}_{\xi}^e}{\partial \overline{\underline{\mathbb{X}}}} \right\rangle_V = \langle \underline{\mathbb{Z}} \rangle_V + \frac{1}{V} \underline{\boldsymbol{L}}^T \, \frac{\partial \Delta \underline{\widetilde{\boldsymbol{D}}}_{\xi}}{\partial \overline{\underline{\mathbb{X}}}} \\ &= \underline{\mathbb{Z}}^{\text{Voigt}} - \frac{1}{V} \underline{\boldsymbol{L}}^T \, \underline{\boldsymbol{K}}^{-1} \, \underline{\boldsymbol{L}} \, . \end{aligned} \tag{31}$$

For a detailed derivation of the effective tangent moduli for magneto-electro-mechanically coupled material response we refer to Schröder et al. (2016).

3 Magneto-Electro-Mechanical Material Models

The two-scale FE^2-method is a general treatment and applicable for all kinds of material models. In order to demonstrate the generality of the method and the different ME responses we performed simulations with several magneto-electric composites. We will start with the description of a linear transversely isotropic model for piezoelectric and piezomagnetic material behaviors. A more precise prediction of the material behavior gives an electrostrictive model, which captures in a specific electric field range the nonlinear properties in a good manner. Finally, we will introduce a ferroelectric model, which considers electric field and stress motivated switchings of remanent microscopic polarizations. This model, based on an orientation distribution function of unit cells, is capable of showing the typical dielectric and butterfly hysteresis curves.

3.1 Linear Piezoelectric and Piezomagnetic Model

For the description of piezoelectric and piezomagnetic material behaviors, we use a generalized coordinate-invariant formulation of a magneto-electro-mechanical enthalpy function ψ for transversely isotropic solids adopted from Schröder and Gross (2004). It is formally given by

$$\psi_1(\alpha) = \psi_1^{mech} + \alpha\psi_1^{pe} + (1-\alpha)\psi_1^{pm} + \psi_1^{diel} + \psi_1^{magn} \tag{32}$$

$$\begin{aligned} \alpha &= 1 \quad \text{for piezoelectric materials} \\ \alpha &= 0 \quad \text{for piezomagnetic materials} \end{aligned}$$

where the individual parts represent functions for the purely mechanical$^{(mech)}$, the piezoelectric$^{(pe)}$, the piezomagnetic$^{(pm)}$, the purely electric$^{(diel)}$ and the purely magnetic$^{(magn)}$ behavior, respectively. Please note that in each phase of the composite either the piezoelectric term ψ_1^{pe} or the piezomagnetic term ψ_1^{pm} will be activated. The individual functions are given explicitly by

$$\begin{aligned} \psi_1^{mech} &= \tfrac{1}{2}\lambda I_1^2 + \mu I_2 + \omega_1 I_5 + \omega_2 I_4^2 + \omega_3 I_1 I_4, \\ \psi_1^{pe} &= \beta_1 I_1 J_2^e + \beta_2 I_4 J_2^e + \beta_3 K_1^e, \\ \psi_1^{pm} &= \kappa_1 I_1 J_2^m + \kappa_2 I_4 J_2^m + \kappa_3 K_1^m, \\ \psi_1^{diel} &= \gamma_1 J_1^e + \gamma_2 (J_2^e)^2, \\ \psi_1^{magn} &= \xi_1 J_1^m + \xi_2 (J_2^m)^2, \end{aligned} \tag{33}$$

formulated in terms of the following invariants

$$\begin{aligned} &I_1 := \mathrm{tr}[\boldsymbol{\varepsilon}], \quad I_2 := \mathrm{tr}[\boldsymbol{\varepsilon}^2], \quad I_4 := \mathrm{tr}[\boldsymbol{\varepsilon m}], \quad I_5 := \mathrm{tr}[\boldsymbol{\varepsilon}^2 \boldsymbol{m}], \\ &J_1^e := \mathrm{tr}[\boldsymbol{E}\otimes\boldsymbol{E}], \quad J_2^e := \mathrm{tr}[\boldsymbol{E}\otimes\boldsymbol{a}], \quad K_1^e := \mathrm{tr}[\boldsymbol{\varepsilon}(\boldsymbol{E}\otimes\boldsymbol{a})], \\ &J_1^m := \mathrm{tr}[\boldsymbol{H}\otimes\boldsymbol{H}]. \quad J_2^m := \mathrm{tr}[\boldsymbol{H}\otimes\boldsymbol{a}], \quad K_1^m := \mathrm{tr}[\boldsymbol{\varepsilon}(\boldsymbol{H}\otimes\boldsymbol{a})], \end{aligned} \tag{34}$$

in which $\boldsymbol{a}$ describes the preferred direction of the respective phase and $\boldsymbol{m} := \boldsymbol{a}\otimes\boldsymbol{a}$ is the associated structural tensor. In this model, the preferred direction is assumed to coincide with the direction of remanent electric polarization or magnetization, depending on the considered phase. The constitutive magneto-electro-mechanical moduli follow consequently as

$$\begin{aligned} \mathbb{C} &= \lambda\mathbf{1}\otimes\mathbf{1} + 2\mu\mathbb{I} + \omega_3[\mathbf{1}\otimes\boldsymbol{m} + \boldsymbol{m}\otimes\mathbf{1}] + 2\omega_2\boldsymbol{m}\otimes\boldsymbol{m} + \omega_1\,\Xi, \\ \boldsymbol{e} &= -\beta_1\boldsymbol{a}\otimes\mathbf{1} - \beta_2\boldsymbol{a}\otimes\boldsymbol{m} - \beta_3\boldsymbol{\theta}, \\ \boldsymbol{q} &= -\kappa_1\boldsymbol{a}\otimes\mathbf{1} - \kappa_2\boldsymbol{a}\otimes\boldsymbol{m} - \kappa_3\boldsymbol{\theta}, \\ \boldsymbol{\epsilon} &= -2\gamma_1\mathbf{1} - 2\gamma_2\boldsymbol{m}, \\ \boldsymbol{\mu} &= -2\xi_1\mathbf{1} - 2\xi_2\boldsymbol{m}, \end{aligned} \tag{35}$$

where we have defined the fourth-order tensors

$$\mathbb{I}_{ijkl} := \frac{1}{2}(\delta_{ik}\delta_{jl} + \delta_{il}\delta_{jk}), \quad \text{and} \quad \Xi_{ijkl} := a_i\delta_{jk}a_l + a_k\delta_{il}a_j, \tag{36}$$

and the third-order tensor

$$\theta_{ijk} := \frac{1}{2}(a_j\delta_{ik} + a_k\delta_{ij}). \tag{37}$$

Note that in each of the two phases the ME-coefficient is zero and either the piezomagnetic or piezoelectric moduli are active. In order to determine the overall ME-coefficient we use the homogenization approach of the FE^2-method. The resulting coupling values are shown in the numerical examples section. The parameters for the coordinate-invariant formulation can be related to the experimental parameters by setting the preferred directions of the phases to $\boldsymbol{a} = \boldsymbol{x}_3$. Then we obtain

$$\begin{aligned}
&\lambda = \mathbb{C}_{1122}, \quad \mu = \tfrac{1}{2}(\mathbb{C}_{1111} - \mathbb{C}_{1122}), \quad \omega_1 = 2\mathbb{C}_{1212} + \mathbb{C}_{1122} - \mathbb{C}_{1111},\\
&\omega_2 = \tfrac{1}{2}(\mathbb{C}_{1111} + \mathbb{C}_{3333}) - 2\mathbb{C}_{1212} - \mathbb{C}_{1133}, \quad \omega_3 = \mathbb{C}_{1133} - \mathbb{C}_{1122},\\
&\beta_1 = -e_{311}, \quad \beta_2 = e_{311} - e_{333} + 2e_{113}, \quad \beta_3 = -2e_{113},\\
&\kappa_1 = -q_{311}, \quad \kappa_2 = q_{311} - q_{333} + 2q_{113}, \quad \kappa_3 = -2q_{113},\\
&\gamma_1 = -\tfrac{1}{2}\epsilon_{11}, \quad \gamma_2 = \tfrac{1}{2}(\epsilon_{11} - \epsilon_{33}), \quad \xi_1 = -\tfrac{1}{2}\mu_{11}, \quad \xi_2 = \tfrac{1}{2}(\mu_{11} - \mu_{33}).
\end{aligned} \tag{38}$$

3.2 *Nonlinear Electrostrictive Model*

As a second example, the electric phase will be modeled as an electrostrictive and paramagnetic solid. Thus, we assume the existence of an isotropic magneto-electric enthalpy function ψ_2 in accordance to the purely electro-mechanical case in Schröder and Keip (2012). The function is decomposed into four parts, which describe the mechanical, the electrostrictive, the dielectric as well as the magnetic behavior as

$$\psi_2 = \psi_2^{mech} + \psi_2^{estr} + \psi_2^{diel} + \psi_2^{magn}, \tag{39}$$

wherein the individual functions are defined in coordinate-invariant form as

$$\begin{aligned}
\psi_2^{mech} &= \tfrac{1}{2}\,\lambda\, I_1^2 + \mu\, I_2\,,\\
\psi_2^{estr} &= \beta_1\, I_1\, J_1 + \beta_2\, K_2\,,\\
\psi_2^{diel} &= \gamma J_1^e\,,\\
\psi_2^{magn} &= \xi J_1^m\,.
\end{aligned} \tag{40}$$

The utilized invariants are given in (34) and by the mixed invariant

$$K_2 = \mathrm{tr}[\varepsilon(\boldsymbol{E} \otimes \boldsymbol{E})] \ . \tag{41}$$

Based on (40), the elastic, dielectric and magnetic moduli follow as

$$\begin{aligned} \mathbb{C} &= \lambda \mathbf{1} \otimes \mathbf{1} + 2\mu \mathbb{I} \ , \\ \epsilon &= -2\gamma \mathbf{1} - 2\,\beta_1 I_1 \mathbf{1} - 2\beta_2 \varepsilon \ , \\ \mu &= -2\,\xi \mathbf{1} \ . \end{aligned} \tag{42}$$

Furthermore, we obtain the third-order electrostrictive coupling modulus

$$\boldsymbol{e}_{estr} = -2\,\beta_1 \boldsymbol{E} \otimes \mathbf{1} - \beta_2(\mathbf{1} \otimes \boldsymbol{E} + \hat{\boldsymbol{e}}) \quad \text{with} \quad \hat{e}_{ijk} = \delta_{ik} E_j \ , \tag{43}$$

which is a function of the electric field and thus only appear when an electric field is applied, see also Maugin et al. (1992), page 109. The symmetric incremental constitutive equations can now be given in matrix form as

$$\begin{bmatrix} \Delta\underline{\sigma} \\ -\Delta\underline{\boldsymbol{D}} \\ \Delta\underline{\boldsymbol{B}} \end{bmatrix} = \begin{bmatrix} \underline{\mathbb{C}} & -\underline{\boldsymbol{e}}^T_{estr} & \underline{\mathbf{0}} \\ -\underline{\boldsymbol{e}}_{estr} & -\underline{\epsilon} & \underline{\mathbf{0}} \\ \underline{\mathbf{0}} & \underline{\mathbf{0}} & \underline{\mu} \end{bmatrix} \begin{bmatrix} \Delta\underline{\varepsilon} \\ \Delta\underline{\boldsymbol{E}} \\ \Delta\underline{\boldsymbol{H}} \end{bmatrix} \ . \tag{44}$$

The identification of the material parameters is defined as

$$\begin{aligned} &\lambda = \mathbb{C}_{1122} \ , \quad \mu = -\tfrac{1}{2}(\mathbb{C}_{1122} - \mathbb{C}_{1111}) \ , \quad \gamma = -\tfrac{1}{2}\epsilon_{11} \ , \\ &\zeta_1 = \mathbb{Q}_{12} \ , \quad \zeta_2 = \mathbb{Q}_{1111} - \mathbb{Q}_{1122} \ , \quad \xi = -\tfrac{1}{2}\mu_{magn} \ , \\ &\beta_1 = -(12\lambda\zeta_1 + 4\lambda\zeta_2)\gamma^2 \ , \quad \beta_2 = -8\mu\zeta_2\gamma^2 \ , \end{aligned} \tag{45}$$

with the electrostrictive moduli $\mathbb{Q} = \zeta_1\, \mathbf{1} \otimes \mathbf{1} + \zeta_2\, II$ which relates the strains to the electric displacements

$$\varepsilon = \mathbb{Q} : (\boldsymbol{D} \otimes \boldsymbol{D}) \ . \tag{46}$$

3.3 Piezoelectric Model with Tetragonal Symmetry

In order to depict the ferroelectric hysteretic behavior the magneto-electric enthalpy function is formulated such that it represents the tetragonal symmetry of barium titanate. The crystal axes are defined by three mutually perpendicular unit vectors: The axis $\hat{\boldsymbol{c}} = \hat{\boldsymbol{a}}_3$ is associated with the direction of the spontaneous polarization. The other two axes $\hat{\boldsymbol{a}}_1$ and $\hat{\boldsymbol{a}}_2$ are oriented perpendicular to $\hat{\boldsymbol{a}}_3$. Based thereon, we define, according to Keip (2012), Schröder et al. (2016), the second-order structural tensors by

$$\hat{\boldsymbol{M}}_{ij} = \hat{\boldsymbol{a}}_i \otimes \hat{\boldsymbol{a}}_j, \quad \hat{\boldsymbol{\Xi}}_1 = (\hat{\boldsymbol{M}}_{13} + \hat{\boldsymbol{M}}_{31}) \quad \text{and} \quad \hat{\boldsymbol{\Xi}}_2 = (\hat{\boldsymbol{M}}_{23} + \hat{\boldsymbol{M}}_{32}) \tag{47}$$

and a third-order structural tensor by

$$\hat{\boldsymbol{\Xi}}_3 = \sum_{i=1}^{3} \left(\hat{\boldsymbol{a}}_i \otimes \hat{\boldsymbol{a}}_i \otimes \hat{\boldsymbol{c}} + \hat{\boldsymbol{a}}_i \otimes \hat{\boldsymbol{c}} \otimes \hat{\boldsymbol{a}}_i\right). \tag{48}$$

The corresponding magneto-electric enthalpy function is formally given as

$$\psi_3 = \psi_3^{mech} + \psi_3^{pe} + \psi_3^{diel} + \psi_3^{magn} \tag{49}$$

with the four parts

$$\begin{aligned}
\psi_3^{mech} &= \tfrac{1}{2}\,\lambda\,\hat{I}_1^2 + \mu\,\hat{I}_2 + \tfrac{1}{2}\,\omega_1\,\hat{I}_5^2 + \omega_2\,\hat{I}_1\,\hat{I}_5 + \tfrac{1}{2}\,\omega_3\,(\hat{I}_6^2 + \hat{I}_7^2) \\
&\quad + \tfrac{1}{2}\,\omega_4\,(\hat{I}_3^2 + \hat{I}_4^2 + \hat{I}_5^2), \\
\psi_3^{pe} &= \beta_1\,\hat{K}_1 + \beta_2\,\hat{I}_1\,\hat{J}_2^e + \beta_3\,\hat{I}_5\,\hat{J}_2^e, \\
\psi_3^{diel} &= \tfrac{1}{2}\,\gamma_1\,\hat{J}_1^e + \tfrac{1}{2}\,(\gamma_2 - \gamma_1)\,(\hat{J}_2^e)^2 - \hat{J}_2^e\,\hat{P}_1, \\
\psi_3^{magn} &= \tfrac{1}{2}\xi_1\,\hat{J}_1^m + \tfrac{1}{2}(\xi_2 - \xi_1)(\hat{J}_2^m)^2.
\end{aligned} \tag{50}$$

The above functions are formulated in terms of the invariants

$$\begin{aligned}
&\hat{I}_1 = \mathrm{tr}[\hat{\varepsilon}_e], \quad \hat{I}_2 = \mathrm{tr}[(\hat{\varepsilon}_e)^2], \quad \hat{I}_3 = \mathrm{tr}[\hat{\varepsilon}_e\hat{\boldsymbol{M}}_{11}], \quad \hat{I}_4 = \mathrm{tr}[\hat{\varepsilon}_e\hat{\boldsymbol{M}}_{22}], \\
&\hat{I}_5 = \mathrm{tr}[\hat{\varepsilon}_e\hat{\boldsymbol{M}}_{33}], \quad \hat{I}_6 = \mathrm{tr}[\hat{\varepsilon}_e\hat{\boldsymbol{\Xi}}_1], \quad \hat{I}_7 = \mathrm{tr}[\hat{\varepsilon}_e\hat{\boldsymbol{\Xi}}_2], \quad \hat{J}_1^e = \mathrm{tr}[\boldsymbol{E} \otimes \boldsymbol{E}], \\
&\hat{J}_1^m = \mathrm{tr}[\boldsymbol{H} \otimes \boldsymbol{H}], \quad \hat{J}_2^e = \mathrm{tr}[\boldsymbol{E} \otimes \hat{\boldsymbol{c}}], \quad \hat{J}_2^m = \mathrm{tr}[\boldsymbol{H} \otimes \hat{\boldsymbol{c}}], \\
&\hat{K}_1 = \mathrm{tr}[\boldsymbol{E} \otimes (\hat{\boldsymbol{\Xi}}_3 : \hat{\varepsilon}_e)], \quad \hat{P}_1 = \mathrm{tr}[\hat{\boldsymbol{P}}_r \otimes \hat{\boldsymbol{c}}].
\end{aligned} \tag{51}$$

Opposed to the previous models, the mechanical and electro-mechanical invariants now depend on the elastic strains $\hat{\varepsilon}_e$. Furthermore, we introduced the remanent polarization $\hat{\boldsymbol{P}}_r$. The reason for this lies in the assumption that the total electric displacement and the total strains can be additively decomposed into a reversible (elastic$_e$) and an irreversible (remanent$_r$) part

$$\hat{\boldsymbol{D}} = \hat{\boldsymbol{D}}_e + \hat{\boldsymbol{P}}_r \quad \text{and} \quad \hat{\varepsilon} = \hat{\varepsilon}_e + \hat{\varepsilon}_r. \tag{52}$$

On a nanoscopic level, the remanent polarization vector as well as the remanent strain tensor can uniquely defined through

$$\hat{\boldsymbol{P}}_r = \hat{P}_s\,\hat{\boldsymbol{c}} \quad \text{and} \quad \hat{\varepsilon}_r = \frac{3}{2}\,\hat{\varepsilon}_s\,\mathrm{dev}\,(\hat{\boldsymbol{c}} \otimes \hat{\boldsymbol{c}}), \tag{53}$$

since in the tetragonal phase each crystallite of barium titanate is spontaneously polarized in the direction of the preferred axis $\hat{\boldsymbol{c}}$. Here, $\hat{P}_s$ denotes the spontaneous polarization and $\hat{\varepsilon}_s$ denotes the spontaneous strain. Based on the magneto-electric enthalpy function, the constitutive moduli follow as

$$\hat{\mathbb{C}} = \lambda \mathbf{1} \otimes \mathbf{1} + 2\mu \mathbb{I} + \omega_1 \hat{M}_{33} \otimes \hat{M}_{33} + \omega_2 (\mathbf{1} \otimes \hat{M}_{33} + \hat{M}_{33} \otimes \mathbf{1})$$
$$+ \omega_3 (\hat{\Xi}_1 \otimes \hat{\Xi}_1 + \hat{\Xi}_2 \otimes \hat{\Xi}_2) + \omega_4 \sum_{i=1}^{3} \hat{M}_{ii} \otimes \hat{M}_{ii},$$
$$\hat{e} = -\beta_1 \hat{\Xi}_3 - \beta_2 \hat{c} \otimes \mathbf{1} - \beta_3 \hat{c} \otimes \hat{c} \otimes \hat{c},$$
$$\hat{\epsilon} = -\gamma_1 \mathbf{1} - (\gamma_2 - \gamma_1)\hat{c} \otimes \hat{c},$$
$$\hat{\mu} = -\xi_1 \mathbf{1} - (\xi_2 - \xi_1)\hat{c} \otimes \hat{c}.$$
(54)

Using the above expressions, the constitutive relations can be evaluated.

Switching criterion. The criterion for switching is based on the fundamental work of Hwang et al. (1995). Therein, the authors motivated energetic criteria for switching of the remanent variables $\hat{P}_r$ and $\hat{\varepsilon}_r$. The model is based on the incremental change of free energy as a result of ferroelectric and/or ferroelastic switching

$$\Delta \psi = \Delta \psi^{elec} + \Delta \psi^{mech} = E \cdot \Delta \hat{P}_r + \sigma : \Delta \hat{\varepsilon}_r. \tag{55}$$

In this equation, the increments of spontaneous polarization and spontaneous strain (in terms of the individual directions of switching) can be identified from the tetragonal crystal lattice, see Fig. 4. Since the amount of remanent strains is independent of the sign of the remanent polarization, only two switching possibilities follow for the spontaneous strain, with each option including two possible directions of polarization

$$\Delta \hat{\varepsilon}_{r,2,4} = \tfrac{3}{2} \hat{\varepsilon}_s \, \text{dev} \, (\hat{a}_1 \otimes \hat{a}_1 - \hat{c} \otimes \hat{c})$$
$$\text{and } \Delta \hat{\varepsilon}_{r,3,5} = \tfrac{3}{2} \hat{\varepsilon}_s \, \text{dev} \, (\hat{a}_2 \otimes \hat{a}_2 - \hat{c} \otimes \hat{c}). \tag{56}$$

According to the model formulation proposed in Hwang et al. (1995), a homogenization approach based on orientation distributions of the crystallographic axes is utilized. Therefore, in each material point a certain number n of tetragonal crystal lattice orientations (triads) is prescribed, for which the material law is evaluated and homogenized over the set of orientations. Then, the corresponding microscopic

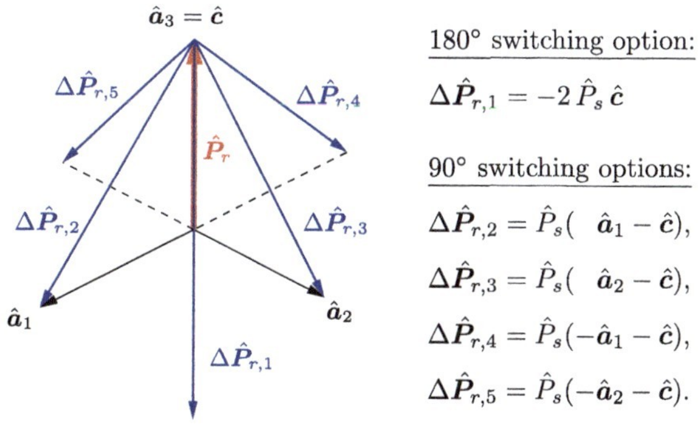

Fig. 4 180° and 90° switching options in tetragonal unit cell. The spontaneous polarization of the unit cell is given by $\hat{P}_r = \hat{P}_s \hat{c}$

180° switching option:
$\Delta \hat{P}_{r,1} = -2\hat{P}_s \hat{c}$

90° switching options:
$\Delta \hat{P}_{r,2} = \hat{P}_s(\hat{a}_1 - \hat{c}),$
$\Delta \hat{P}_{r,3} = \hat{P}_s(\hat{a}_2 - \hat{c}),$
$\Delta \hat{P}_{r,4} = \hat{P}_s(-\hat{a}_1 - \hat{c}),$
$\Delta \hat{P}_{r,5} = \hat{P}_s(-\hat{a}_2 - \hat{c}).$

counterparts based on n nanoscopic orientations are determined as the homogenized quantities

$$\boldsymbol{\sigma} = \frac{1}{n}\sum_{i=1}^{n} \hat{\boldsymbol{\sigma}}_i \quad \text{and} \quad \boldsymbol{D} = \frac{1}{n}\sum_{i=1}^{n} \hat{\boldsymbol{D}}_i \,. \tag{57}$$

Analogously, the overall microscopic mechanical, piezoelectric and dielectric moduli are computed by

$$\mathbb{C} = \frac{1}{n}\sum_{i=1}^{n} \hat{\mathbb{C}}_i \,, \quad \boldsymbol{e} = \frac{1}{n}\sum_{i=1}^{n} \hat{\boldsymbol{e}}_i \,, \quad \boldsymbol{\mu} = \frac{1}{n}\sum_{i=1}^{n} \hat{\boldsymbol{\mu}}_i \,, \quad \text{and} \quad \boldsymbol{\epsilon} = \frac{1}{n}\sum_{i=1}^{n} \hat{\boldsymbol{\epsilon}}_i \,, \tag{58}$$

where the nanoscopic moduli are determined from (54). In order to specify the critical energy barrier at which ferroelectric and/or ferroelastic switching is induced, the amount of dissipated work during switching is considered. We approximate the work dissipated during a discrete switching process by

$$\mathcal{W}^{diss}_{e,180^\circ} = 2\,\hat{P}_s\,\hat{E}_c \,, \quad \mathcal{W}^{diss}_{e,90^\circ} = \hat{P}_s\,\hat{E}_c \quad \text{and} \quad \mathcal{W}^{diss}_{m,90^\circ} = \frac{3}{2}\hat{\varepsilon}_s\,\hat{\sigma}_c \tag{59}$$

for 180° switching and 90° switching, respectively. This motivates the formulation of a normalized switching criterion of the form

$$\frac{\boldsymbol{E}\cdot\Delta\hat{\boldsymbol{P}}_{r,1}}{\mathcal{W}^{diss}_{e,180^\circ}} \geq 1 \quad \text{and} \quad \frac{\boldsymbol{E}\cdot\Delta\hat{\boldsymbol{P}}_{r,k}}{\mathcal{W}^{diss}_{e,90^\circ}} + \frac{\boldsymbol{\sigma}:\Delta\hat{\boldsymbol{\varepsilon}}_{r,k}}{\mathcal{W}^{diss}_{m,90^\circ}} \geq 1 \quad \text{for} \quad k = 2, ..., 5. \tag{60}$$

In detail, we evaluate the switching criteria in consideration of the microscopic electric field and stress (Reuss bound), which is in accordance with Hwang et al. (1995). Finally, it should be noted that if two or more switching criteria are met simultaneously, the one with greatest dissipation drives the switching process. If two or more criteria are met with equal amount of dissipation, the choice is taken randomly between the associated directions. The switching process involves a corresponding rotation of the crystallographic axes of the unit cells, which results in an update of the constitutive mechanical, piezoelectric and dielectric moduli, cf. (54). However, a simultaneous switching of many orientations in one load step may cause repeated back-and-forth switching of spontaneous polarizations. In order to avoid such behavior, we allow a maximum number n_{swit} of orientation switchings, with the highest dissipation, per Gauss point to switch. After solving the microscopic boundary value problem and homogenizing the microscopic fields, the macroscopic problem is solved. Based on the new solution, we localize the magneto-electro-mechanical fields to the microscale and the procedure starts again and is repeated until for all polarization vectors the switching criterion is not fulfilled any longer. The algorithmic step-by-step documentation can be found in the next Table.

(i) Next macroscopic load step

(ii) While macroscopic equilibrium is not found do

- (a) In each macroscopic Gauss point: compute electric field $\overline{\boldsymbol{E}}$ and strain $\overline{\varepsilon}$.
- (b) Update boundary conditions along $\partial\mathcal{RVE}$.
- (c) In each microscopic Gauss point: compute microscopic quantities $\boldsymbol{\sigma}$, $\boldsymbol{D}$, $\boldsymbol{B}$, $\mathbb{C}$, $\boldsymbol{e}$, and $\boldsymbol{\epsilon}$ see (58).
- (d) Solve microscopic boundary value problem and determine homogenized quantities.
- (e) Solve macroscopic boundary value problem.

Else (equilibrium is found): Go to (iii) and check if switching is possible.

(iii) In each microscopic Gauss point: compute microscopic stress $\boldsymbol{\sigma}$ and microscopic electric displacement $\boldsymbol{D}$.

(iv) In each microscopic Gauss point and for each of the attached crystallographic triads $\{\hat{\boldsymbol{a}}_1, \hat{\boldsymbol{a}}_2, \hat{\boldsymbol{c}}\}^n$, where n is the number of triads: compute all possible changes in remanent polarization $\Delta\hat{\boldsymbol{P}}_i^n$ $(i = 1, ..., 5)$ and remanent strain $\Delta\hat{\varepsilon}_i^n$ $(i = 2, 3)$.

(v) Evaluate switching criteria (60) in consideration of all possible changes of remanent quantities $\Delta\hat{\boldsymbol{P}}_i^n$ $(i = 1, ..., 5)$ and $\Delta\hat{\varepsilon}_i^n$ $(i = 2, 3)$ for all attached triads $\{\hat{\boldsymbol{a}}_1, \hat{\boldsymbol{a}}_2, \hat{\boldsymbol{c}}\}^n$.

(vi) Store evaluated switching criteria per Gauss point which meet the criteria (60) from highest to lowest dissipation in `list`. Store its number of entries in $n_{\texttt{list}}$.

(vii) If $n_{\texttt{list}} > 0$ (dissipative step) then

- (a) Choose number $n_{\text{swit}} \leq n_{\texttt{list}}$ of spontaneous polarizations allowed to switch simultaneously.
- (b) Perform update of the first n_{swit} polarization directions associated with the entries stored in `list`.
- (c) Save new orientation directions and compute new material tangents $\hat{\mathbb{C}}$, $\hat{\boldsymbol{e}}$, and $\hat{\boldsymbol{\epsilon}}$ for the new state of crystallographic triads $\{\hat{\boldsymbol{a}}_1, \hat{\boldsymbol{a}}_2, \hat{\boldsymbol{c}}\}^n$.
- (d) Go to (ii).

Else (switching criterion is no longer fulfilled): Go to (i).

Ferroelectric Response of Different Orientation Distributions. In order to represent the initially unpolar character of the matrix material, we consider an initially isotropic distribution of the spontaneous polarization vectors. This is achieved on the one hand by an error function which minimizes the difference between an average of distributed transversely isotropic properties with isotropic properties, and on the other hand through the use of a triangular collocation on a geodesic dome (Fuller 1965), whose surface is uniformly subdivided into multiple triangles, see for example Kurzhöfer et al. (2005) or Schröder et al. (2016). Subsequently, the orientations, characterized by arrows, then point from the center of the geodesic sphere into the

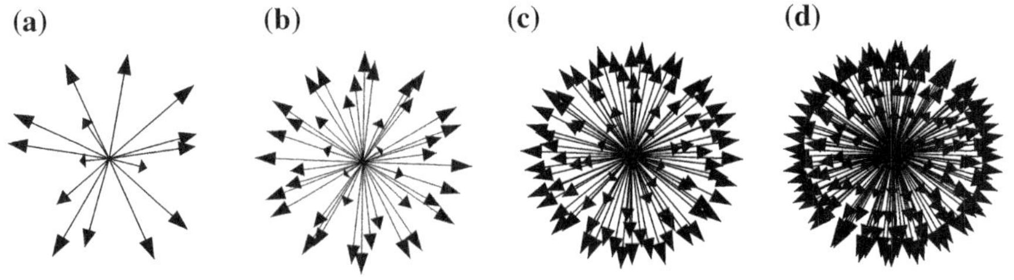

Fig. 5 Distributions of **a** 15, **b** 42, **c** 92 and **d** 162 orientations

Table 3 Material parameters of single crystal $BaTiO_3$ (Zgonik et al. 1994)

Modulus	Unit	$BaTiO_3$	Modulus	Unit	$BaTiO_3$
$\hat{\mathbb{C}}_{1111}$	N/mm^2	222000	$\hat{\epsilon}_{11}$ ($\hat{\epsilon}^r_{11}$)	mC/kVm (1)	0.019 (2146)
$\hat{\mathbb{C}}_{1122}$	N/mm^2	108000	$\hat{\epsilon}_{33}$ ($\hat{\epsilon}^r_{33}$)	mC/kVm (1)	0.000496 (56)
$\hat{\mathbb{C}}_{1133}$	N/mm^2	111000	$\hat{\mu}_{11}$ ($\hat{\mu}^r_{11}$)	N/kA^2 (1)	1.26 (1.0)
$\hat{\mathbb{C}}_{3333}$	N/mm^2	151000	$\hat{\mu}_{33}$ ($\hat{\mu}^r_{33}$)	N/kA^2 (1)	1.26 (1.0)
$\hat{\mathbb{C}}_{1212}$	N/mm^2	134000	$\hat{E}_c$	kV/mm	1.0
$\hat{\mathbb{C}}_{1313}$	N/mm^2	61000	$\hat{\sigma}_c$	N/mm^2	100
$\hat{e}_{311}$	C/m^2	−0.7	$\hat{\varepsilon}_s$	1	0.00834
$\hat{e}_{333}$	C/m^2	6.7	$\hat{P}_s$	C/m^2	0.26
$\hat{e}_{131}$	C/m^2	34.2			

direction of a corresponding corner point of the surface triangles. By subdividing each triangular edge into multiple subsections, which is denoted by frequencies, the number of orientations can be increased. Typically, this construction principle yields 42, 92, 162 or 1002 orientations for a frequency of 2, 3, 4 or 10. Figure 5 shows the associated preferred directions of the error function for 15 orientations and of the approximated geodesic domes for 42, 92 and 162 orientations. In the following example, we consider different numbers of preferred directions in the Gauss points of the electric material on the microscale, each belonging to one tetragonal crystallite. In order to generate an initially isotropic and unpolar behavior, the preferred directions are oriented homogeneously in the three-dimensional space. These orientations can switch its directions if the ferroelectroelastic switching criterion (60) due to the microscopic electric field and stresses is fulfilled. In order to investigate the material response of a ferroelectric single-phase material, we performed simulations with different numbers of orientations at each microscopic gauss point, which represent a set of barium titanate unit cells. The set of material parameters used to describe tetragonal barium titanate is listed in Table 3.

In the initial state the orientations are distributed uniformly in the three-dimensional space to obtain an unpoled ferroelectric bulk material, in order to approximate a real material without a pre-polarization. Figure 6 shows the resulting butterfly and dielectric hysteresis curves for an applied alternating vertical electric field $\overline{E}_3$,

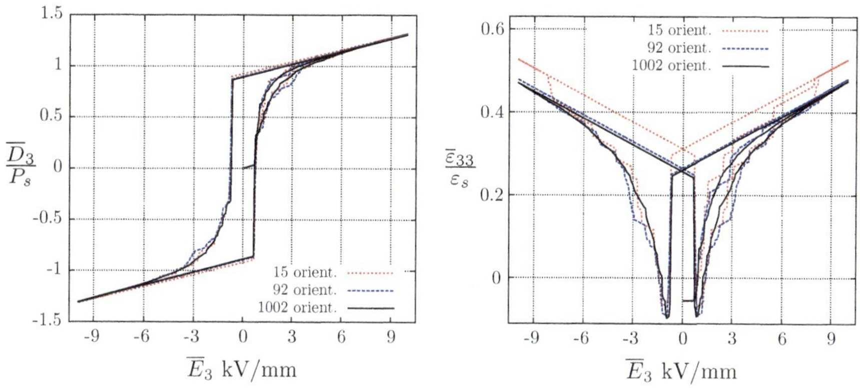

Fig. 6 Macroscopic dielectric and butterfly hysteresis loop using 15, 92 and 1002 orientations

with 15, 92 and 1002 orientations. The maximum applied electric field is increased to 10 kV/mm to ensure that each orientation is switched to its most efficient direction as well as to obtain a perfect symmetry of the hysteresis loops. As shown in Fig. 6 the macroscopic response converges to smoother hysteresis loops for a larger number of orientations. We state that a number of 92 orientations obtained by the construction of a geodesic dome is a good approximation to depict the hysteresis loops.

4 Numerical Examples

In the following we investigate three different material models for the simulation of magneto-electric composites. First, we use an electrostrictive model for the electro-active phase with piezomagnetic inclusions and take a look at the convergence behavior of the Newton method on both scales. In the second example we consider the response of a two-phase composite consisting of linear piezoelectric and piezomagnetic phases. In the third and fourth example, we are capable of simulating the ferroelectric hysteretic behavior and consider on the one hand the ability to describe transversely isotropic properties with the used model and on the other hand the response of a complex microstructure with two ellipsoidal inclusions in a cubic shaped matrix.

4.1 Electrostrictive/Piezomagnetic Cantilever Beam

In the case of ferroelectric solids the macroscopic overall deformations and electric polarization show a hysteretic behavior. In order to capture such nonlinear properties we consider in this numerical example an electrostrictive model, which is described

Table 4 Material parameters of lead magnesium niobate [$Pb(Mg_{1/3}Nb_{2/3})O_3$] taken from Sundar and Newnham (1992) ($[\mathbb{C}] = \text{GPa}$, $[\mathbb{Q}] = 10^{-2}\frac{\text{m}^4}{\text{C}^2}$, $[\epsilon_{diel}] = 10^{-9}\frac{\text{C}}{\text{Vm}}$, $[\mu_{magn}] = \frac{\text{N}}{\text{kA}^2}$)

$\mathbb{C}_{11}$	$\mathbb{C}_{12}$	$\mathbb{Q}_{11}$	$\mathbb{Q}_{12}$	ϵ_{diel} (ϵ^r)	μ_{magn} (μ^r)
137.04	48.148	2.60	−0.96	$66.375 \cdot 10^{-2}$ (74.97)	1.26 (1)

in Sect. 3.2. Especially in low electric field regions, where an increasing number of microscopic remanent polarizations begin to reorientate, an electrostrictive model can capture such nonlinear behavior in a very good manner. Additionally, we take a closer look on the macroscopic and microscopic convergence characteristic within the Newton-Raphson iteration scheme. Due to the consistent linearization of the macroscopic equations, see Sect. 2.3, we obtain a quadratic convergence on both scales. Therefore, we analyze the behavior of a composite with a nonlinear electrostrictive matrix and piezomagnetic inclusions. On the macroscale, we consider a two-dimensional (plane-strain) coupled boundary value problem, where a cantilever beam is loaded with a horizontal electric field $\overline{E}_1$ and a mechanical load $\overline{F}_2$ at its right end, see Fig. 7. At each macroscopic integration point, a three-dimensional $\mathcal{RVE}$ is attached. It consists of a brick-shaped electrostrictive matrix with a spherical piezomagnetic inclusion with 20% volume fraction. For the electrostrictive phase, we choose material parameters of lead magnesium niobate (PMN) adopted from Sundar and Newnham (1992), see Table 4. The material parameters for the piezomagnetic inclusions ($CoFe_2O_4$) are given in Table 5. In the present example, we assume the preferred direction of the piezomagnetic phase pointing in positive horizontal direction. The response of the cantilever beam as a whole and of several $\mathcal{RVE}$s at different macroscopic integration points are also depicted in Fig. 7. In detail, we see the horizontal component of the magnetic field on both scales ($\overline{H}_1$ and H_1). The inhomogeneous distribution at the left boundary is a result of the vertical force $\overline{F}_2$, which induces compressive and tensile stresses in this region. The heterogeneity of the field can also be observed in the attached $\mathcal{RVE}$s located at Gauss points close to the left boundary of the cantilever beam. In Fig. 8, we take a look on the numerical convergence of the Newton-Raphson iteration schemes on both scales. Between each macroscopic iteration step, the microscopic boundary value problems of all $\mathcal{RVE}$s are solved and the effective quantities are given back to the macroscale. As can be seen in Fig. 8, quadratic convergences on both scales, the micro- and the macroscale, are obtained. This shows the accurateness of the computation of the effective algorithmic tangents.

4.2 *Piezoelectric/Piezomagnetic Composites*

As described in Sect. 3.1 we used a transversely isotropic approach to simulate the piezoelectric/-magnetic behavior. In the case of the investigated barium titanate, the

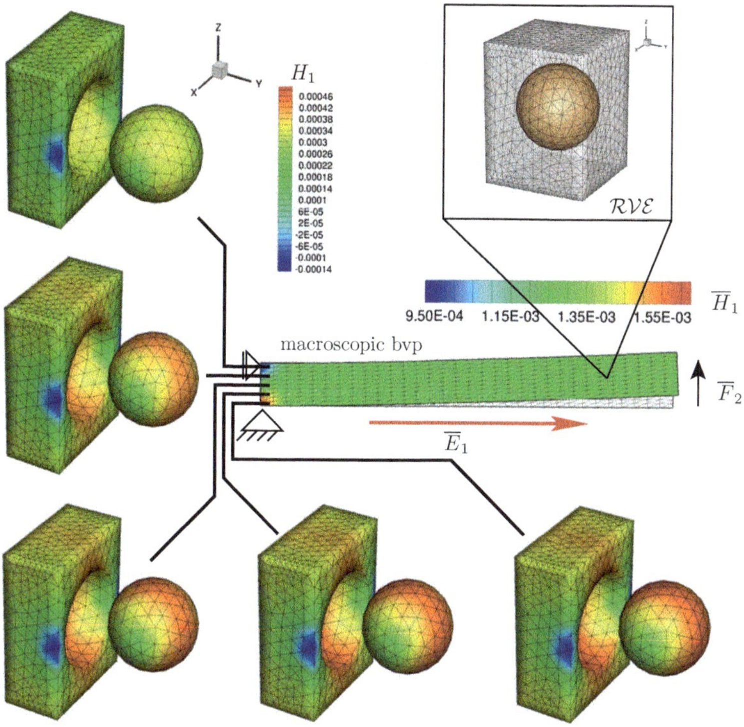

Fig. 7 Two-dimensional macroscopic cantilever beam with attached three-dimensional microstructures

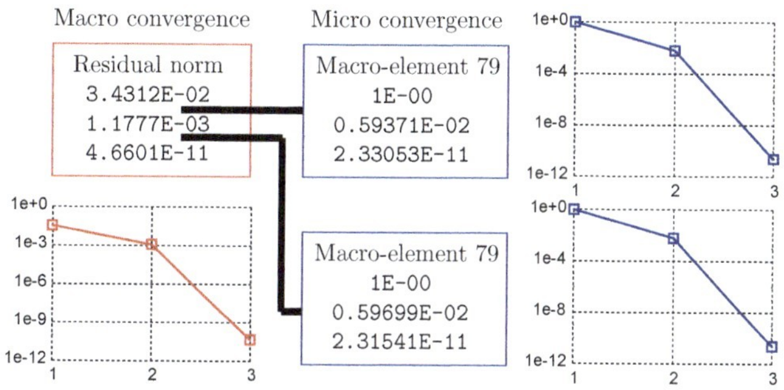

Fig. 8 Convergence of the Newton method on both scales. The macroscopic and relative microscopic residuums of one integration point of an arbitrary macroscopic element (here macro-element 79) are described with the red and blue curve, respectively

most reasonable approach using this model is to describe the real material behavior after a polarization process, whereby such materials are approximated by a transversely isotropic symmetry. The used material parameters are listed in Table 5. An extension to this model, where different polarization states are simulated, is given in Labusch et al. (2014). In this contribution, we take a closer look on the influence

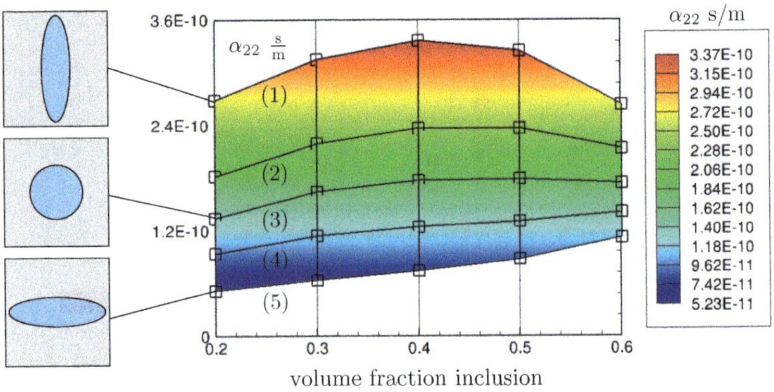

Fig. 9 Influence of the volume fraction (x-axis) and shape (different curves) of the inclusions on the magneto-electric coefficient (y-axis)

Table 5 Material parameters of polycrystalline $BaTiO_3/CoFe_2O_4$, transversely isotropic approximation

Parameter	Unit	$BaTiO_3$	$CoFe_2O_4$
$\mathbb{C}_{1111}$	N/mm^2	$16.6 \cdot 10^4$	$21.21 \cdot 10^4$
$\mathbb{C}_{1122}$	N/mm^2	$7.66 \cdot 10^4$	$7.45 \cdot 10^4$
$\mathbb{C}_{1133}$	N/mm^2	$7.75 \cdot 10^4$	$7.45 \cdot 10^4$
$\mathbb{C}_{3333}$	N/mm^2	$16.2 \cdot 10^4$	$21.21 \cdot 10^4$
$\mathbb{C}_{1212}$	N/mm^2	$4.29 \cdot 10^4$	$6.88 \cdot 10^4$
ϵ_{11} (ϵ^r_{11})	mC/kVm (1)	0.0112 (1264.9)	$8 \cdot 10^{-5}$ (9.04)
ϵ_{33} (ϵ^r_{33})	mC/kVm (1)	0.0126 (1423.1)	$9.3 \cdot 10^{-5}$ (10.5)
μ_{11} (μ^r_{11})	N/kA^2 (1)	1.26 (1)	157.0 (124.9)
μ_{33} (μ^r_{33})	N/kA^2 (1)	1.26 (1)	157.0 (124.9)
e_{311}	C/m^2	−4.4	0
e_{333}	C/m^2	18.6	0
e_{113}	C/m^2	11.6	0
q_{311}	N/Am	0	580.3
q_{333}	N/Am	0	−699.7
q_{113}	N/Am	0	550.0

of different microscopic morphologies. Due to the fact, that the ME coefficient in composites materials is a strain-induced property, the microstructure has a significant impact on this product properties. Therefore, we consider two-phase composites with a piezoelectric matrix and piezomagnetic inclusions and increase the volume fractions and modify the geometry and orientation of the inclusions. Figure 9 shows on the left hand side different microstructures, which are loaded with an applied macroscopic electric field $\overline{E}_2$ in vertical direction.

The diagram shows five curves of the resulting magneto-electric coefficient, which depends on the volume fraction of the inclusion material (x-axis) as well as on its geometry. In this study we considered five different geometries with different proportions of the radii of the ellipsoidal inclusion. The topmost curve (1) shows the ME coefficients of vertical ellipsoidal inclusions with increasing volume fractions, whose long major axis points into the direction of the applied electric field. The curve in the middle (3) investigates the behavior of circular inclusions and the bottommost curve (5) the response of an ellipsoidal inclusions with horizontaly aligned major axis. Less stretched ellipsoids are considered in the curves (2) and (4). We can clearly observe that the microscopic morphology has a large influence on the effective ME properties and can constitute this effect by microscopic electric field concentrations. At the vertical boundaries between the electric and magnetic phase, electric field concentrations appear, where the deformations of the matrix material, produced by the piezoelectric effect, are transferred to the inclusion. In the case of vertical aligned ellipsoidal inclusions the strains can be transferred at a larger surface of the inclusions, such that we obtain the highest magneto-electric coefficient for a composite with vertical inclusions with a volume fraction of 40 percent. However, this model is not capable of capturing different polarization states, such that a comparison with experimentally measured ME coefficients shows large deviations. Therefore, an extension of the model is necessary, see Labusch et al. (2014), or nonlinear models have to be considered.

4.3 *Ferroelectric Matrix with Cylindrical Magnetic Inclusions*

In the third numerical example we consider a two-phase composite with a ferroelectric matrix with piezomagnetic inclusions. Here, we used a nonlinear ferroelectric/-elastic model for the electric phase, which was described in Sect. 3.3. We will take a closer look at the resulting strain-induced ME coefficient. In detail, the composite under consideration consists of an electro-active matrix with magneto-active fibers, which are aligned in vertical direction. The used material parameters for the piezomagnetic material are listed in Table 5. The proposed ferroelectric/-elastic model has two advantages: First, it is capable to describe the typical dielectric and butterfly hysteresis loops of the investigated barium titanate. Figure 10a–c shows the effective dielectric and butterfly hysteresis loop as well as the magneto-electric coefficient for a composite with 30% volume fraction of the inclusion material. Since it is a strain-induced property, the magnitude of the coefficient depends on the transferred strains, which can be seen in the butterfly hysteresis loop. Once all orientations are aligned in their most preferred direction for high electric fields, the properties of the matrix material are piezoelectric, with the result of a saturated ME coefficient. Furthermore, it is possible to investigate the influence of different polarization states of the matrix material on the ME coefficient, which is done with a simplified model, proposed in Labusch et al. (2014). The second aspect of this example is to show that transversely isotropic material properties are obtained with the ferroelectric model after a polarization process. We used two different kinds of material parameters for barium titanate in the numerical results which are shown in Fig. 10d. On the one hand we used a linear piezoelectric model, described in Sect. 3.1, with material parameters adjusted to polarized samples with transversely isotropic properties. On the other hand, we considered parameters adjusted to the behavior of barium titanate unit cells, which are used for the ferroelectric/-elastic model, described in Sect. 3.3. By using the latter model, we attached an orientation distribution of unit cells in each microscopic integration point. After a polarization process the orientations are aligned in the direction of the applied macroscopic electric field, such that transversely isotropic properties are obtained. In order to compare both models, we depict the ME coefficients of the nonlinear model after reducing the applied electric field after the polarization process of the ferroelectric matrix. We start by considering the ME coefficients for different volume fractions in order to investigate its dependency on the ME coefficient. We can observe that the highest coefficient is obtained at a volume fraction of 50% for both phases and decreases for higher volume fractions of the magnetic inclusions and drops down to zero for purely piezoelectric materials. The curve reaches the value of 78% on the x-axis, which is the maximum volume fraction for cylindrical inclusions. However, we have to consider the maximum applied electric field and the consequently received polarization state. By comparing the results of the ME coefficients for the linear and nonlinear material model for a maximum applied electric field of 6 kV/mm, we observe slightly different results. At 30% volume fraction of the inclusion material the ME coefficient computed with the nonlinear model reaches the value of $\overline{\alpha}_{33} \approx 2.63 \cdot 10^{-9}$ s/m, which is 6.74% lower than the value in case of the linear approach. At a volume fraction of 50% inclusion material the coefficient for the nonlinear model is 5.86% lower and for 70% inclusion material we observe an ME coefficient 10.12% higher than the value for the linear approach. These deviations have the following reasons. First, we use different material parameters for both simulations, describing the transversely isotropic behavior and respectively the behavior of a tetragonal unit cell. Second, the maximum value of the alternating electric field is not high enough to perfectly align all orientations in the direction of the applied field and to reach the maximum saturation polarization. However, by increasing the maximum electric field up to 12 kV/mm an increased number of orientations have changed their directions, such that we obtain a higher polarization state and therefore increased piezoelectric properties. The deviations of the ME coefficients for 30%, 50% and 70% volume fraction of the inclusion, reduce to only 1.00%, 6.96% and 3.18%, respectively. The still larger difference for 50% could be caused by the ferroelastic switching processes due to internal stresses.

Obviously, the maximum coupling coefficient is not obtained at the vanishing electric field after the polarization process. Instead, the highest ME coupling values are achieved when the butterfly hysteresis loop reaches its highest slope. In this case most of the orientations change their directions resulting into high strains and therefore high strain-induced ME coefficients, determined by $\overline{\boldsymbol{\alpha}} = \partial\overline{\boldsymbol{B}}/\partial\overline{\boldsymbol{E}}$. Thus,

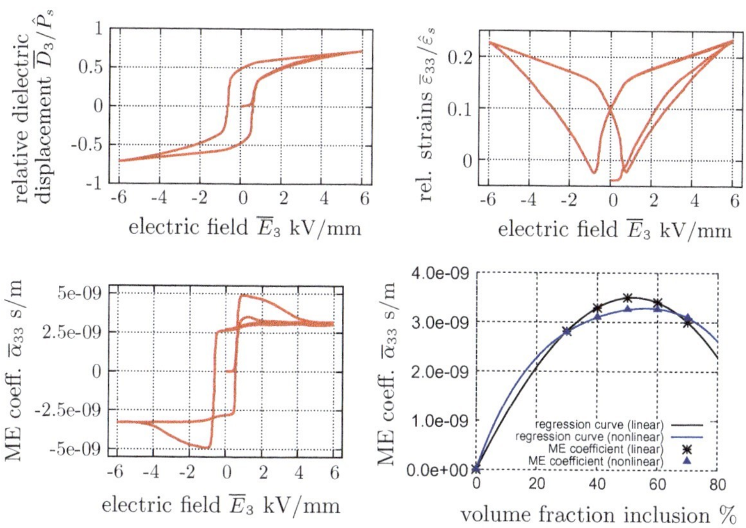

Fig. 10 **a** Effective dielectric $\overline{D}_3/\hat{P}_s$ and **b** butterfly $\overline{\varepsilon}_{33}/\hat{\varepsilon}_s$ hysteresis loop as well as **c** ME coefficient $\overline{\alpha}_{33}$ in s/m, depending on an alternating applied electric field $\overline{E}_3$ in kV/mm for 30% volume fraction of the inclusion. **d** Comparison of the results using linear and nonlinear models for the ferroelectric matrix material

the nonlinear ferroelectric model seems to be meaningful, particularly to capture the behavior of the ME coefficients in the case of a reorientation of the unit cells. However, such nonlinearities are difficult to exploit in an actual ME device.

4.4 Ferroelectric Matrix with Ellipsoidal Magnetic Inclusions

In the last numerical example, we consider a complex three dimensional microstructure consisting of a ferroelectric matrix and piezomagnetic inclusions with two different ellipsoidal shapes. The matrix material is characterized by the ferroelectric/-elastic model as in the previous example. As already shown in Sect. 4.2 the morphology of the microstructure has a significant influence on the magneto-electric coefficient. Thus, we take a closer look on the microscopic field distributions in a cross section of the attached $\mathcal{RVE}$s. Figure 11 depicts the macroscopic boundary value problem with the attached $\mathcal{RVE}$. The macroscopic cube is loaded with an alternating electric field with a maximum value of 6 kV/mm. We take a closer look at the distributions of the microscopic electric fields, magnetic fields as well as the stresses in vertical direction, which are shown in Fig. 12 at a maximum applied field of $\overline{E}_3 = 6$ kV/mm. It can clearly be observed that at the vertical boundaries of the inclusions microscopic electric field concentrations occur, see Fig. 12a, which cause an alignment of the attached orientations in field direction. As a consequence the

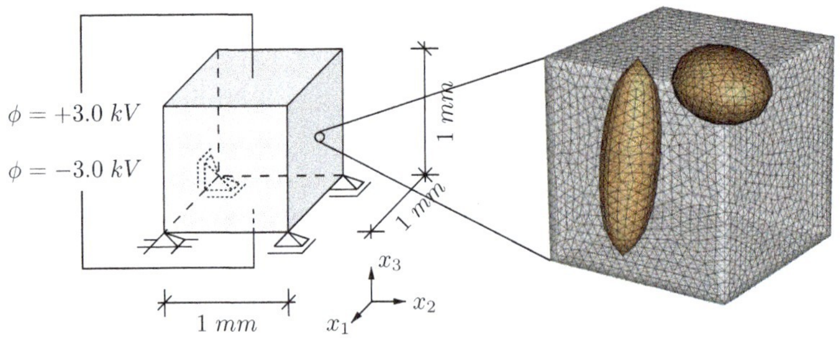

Fig. 11 Macroscopic bvp ($\overline{E}_3^{\max} = 6$ kV/mm) with attached $\mathcal{RVE}$

piezoelectric properties in these regions increase, such that the piezoelectric effect yield higher strains, resulting in strain-induced ME properties. Due to the large interface between the matrix and the inclusion at the vertical ellipsoid, higher magnetic fields can be observed in Fig. 12c. However, microscopic vertical stresses could lead to a ferroelastic switching of the orientations perpendicular to the direction of the applied electric field. This switching could decrease the piezoelectric properties in field direction and may yield a negative effect on the ME coefficient. The σ_{33}-stress distribution is depicted in Fig. 12b. Furthermore, attention has to be taken on the microscopic stress distribution in the inclusions. In this contribution a linear piezomagnetic behavior is assumed for the magnetic phase, however, such stresses could also influence a reorientation of microscopic remanent magnetizations. In future works, the influence of nonlinear model for the magnetostrictive behavior has to be analyzed. The overall effective properties of the composite are shown in Fig. 13, where the dielectric and butterfly hysteresis as well as the ME coefficient are depicted.

5 Summary

In this contribution, we proposed a two-scale homogenization approach for magneto-electro-mechanically coupled boundary value problems. The procedure was implemented into the FE^2-method, which allows to compute macroscopic boundary value problems in consideration of attached microscopic representative volume elements. The major focus by using this method was to analyze the strain-induced magneto-electric coefficient of such composites with different morphologies and different material models. We demonstrated that suitable approaches for the ferroelectric material behavior are capable for the description of typical hysteresis curves. However, the assumption of a linear material behavior for the magnetic inclusions is a simplification which has to be replaced by an appropriate magnetostrictive material model. Therefore, future works will concentrate on the extention of the microscopic model to a nonlinear magnetostrictive behavior. This will allow to capture the whole range of ferroic nonlinearities of ME composites.

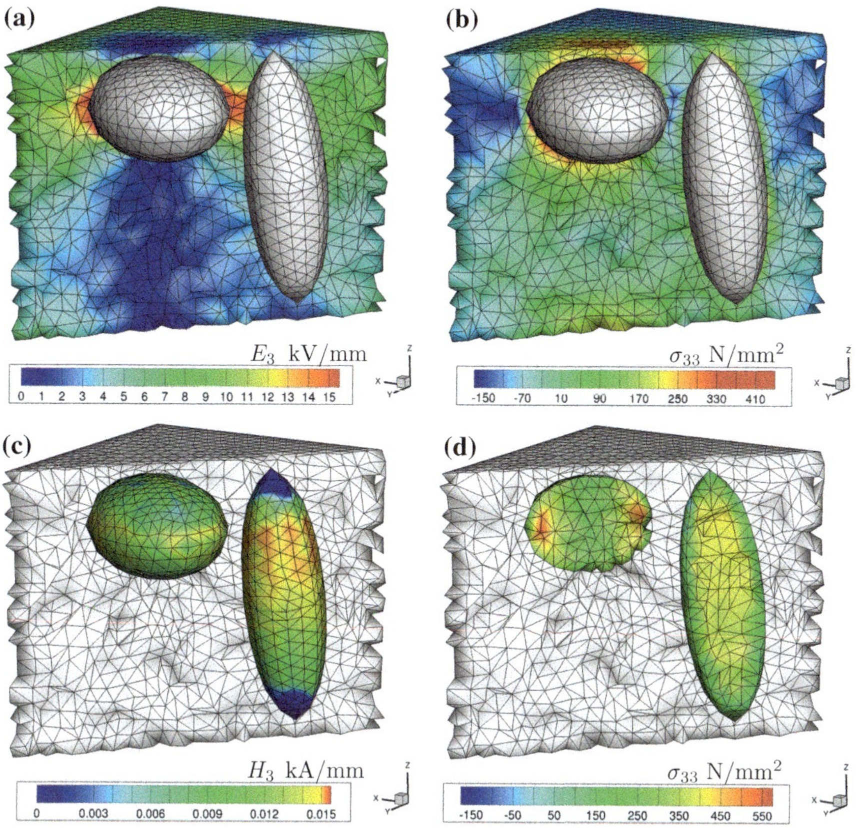

Fig. 12 Distributions of the microscopic vertical **a** electric field E_3 and **b** stresses σ_{33} in the matrix material, as well as **c** magnetic field H_3 and **d** stresses σ_{33} in the inclusion material

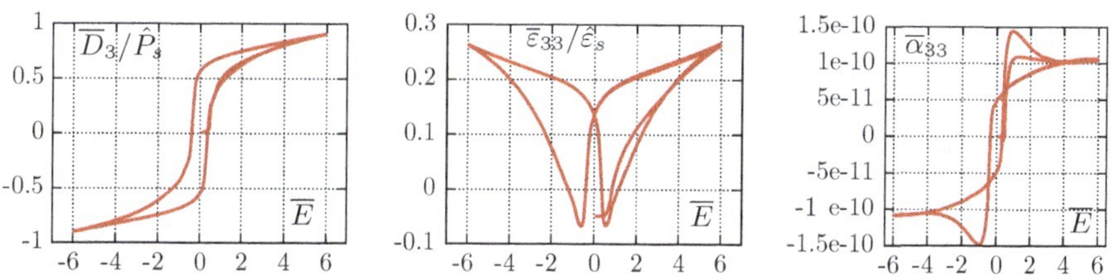

Fig. 13 Effective dielectric $\overline{D}_3/\hat{P}_s$ and butterfly $\overline{\varepsilon}_{33}/\hat{\varepsilon}_s$ hysteresis and ME coefficient $\overline{\alpha}_{33}$ in s/m, depending on an electric field $\overline{E}$ in kV/mm

References

Al'shin, B. I., & Astrov, D. N. (1963). Magnetoelectric effect in titanium oxide Ti_2O_3. *Soviet Physics JETP*, *17*, 809–811.

Ascher, E., Rieder, H., Schmid, H., & Stössel, H. (1966). Some properties of ferromagnetoelectric nickel-iodine boracite, $Ni_3B_7O_{13}I$. *Journal of Applied Physics*, *37*, 1404–1405.

Astrov, D. N. (1960). The magnetoelectric effect in antiferromagnetics. *Soviet Physics JETP*, *38*, 984–985.

Astrov, D. N. (1961). Magnetoelectric effect in chromium oxide. *Journal of Experimental and Theoretical Physics*, *40*, 1035–1041.

Bibes, M., & Barthélémy, A. (2008). Multiferroics: Towards a magnetoelectric memory. *Nature Materials*, *7*(6), 425–426. ISSN 1476-1122.

Bichurin, M. I., Petrov, V. M., & Srinivasan, G. (2003). Theory of low-frequency magnetoelectric coupling in magnetostrictive-piezoelectric bilayers. *Physical Review B*, *68*, 054402.

Bichurin, M. I., Petrov, V. M., Averkin, S. V., & Liverts, E. (2010). Present status of theoretical modeling the magnetoelectric effect in magnetostrictive-piezoelectric nanostructures. Part I: Low frequency and electromechanical resonance ranges. *Journal of Applied Physics*, *107*(5), 053904.

Cheong, S.-W., & Mostovoy, M. (2007). Multiferroics: A magnetic twist for ferroelectricity. *Nature Materials*, *6*(1), 13–20. ISSN 1476-1122.

Crottaz, O., Rivera, J.-P., Revaz, B., & Schmid, H. (1997). Magnetoelectric effect of $Mn_3B_7O_{13}I$ boracite. *Ferroelectrics*, *204*, 125–133.

Debye, P. (1926). Bemerkung zu einigen neuen versuchen über einen magneto-elektrischen richteffekt. *Zeitschrift für Physik*, *36*, 300–301.

Dzyaloshinskii, I. E. (1959). On the magneto-electrical effect in antiferromagnets. *Soviet Physics Jetp-Ussr*, *37*, 881–882.

Eerenstein, W., Mathur, N. D., & Scott, J. F. (2006). Multiferroic and magnetoelectric materials. *Nature*, *442*(7104), 759–765.

Eerenstein, W., Wiora, M., Prieto, J. L., Scott, J. F., & Mathur, N. D. (2007). Giant sharp and persistent converse magnetoelectric effects in multiferroic epitaxial heterostructures. *Nature Materials*, *6*(5), 348–351.

Etier, M., Shvartsman, V. V., Gao, Y., Landers, J., Wende, H., & Lupascu, D. C. (2013). Magnetoelectric effect in (0–3) CoFe2O4-BaTiO3 (20/80) composite ceramics prepared by the organosol route. *Ferroelectrics*, *448*, 77–85.

Feyel, F., & Chaboche, J.-L. (2000). FE^2 multiscale approach for modelling the elastoviscoplastic behavior of long fibre SiC/Ti composite materials. *Computer Methods in Applied Mechanics and Engineering*, *183*, 309–330.

Fiebig, M. (2005). Revival of the magnetoelectric effect. *Journal of Physics D: Applied Physics*, *38*, R123–R152.

Folen, V. J., Rado, G. T., & Stalder, E. W. (1961). Anisotropy of the magnetoelectric effect in Cr_2O_3. *Physical Review Letters*, *6*, 607–608.

Fuller, R. B. (1965). Geodesic Structures.

Harshé, G., Dougherty, J. P., & Newnham, R. E. (1993a). Theoretical modeling of multilayer magnetoelectric composites. *International Journal of Applied Electromagnetics in Materials*, *4*, 145–159.

Harshé, G., Dougherty, J. P., & Newnham, R. E. (1993b). Theoretical modeling of 3–0/0-3 magnetoelectric composites. *International Journal of Applied Electromagnetics in Materials*, *4*, 161–171.

Hill, N. A. (2000). Why are there so few magnetic ferroelectrics? *Journal of Physical Chemistry B*, *104*, 6694–6709.

Hill, R. (1963). Elastic properties of reinforced solids - some theoretical principles. *Journal of the Mechanics and Physics of Solids*, *11*, 357–372.

Howes, B., Pelizzone, M., Fischer, P., Tabares-Munoz, C., Rivera, J.-P., & Schmid, H. (1984). Characterization of some magnetic and magnetoelectric properties of ferroelectric $Pb(Fe_{0.5}Nb_{0.5})O_3$. *Ferroelectrics*, *54*, 317–320.

Huang, J. H., & Kuo, W.-S. (1997). The analysis of piezoelectric/piezomagnetic composite materials containing ellipsoidal inclusions. *Journal of Applied Physics*, *81*, 1378–1386.

Hwang, S. C., Lynch, C. S., & McMeeking, R. M. (1995). Ferroelectric/ferroelastic interaction and a polarization switching model. *Acta Metallurgica et Materialia*, *43*, 2073–2084.

Javili, A., Chatzigeorgiou, G., & Steinmann, P. (2013). Computational homogenization in magneto-mechanics. *International Journal of Solids and Structures*, *50*, 4197–4216.

Jayachandran, K. P., Guedes, J. M., & Rodrigues, H. C. (2013). A generic homogenization model for magnetoelectric multiferroics. *Journal of Intelligent Material Systems and Structures*.

Jin, K., & Aboudi, J. (2015). Macroscopic behavior prediction of multiferroic composites. *International Journal of Engineering Science*, *94*, 226–241.

Keip, M.-A. (2012). *Modeling of electro-mechanically coupled materials on multiple scales*. Ph.D. thesis, University of Duisburg-Essen.

Keip, M.-A., & Rambausek, M. (2016). A multiscale approach to the computational characterization of magnetotheological elastomers. *International Journal For Numerical Methods In Engineering*, *107*, 338–360.

Keip, M.-A., & Rambausek, M. (2017). Computational and analytical investigation of shape effects in the experimental characterization of magnetorheological elastomers. *International Journal of Solids and Structures*.

Keip, M.-A., Steinmann, P., & Schröder, J. (2014). Two-scale computational homogenization of electro-elasticity at finite strains. *Computer Methods in Applied Mechanics and Engineering*, *278*, 62–79. ISSN 0045-7825.

Khomskii, D. (2009). Classifying multiferroics: Mechanisms and effects. *Physics*, *2*, 20.

Kouznetsova, V., Brekelmans, W. A. M., & Baaijens, F. P. T. (2001). An approach to micro-macro modeling of heterogeneous materials. *Computational Mechanics*, *27*, 37–48.

Kouznetsova, V., Geers, M. G. D., & Brekelmans, W. A. M. (2002). Multi-scale constitutive modelling of heterogeneous materials with a gradient-enhanced computational homogenization scheme. *International Journal for Numerical Methods in Engineering*, *54*(8), 1235–1260.

Kuo, H.-Y., & Wang, Y.-L. (2012). Optimization of magnetoelectricity in multiferroic fibrous composites. *Mechanics of Materials*, *50*, 88–99.

Kurzhöfer, I., Schröder, J., & Romanowski, H. (2005). Simulation of polycrystalline ferroelectrics based on discrete orientation distribution functions. *Proceedings in Applied Mathematics and Mechanics*, *5*, 307–308.

Labusch, M., Etier, M., Lupascu, D. C., Schröder, J., & Keip, M.-A. (2014). Product properties of a two-phase magneto-electric composite: Synthesis and numerical modeling. *Computational Mechanics*, *54*, 71–83.

Landau, L. D., & Lifshitz, E. M. (1960). *Electrodynamics of continuous media*. Oxford: Pergamon Press.

Lee, J. S., Boyd, J. G., & Lagoudas, D. C. (2005). Effective properties of three-phase electro-magneto-elastic composites. *International Journal Of Engineering Science*, *43*(10), 790–825.

Mandel, J., & Dantu, P. (1963). Conribution à l'étude théorique et expérimentale du coefficient d'élasticité d'un milieu hétérogène mais statistiquement homogène. *Annales des Ponts et Chaussées*.

Martin, L. W., Chu, Y.-H., & Ramesh, R. (2010). Advances in the growth and characterization of magnetic, ferroelectric, and multiferroic oxide thin films. *Materials Science and Engineering: R: Reports*, *68*(46), 89–133. ISSN 0927-796X.

Maugin, G. A., Pouget, J., Drouot, R., & Collet, B. (1992). *Nonlinear electromechanical couplings*. New York: Wiley.

Miehe, C. (2003). Computational micro-to-macro transitions for discretized micro-structures of heterogeneous materials at finite strains based on the minimization of averaged incremental energy. *Computer Methods in Applied Mechanics and Engineering*, *192*, 559–591.

Miehe, C., & Bayreuther, C. G. (2007). On mutiscale FE analyses of heterogeneous structures: from homogenization to multigrid solvers. *International Journal for Numerical Methods in Engineering*, *71*, 1135–1180.

Miehe, C., & Koch, A. (2002). Computational micro-to-macro transitions of discretized microstructures undergoing small strains. *Archive of Applied Mechanics*, *72*, 300–317.
Miehe, C., Schotte, J., & Schröder, J. (1999a). Computational micro-macro transitions and overall moduli in the analysis of polycrystals at large strains. *Computational Materials Science*, *16*(1–4), 372–382.
Miehe, C., Schröder, J., & Schotte, J. (1999b). Computational homogenization analysis in finite plasticity. Simulation of texture development in polycrystalline materials. *Computer Methods in Applied Mechanics and Engineering*, *171*, 387–418.
Miehe, C., Rosato, D., & Kiefer, B. (2011). Variational principles in dissipative electro-magneto-mechanics: A framework for the macro-modeling of functional materials. *International Journal For Numerical Methods In Engineering*, *86*, 1225–1276.
Nan, C.-W. (1994). Magnetoelectric effect in composites of piezoelectric and piezomagnetic phases. *Physical Review B*, *50*, 6082–6088.
Nan, C.-W., Liu, G., Lin, Y., & Chen, H. (2005). Magnetic-field-induced electric polarization in multiferroic nanostructures. *Physical Review Letters*, *94*(19), 197203. 1–4 May 2005.
Nan, C.-W., Bichurin, M. I., Dong, S., Viehland, D., & Srinivasan, G. (2008). Multiferroic magnetoelectric composites: Historical perspective, status, and future directions. *Journal of Applied Physics*, *103*(3), 031101.
O'Dell, T. H. (1967). An induced magneto-electric effect in yttrium iron garnet. *Philosophical Magazine*, *16*, 487–494.
Özdemir, I., Brekelmans, W. A. M., & Geers, M. G. D. (2008). Computational homogenization for heat conduction in heterogeneous solids. *International Journal for Numerical Methods in Engineering*, *73*, 185–204.
Perić, D., de Souza Neto, E. A., Feijóo, R. A., Partovi, M., & Carneiro Molina, A. J. (2011). On micro-to-macro transitions for multi-scale analysis of non-linear heterogeneous materials. *International Journal for Numerical Methods in Engineering*, *87*, 149–170.
Priya, S., Islam, R., Dong, S. X., & Viehland, D. (2007). Recent advancements in magnetoelectric particulate and laminate composites. *Journal of Electroceramics*, *19*(1), 147–164.
Rado, G. T. (1964). Observation and possible mechanisms of magnetoelectric effects in a ferromagnet. *Physical Review Letters*, *13*, 335–337.
Rado, G. T., & Folen, V. J. (1961). Observation of the magentically induced magnetoelectric effect and evidence for antiferromagnetic domains. *Physical Review Letters*, *7*, 310–311.
Rado, G. T., Ferrari, J. M., & Maisch, W. G. (1984). Magnetoelectric susceptibility and magnetic symmetry of magnetoelectrically annealed $TbPO_4$. *Physical Review B*, *29*, 4041–4048.
Ramesh, R., & Spaldin, N. A. (2007). Multiferroics: Progress and prospects in thin films. *Nature Materials*, *6*(1), 21–29. ISSN 1476-1122.
Rivera, J.-P. (1994a). On definitions, units, measurements, tensor forms of the linear magnetoelectric effect and on a new dynamic method applied to Cr-Cl boracite. *Ferroelectrics*, *161*, 165–180.
Rivera, J.-P. (1994b). The linear magnetoelectric effect in $LiCoPO_4$ revisited. *Ferroelectrics*, *161*, 147–164.
Rivera, J.-P., & Schmid, H. (1997). On the birefringence of magnetoelectric $BiFeO_3$. *Ferroelectrics*, *204*, 23–33.
Röntgen, W. C. (1888). Ueber die durch bewegung eines im homogenen electrischen felde befindlichen dielectricums hervorgerufene electrodynamische kraft. *Annalen der Physik*, *271*, 264–270.
Ryu, J., Priya, S., Uchino, K., & Kim, H. E. (2002). Magnetoelectric effect in composites of magnetostrictive and piezoelectric materials. *Journal of Electroceramics*, *8*, 107–119.
Schmid, H. (1994). Multi-ferroic magnetoelectrics. *Ferroelectrics*, *162*, 317–338.
Schröder, J. (2009). Derivation of the localization and homogenization conditions for electro-mechanically coupled problems. *Computational Materials Science*, *46*(3), 595–599.
Schröder, J. (2014). A numerical two-scale homogenization scheme: The FE^2–method. In J. Schröder, & K. Hackl (Eds.), *Plasticity and Beyond* (Vol. 550, pp. 1–64). CISM Courses and Lectures. Berlin: Springer.

Schröder, J., & Gross, D. (2004). Invariant formulation of the electromechanical enthalpy function of transversely isotropic piezoelectric materials. *Archive of Applied Mechanics, 73*, 533–552.

Schröder, J., & Keip, M.-A. (2012). Two-scale homogenization of electromechanically coupled boundary value problems. *Computational Mechanics, 50*, 229–244.

Schröder, J., Labusch, M., & Keip, M.-A. (2016). Algorithmic two-scale transition for magneto-electro-mechanically coupled problems - FE^2-scheme: Localization and homogenization. *Computer Methods in Applied Mechanics and Engineering, 302*, 253–280.

Shi, Y., & Gao, Y. (2014). A nonlinear magnetoelectric model for magnetoelectric layered composite with coupling stress. *Journal of Magnetism and Magnetic Materials, 360*, 131–136.

Shtrikman, S., & Treves, D. (1963). Observation of the magnetoelectric effect in Cr_2O_3 powders. *Physical Review, 130*, 986–988.

Shvartsman, V. V., Alawneh, F., Borisov, P., Kozodaev, D., & Lupascu, D. C. (2011). Converse magnetoelectric effect in $CoFe_2O_4$-$BaTiO_3$ composites with a core-shell structure. *Smart Materials and Structures, 20*.

Smit, R. J. M., Brekelmans, W. A. M., & Meijer, H. E. H. (1998). Prediction of the mechanical behavior of nonlinear heterogeneous systems by multi-level finite element modeling. *Computer Methods in Applied Mechanics and Engineering, 155*, 181–192.

Spaldin, N. A., & Fiebig, M. (2005). The renaissance of magnetoelectric multiferroics. *Science, 309*, 391–392.

Srinivas, S., Li, J. Y., Zhou, Y. C., & Soh, A. K. (2006). The effective magnetoelectroelastic moduli of matrix-based multiferroic composites. *Journal of Applied Physics, 99*(4), 043905.

Srinivasan, G. (2010). Magnetoelectric composites. *Annual Review of Materials Research, 40*, 1–26.

Sundar, V., & Newnham, R. E. (1992). Electrostriction and polarization. *Ferroelectrics, 135*, 431–446.

Temizer, I., & Wriggers, P. (2008). On the computation of the macroscopic tangent for multiscale volumetric homogenization problems. *Computer Methods in Applied Mechanics and Engineering, 198*, 495–510.

Terada, K., Hori, M., Kyoya, T., & Kikuchi, N. (2000). Simulation of the multi-scale convergence in computational homogenization approach. *International Journal of Solids and Structures, 37*, 2285–2311.

Terada, K., Saiki, I., Matsui, K., & Yamakawa, Y. (2003). Two-scale kinematics and linearization for simultaneous two-scale analysis of periodic heterogeneous solids at finite strain. *Computer Methods in Applied Mechanics and Engineering, 192*(31–32), 3531–3563.

van der Sluis, O., Schreurs, P. J. G., Brekelmans, W. A. M., & Meijer, H. E. H. (2000). Overall behavior of heterogeneous elastoviscoplastic materials: effect of microstructural modelling. *Mechanics of Materials, 32*, 449–462.

van Suchtelen, J. (1972). Product properties: A new application of composite materials. *Philips Research Reports, 27*, 28–37.

Vaz, C. A. F., Hoffman, J., Ahn, C. H., & Ramesh, R. (2010). Magnetoelectric coupling effects in multiferroic complex oxide composite structures. *Advanced Materials, 22*, 2900–2918.

Wang, Y., Hu, J., Lin, Y., & Nan, C.-W. (2010). Multiferroic magnetoelectric composite nanostructures. *NPG asia materials, 2*(2), 61–68.

Watanabe, T., & Kohn, K. (1989). Magnetoelectric effect and low temperature transition of $PbFe_{0.5}Nb_{0.5}O_3$ single crystal. *Phase Transitions, 15*, 57–68.

Wilson, H. A. (1905). On the electric effect of rotating a dielectric in a magnetic field. *Philosophical Transactions of the Royal Society of London, 204*, 121–137.

Ye, Z.-G., Rivera, J.-P., Schmid, H., Haida, M., & Kohn, K. (1994). Magnetoelectric effect and magnetic torque of chromium chlorine boracite $Cr_3B_7O_{13}Cl$. *Ferroelectrics, 161*, 99–110.

Zgonik, M., Bernasconi, P., Duelli, M., Schlesser, R., & Günter, P. (1994). Dielectric, elastic, piezoelectric, electro-optic, and elasto-optic tensors of $BaTiO_3$ crystals. *Physical Review B, 50*(9), 5941–5949.

Zohdi, T. (2012). *Electromagnetic properties of multiphase dielectrics - a primer on modeling, theory and computation*. Berlin: Springer.

Zohdi, T. I. (2010). Simulation of coupled microscale multiphysical-fields in particulate-doped dielectrics with staggered adaptive fdtd. *Computer Methods in Applied Mechanics and Engineering*, *199*, 3250–3269.

Multiscale Modeling of Electroactive Polymer Composites

Marc-André Keip and Jörg Schröder

Abstract Electroactive polymer composites are materials that consist of an elastomeric matrix and dispersed high-dielectric-modulus or metallic inclusions. The addition of the inclusions generally leads to a significant enhancement of the electrostatic actuation or, more generally, of the overall electro-mechanical coupling. This enhancement is mainly due to the contrast of dielectric moduli of the individual phases, which induces fluctuations of the electric field in the matrix material. The present contribution aims at the derivation and implementation of a multiscale homogenization framework for the macroscopic simulation of electroactive polymer composites with explicit consideration of their microscopic structure. This is achieved through the development of a two-scale computational homogenization approach for electro-mechanically coupled solids at finite deformations. The microscopic part of the problem is defined on a representative volume element that is attached at each integration point of the macroscopic domain. In order to derive energetically consistent transition conditions between the scales a generalized form of the Hill-Mandel condition extended to electro-elastic phenomena at large deformations is exploited. An efficient solution of the macroscopic boundary value problem is guaranteed by means of an algorithmically consistent tangent. The method is applied to the simulation of different dielectric polymer-ceramic composites, which are analzyed with regard to their effective actuation properties. In addition to that, an example of a multiscale electro-mechanical actuator at large deformations is presented.

M.-A. Keip (✉)
Chair of Material Theory, Institute of Applied Mechanics, University of Stuttgart, Pfaffenwaldring 7, 70569 Stuttgart, Germany
e-mail: marc-andre.keip@mechbau.uni-stuttgart.de

J. Schröder
Department of Civil Engineering, Faculty of Engineering, Institute of Mechanics, University of Duisburg-Essen, Universitätsstraße 15, 45141 Essen, Germany

J. Schröder and Doru C. Lupascu (eds.), *Ferroic Functional Materials*,
CISM International Centre for Mechanical Sciences 581,
https://doi.org/10.1007/978-3-319-68883-1_6

1 Introduction

Electroactive polymers (EAP) are dielectric materials that show mechanical deformations under applied electric fields. While essentially all dielectric materials have this behavior in common, elastomeric EAPs obtain their pronounced actuation properties from their *softness*. Most roughly, electro-mechanically coupled solids can be subdivided into two main classes: those of the hard and those of the soft type. While hard materials (such as piezo- and ferroelectrics) show small deformations and high actuation forces, soft materials (such as dielectric and ionic EAPs) show large deformations and low actuation forces. The large electro-mechanically induced deformations of the latter make them attractive for the development of smart devices in the field of, for example, artificial muscles and haptic displays (Pelrine et al. 2000; Bar-Cohen 2001; Kim and Tadokoro 2007; Kovacs et al. 2007; Carpi et al. 2011).

The present contribution focusses on the continuum-mechanical modeling of *soft dielectric EAPs* and covers ingredients of their continuum-mechanical description and aspects of multiscale numerical treatments (see also Keip et al. 2014). The electro-mechanical theory employed in this paper traces back to the fundmental works on finite electro-elasticity provided by Toupin (1956), Eringen (1963), Maugin (1988), Eringen and Maugin (1990), Kovetz (2000) as well as more recent approaches to material modeling documented in Bhattacharya et al. (2001), McMeeking and Landis (2005), Dorfmann and Ogden (2005), Dorfmann and Ogden (2006), Goulbourne et al. (2007), Suo et al. (2008), Bustamante et al. (2009), Cohen (2014), Zäh and Miehe (2015), Cohen et al. (2016). For numerical implementations related to the current work we refer to, for example, Vu et al. (2007), Vu and Steinmann (2007), Skatulla et al. (2009), Vu and Steinmann (2010), Steinmann (2011), Skatulla et al. (2012), Ask et al. (2012), Klinkel et al. (2013) and Pelteret et al. (2016).

In that respect, the present contribution aims at providing a *computational multiscale view on soft dielectric EAP composites*. Our motivation for studying soft EAP composites is rooted in experimental investigations that revealed that the actuation performance of dielectrics can drastically be enhanced through the design of composite structures. In fact, the addition of high-electric-permittivity (high-ϵ) particles into the dielectric matrix gives a strong enhancement of the electro-mechanical response (Zhang et al. 2002; Huang et al. 2003, 2004, 2005; Li et al. 2004; Carpi and Rossi 2005; Zhang et al. 2007). This phenomenon can roughly be explained by the interplay between the elastomeric matrix and the inclusions. Here one notes two distinct contributions: First, the addition of high-ϵ particles increases the overall electric permittivity of the composite (Zhang et al. 2002). Second, the high contrast between the composite's phases gives rise to electric-field fluctuations (Tian et al. 2012). A third driving mechanism can be activated by the embedding metal particles: Then, the forces between the free-charge carriers that reside in the metal, are able to further enhance the coupling properties (Lopez-Pamies 2014; Lopez-Pamies et al. 2014).

Such multiscale phenomena are in the focus of homogenization theory, both in the analytical and computational area. For analytical approaches we refer to DeBotton et al. (2007), Ponte Castañeda and Siboni (2012), Cao and Zhao (2013), Siboni and Ponte Castañeda (2013), Rudykh et al. (2013), Siboni and Ponte Castañeda (2014), Lefèvre and Lopez-Pamies (2017). Computational methods are given, for example, by Müller et al. (2010), Klassen et al. (2012).

Whenever it comes to the action of large electric actuation fields inducing large mechanical displacements, instability phenomena may arise. In this context, we can distinguish two important effects associated with electro-mechanical stability. First, instabilities may occur when electrostatic forces can no longer be compensated by mechanical forces. Second, there is the danger of electric breakdown. While the latter is of course possible in any electrically charged dielectric the former is of particular relevance in soft dielectrics. Phenomena associated with stability are highly nonlinear and may arise on different length-scales (Bertoldi and Gei 2011; Li and Landis 2012; Rudykh et al. 2014; Siboni et al. 2015). For the case of homogeneous materials stability issues are addressed, for example, in Zhao and Suo (2007), Xu et al. (2010), Xu et al. (2012) for static and dynamic conditions. Experimental observations are documented in, for example, Plante and Dubowsky (2006) and Zhao and Wang (2014). Recent computational investigations of stability issues can be found in the works Miehe et al. (2015a), Miehe et al. (2015b), Miehe et al. (2016) and Goshkoderia and Rudykh (2017). At this point we note that the investigation of stability phenomena is not in the focus of the present contribution. So we instead refer the interestest reader to the above literature and the works cited therein.

The aim of the present contribution is to provide a multi-scale computational homogenization framework for electro-elastic boundary value problems at large deformations, which can be applied to the simulation, characterization and optimization of soft electro-elastic bodies. The framework is rooted in the FE^2-method, which solves a macroscopic boundary value problem in consideration of microscopic representative volume elements (RVE). This method has a long tradition in mechanics (Smit et al. 1998; Miehe et al. 1999a, b; Michel et al. 1999; Schröder 2000; Terada and Kikuchi 2001; Miehe and Koch 2002; Kouznetsova et al. 2002; Terada et al. 2003; Markovic et al. 2005; Somer et al. 2009; Schröder 2014) and has recently been extended to physically coupled materials like thermo-elastic solids (Özdemir et al. 2008a, b) electro-elastic solids (Schröder 2009; Schröder and Keip 2010; 2012; Keip et al. 2014) and magneto-elastic solids (Javili et al. 2013; Keip and Rambausek 2016, 2017). Furthermore, its application to the analysis of magneto-electric multiferroics was shown in Labusch et al. (2014), Schröder et al. (2015), Schröder et al. (2016). At this point it is also worth to mention the recent approaches to the computational homogenization of gradient-extended models in the area of electro- and magneto-mechanics by Zäh and Miehe (2013), Keip et al. (2015), Sridhar et al. (2016).

The outline of the present contribution is as follows. In Sect. 2 we provide the fundamental kinematical relations and balance equations of finite electro-elasto-statics. Section 3 then outlines a first-order two-scale homogenization approach of finite electro-elasto-statics. Here, the emphasis will be on the formulation of the coupled BVPs at macro- and micro-scale as well as on the related transitions between these scales. We also provide an expression of the macroscopic tangent moduli. Chapter 4 shows numerical examples with applications in the area of polymer-ceramic composites. In detail, the effective electro-elastic coupling of selected representative micro-structures will be analyzed. Furthermore, we document the functionality of the method in a full two-scale simulation of an electric actuator. Finally, we close the contribution with a short summary and a conclusion in Chap. 5.

2 Governing Equations of Electro-Elasto-Statics at Finite Strains

In this section, we briefly summarize the fundamental equations of the continuum theory of electro-elasto-statics at finite strains. For a more detailed discussion we would like to refer to Dorfmann and Ogden (2005), Dorfmann and Ogden (2006) or Vu et al. (2007).

In the undeformed (reference) configuration the electro-elastic body of interest is denoted by $\mathcal{B} \subset \mathbb{R}^3$ and parameterized in the referential coordinates $\boldsymbol{X}$. In its deformed configuration the same body is denoted by $\mathcal{S} \subset \mathbb{R}^3$ and parameterized in the current coordinates $\boldsymbol{x}$.

Under the classical assumptions of electro-statics (absence of magnetic fields and free currents) the electric field in the current configuration is governed by Faraday's law of electro-statics

$$\operatorname{curl} \boldsymbol{e} = \boldsymbol{0}. \tag{1}$$

In order to automatically fulfill this equation, the electric field is expressed as the gradient of some scalar electric potential ϕ with respect to the current coordinates as

$$\boldsymbol{e} := -\operatorname{grad} \phi \tag{2}$$

Furthermore, we introduce the electric displacement in the current configuration $\boldsymbol{d}$ as

$$\boldsymbol{d} = \epsilon_0 \boldsymbol{e} + \boldsymbol{p}, \tag{3}$$

in which ϵ_0 is the permittivity of free space and $\boldsymbol{p}$ is the polarization. Based on that, the electric balance equation is given by Gauss's law

$$\operatorname{div} \boldsymbol{d} = \rho, \tag{4}$$

where ρ is the density of free electric charge carriers. In what follows, we will always assume that $\rho = 0$.

When we want to write down the mechanical balance equation we have to take into account that the electric field interacts with the considered body. According to, for example, Pao (1978) such electro-static force can be accounted for through the introduction of a electro-static body force $\boldsymbol{f}^{\text{elec}}$ given by

$$f^{\text{elec}} = \text{grad } \boldsymbol{e} \cdot \boldsymbol{p}. \tag{5}$$

Based on that, we give the balance of linear momentum in the absence of mechanical body forces in the current configuration as

$$\text{div } \boldsymbol{\sigma}^{\text{mech}} + f^{\text{elec}} = \mathbf{0}, \tag{6}$$

in which $\boldsymbol{\sigma}^{\text{mech}}$ denotes the mechanical part of the Cauchy stress tensor. In the latter equation, we are able to express the electro-static body force by the help of the second-order Maxwell stress tensor

$$\boldsymbol{\sigma}^{\text{elec}} = \boldsymbol{e} \otimes \boldsymbol{d} - \frac{1}{2}\epsilon_0(\boldsymbol{e} \cdot \boldsymbol{e})\mathbf{1} \tag{7}$$

through

$$f^{\text{elec}} = \text{div } \boldsymbol{\sigma}^{\text{elec}}. \tag{8}$$

Above, $\mathbf{1}$ denotes the second-order identity tensor. Based on that, the balance of linear momentum can be reformulated to

$$\text{div } \boldsymbol{\sigma} = \mathbf{0} \quad \text{with} \quad \boldsymbol{\sigma} = \boldsymbol{\sigma}^{\text{mech}} + \boldsymbol{\sigma}^{\text{elec}}. \tag{9}$$

Here, we have introduced the total Cauchy stress tensor $\boldsymbol{\sigma}$ as the sum of the mechanical Cauchy stress $\boldsymbol{\sigma}^{\text{mech}}$ and the electric Maxwell stress $\boldsymbol{\sigma}^{\text{elec}}$.

Now that we have assembled all fundamental quantities in their natural (current) configuration, we formulate associated expressions with respect to the reference configuration. In order to do so, we use the deformation map $\varphi : \mathcal{B} \to \mathcal{S}$ which maps points $\boldsymbol{X} \in \mathcal{B}$ onto points $\boldsymbol{x} \in \mathcal{S}$. The deformation gradient $\boldsymbol{F}$ can be defined in a classical way by

$$\boldsymbol{F} = \text{Grad } \varphi \tag{10}$$

with its determinant $J := \det \boldsymbol{F} > 0$. Based on that, we are able to define the electric field with respect to the reference configuration $\boldsymbol{E}$ and the right Cauchy-Green tensor $\boldsymbol{C}$ by

$$\boldsymbol{E} = \boldsymbol{F}^T \cdot \boldsymbol{e} \quad \text{and} \quad \boldsymbol{C} = \boldsymbol{F}^T \cdot \boldsymbol{F}. \tag{11}$$

Following Dorfmann and Ogden (2005) we now assume the existence of a potential Ψ per unit reference volume, from which the total stresses and the electric displacements can be calculated. In order to obtain constitutive equations that satisfy the principle of material objectivity a priori, we postulate the functional dependence $\Psi := \hat{\Psi}(\boldsymbol{E}, \boldsymbol{C})$. The function Ψ denotes the total electric Gibbs energy,[1] from which

[1]Formally, the electric Gibbs energy is obtained by a Legendre transform of the internal energy $\mathcal{E} := \hat{\mathcal{E}}(\boldsymbol{D}, \boldsymbol{C})$ via $\Psi := \mathcal{E} - \boldsymbol{E} \cdot \boldsymbol{D}$. Note that in an isothermal process, the electric Gibbs energy is equivalent to the electric enthalpy, which is frequently introduced in electro-mechanical problems.

we can compute the total first Piola-Kirchhoff stresses and the electric displacement associated with the reference configuration as

$$P = 2F \cdot \frac{\partial \Psi}{\partial C} \quad \text{and} \quad D = -\frac{\partial \Psi}{\partial E}. \tag{12}$$

Based on that, the quantities with respect to the current configuration follow through the classical push-forward operations

$$\sigma = \frac{1}{J} P \cdot F^T \quad \text{and} \quad d = \frac{1}{J} F \cdot D. \tag{13}$$

3 Electro-Elasto-Static Boundary Value Problems on the Macro- and the Micro-scale

Within this contribution, we want to describe macroscopic bodies that have a periodic heterogeneous micro-structure. In this context, we assume that the typical length scales of the periodic micro-stucture is very much smaller than the typical dimensions of the macroscopic body. In other words, we assume perfect scale separation of the macro- and the micro-scale. From this follows that microsopic volumes degenerate to points at the macroscale, and macroscopic quantities can be assumed constant at micro-level. These classical assumptions lay the ground for the application of the well-established FE^2-method (see refences in the introduction).

In the FE^2-method BVPs are solved on two separated scales, in the following called the macroscopic and the microscopic scale. Both BVPs are connected through homogenization and localization conditions, which will be outlined at a later stage of this contribution. At this point, we want to start with the introduction of the coupled BVPs at the macro- and the micro-scale. Both are based on the fundamental equations provided in Sect. 2.

The procedures introduced below closely follow the ideas of computational homogenization for mechanical problems provided by Miehe and coworkers, see for example Miehe et al. (1999a, b) and Miehe and Koch (2002) For a recent review article we refer to Schröder (2014). Previous extensions of the FE^2-method to electro-elastic problems are given by Schröder (2009), Schröder and Keip (2012), Keip et al. (2014).

3.1 Boundary Value Problem on the Macroscopic Scale

In analogy to the basic equations given in Sect. 2, we denote the body of interest on the macroscopic scale as[2] $\overline{\mathcal{B}} \subset \mathbb{R}^3$ and parameterize it in the coordinates $\overline{X}$.

[2]In the following, all quantities associated with the macro-scale will be labelled with an overline.

Furthermore, the macroscopic deformation map and electric potential are written as $\overline{\varphi}$ and $\overline{\phi}$, respectively. Based on that, we define the macroscopic deformation gradient and electric field as

$$\overline{\boldsymbol{F}} := \overline{\mathrm{Grad}}\,\overline{\varphi} \quad \text{and} \quad \overline{\boldsymbol{E}} := -\overline{\mathrm{Grad}}\,\overline{\phi} \quad \text{in } \overline{\mathcal{B}}. \tag{14}$$

The balance laws on the macroscopic scale consequently appear as

$$\overline{\mathrm{Div}}\,\overline{\boldsymbol{P}} = \boldsymbol{0} \quad \text{and} \quad \overline{\mathrm{Div}}\,\overline{\boldsymbol{D}} = 0 \quad \text{in } \overline{\mathcal{B}}, \tag{15}$$

We prescribe the Dirichlet-type boundary conditions

$$\overline{\varphi} = \overline{\varphi}_b \quad \text{on } \partial\overline{\mathcal{B}}_{\overline{\varphi}} \qquad \text{and} \qquad \overline{\phi} = \overline{\phi}_b \quad \text{on } \partial\overline{\mathcal{B}}_{\overline{\phi}} \tag{16}$$

as well as boundary condition of the Neumann-type

$$\overline{\boldsymbol{t}} = [\![\overline{\boldsymbol{P}}]\!] \cdot \overline{\boldsymbol{N}} \quad \text{on } \partial\overline{\mathcal{B}}_{\overline{t}} \qquad \text{and} \qquad -\overline{Q} = [\![\overline{\boldsymbol{D}}]\!] \cdot \overline{\boldsymbol{N}} \quad \text{on } \partial\overline{\mathcal{B}}_{\overline{Q}}. \tag{17}$$

where $\overline{\boldsymbol{t}}$ are mechanical surface tractions and $\overline{Q}$ are electric surface charges. The vector $\overline{\boldsymbol{N}}$ is a unit normal vector pointing outwards from the body. In the above relations it is crucial to note that all Neumann boundary conditions are formulated in terms of the jumps $[\![\bullet]\!] := \bullet_{\text{inside}} - \bullet_{\text{outside}}$ across the boundary of the body. While such definition is rather usual in electro-static problems, the jump condition in the mechanical part may appear unusual on the first view. It however originates from the fact that even in the free space (vacuum) there exists a Maxwell stress

$$\overline{\boldsymbol{\sigma}}^{\text{elec}}_{\text{vacuum}} = \epsilon_0[\overline{\boldsymbol{e}} \otimes \overline{\boldsymbol{e}} - \frac{1}{2}(\overline{\boldsymbol{e}} \cdot \overline{\boldsymbol{e}})\boldsymbol{1}], \tag{18}$$

see, e.g., Steinmann (2011), Pelteret et al. (2016) for further insights.

3.2 *Definition of Macroscopic Quantities via Homogenization*

In the above equations, we postulate that the macro-fields $\{\overline{\boldsymbol{F}}, \overline{\boldsymbol{P}}, \overline{\boldsymbol{E}}, \overline{\boldsymbol{D}}\}$ can be computed as volume averages along the boundary of a microscopic volume element (RVE).

In its reference state the RVE is described on a domain $\mathcal{B} \subset \mathbb{R}^3$ with boundary $\partial\mathcal{B}$ and parameterized in the microscopic coordinates[3] $\boldsymbol{X}$. Then, the volume average of the surface integral of a quantity $\bullet$ is defined by

[3]All quantities associated with the mirco-scale will have no special labeling.

$$\langle \bullet \rangle := \frac{1}{V} \int_{\partial \mathcal{B}} \bullet \, \mathrm{d}A, \tag{19}$$

where $V = \int_{\mathcal{B}} \mathrm{d}V$ is the volume of the RVE. Then, the macroscopic electro-mechanical fields are defined by

$$\overline{\boldsymbol{F}} := \langle \boldsymbol{\varphi} \otimes \boldsymbol{N} \rangle, \quad \overline{\boldsymbol{E}} := -\langle \phi \boldsymbol{N} \rangle, \quad \overline{\boldsymbol{P}} := \langle \boldsymbol{t} \otimes \boldsymbol{X} \rangle, \quad \overline{\boldsymbol{D}} := -\langle Q \boldsymbol{X} \rangle, \tag{20}$$

where $\boldsymbol{N}$ is a unit normal vector pointing outwards from the RVE.

By assuming continuity of the displacements and the electric potential across the RVE as well as by neglecting mechanical body forces and free electric charges on the RVE, we are able to transform the above surface integrals into volume integrals

$$\langle\langle \bullet \rangle\rangle := \frac{1}{V} \int_{\mathcal{B}} \bullet \, \mathrm{d}V. \tag{21}$$

Thus, we can alternatively define the macroscopic electro-mechanical fields through

$$\overline{\boldsymbol{F}} := \langle\langle \boldsymbol{F} \rangle\rangle, \quad \overline{\boldsymbol{E}} := \langle\langle \boldsymbol{E} \rangle\rangle, \quad \overline{\boldsymbol{P}} := \langle\langle \boldsymbol{P} \rangle\rangle, \quad \overline{\boldsymbol{D}} := \langle\langle \boldsymbol{D} \rangle\rangle. \tag{22}$$

For a detailed derivation of small-strain analogues see Keip (2012).

3.3 Boundary Value Problem on the Microscopic Scale

Above, we have introduced the microscopic deformation gradient $\boldsymbol{F}$ and the microscopic electric field $\boldsymbol{E}$, which we can define in analogy to (14) via

$$\boldsymbol{F} := \operatorname{Grad} \boldsymbol{\varphi} \quad \text{and} \quad \boldsymbol{E} := -\operatorname{Grad} \phi. \tag{23}$$

Furthermore, under the assumptions made above (no mechanical body forces and no free electric charges on the micro-level), the balance equations can be written down as

$$\operatorname{Div} \boldsymbol{P} = \boldsymbol{0} \quad \text{and} \quad \operatorname{Div} \boldsymbol{D} = 0. \tag{24}$$

In view of the considered scale separation we assume that the above microscopic fields $\{\boldsymbol{F}, \boldsymbol{P}, \boldsymbol{E}, \boldsymbol{D}\} =: \Xi$ decompose additively into

$$\Xi := \overline{\Xi} + \widetilde{\Xi} \quad \text{with} \quad \langle\langle \widetilde{\Xi} \rangle\rangle = \boldsymbol{0}. \tag{25}$$

Correspondingly, the primary fields decompose according to

$$\boldsymbol{\varphi} = \overline{\boldsymbol{F}} \cdot \boldsymbol{X} + \widetilde{\boldsymbol{\varphi}} \quad \text{and} \quad \phi = -\overline{\boldsymbol{E}} \cdot \boldsymbol{X} + \widetilde{\phi}. \tag{26}$$

The goal is now to derive energetically consistent boundary conditions along $\partial\mathcal{B}$. This can be achieved by the use of a generalized form of the classical Hill-Mandel macro-homogeneity condition (Hill 1963). Applied to the given electro-mechanical problem (Schröder 2009; Schröder and Keip 2012) the Hill condition requires that the macroscopic rate of electric Gibbs energy Ψ must be equal to the averaged rate of microscopic electric Gibbs energy $\overline{\Psi}$, i.e.

$$\dot{\overline{\Psi}} = \langle\langle \dot{\Psi} \rangle\rangle. \tag{27}$$

For $\Psi = \hat{\Psi}(\boldsymbol{F}, \boldsymbol{E})$ this formally gives

$$\overline{\boldsymbol{P}} : \dot{\overline{\boldsymbol{F}}} - \overline{\boldsymbol{D}} \cdot \dot{\overline{\boldsymbol{E}}} = \langle\langle \boldsymbol{P} : \dot{\boldsymbol{F}} - \boldsymbol{D} \cdot \dot{\boldsymbol{E}} \rangle\rangle, \tag{28}$$

which can be reformulated to[4]

$$\underbrace{\langle\langle \boldsymbol{P} : \dot{\boldsymbol{F}} \rangle\rangle - \overline{\boldsymbol{P}} : \dot{\overline{\boldsymbol{F}}}}_{\mathcal{P}^{\mathrm{mech}}} + \underbrace{\overline{\boldsymbol{D}} \cdot \dot{\overline{\boldsymbol{E}}} - \langle\langle \boldsymbol{D} \cdot \dot{\boldsymbol{E}} \rangle\rangle}_{\mathcal{P}^{\mathrm{elec}}} = 0, \tag{29}$$

A rather simple way to fulfill the above condition is given by the constraints $\boldsymbol{P} := \overline{\boldsymbol{P}}$ or $\dot{\boldsymbol{F}} := \dot{\overline{\boldsymbol{F}}}$ and $\boldsymbol{D} := \overline{\boldsymbol{D}}$ or $\dot{\boldsymbol{E}} := \dot{\overline{\boldsymbol{E}}}$. These resemble the famous Sachs-Reuss and Voigt-Taylor bounds, respectively.

We now want to derive some more general boundary conditions for the microstructure. In order to do so, we have to reformulate the Hill condition into a more suitable format. Since the electric and mechanical part of the Hill condition can be treated separately, we perform the reformulations for both parts individually. In order to outline the basic operations, we exemplarily document the transformations of the mechanical part, see Miehe et al. (1999a, b), Miehe and Koch (2002), Schröder (2000) for further details.

First, we note that the deformation gradient is defined as the material gradient of the current coordinates, so that the first volume integral in (29) can be transformed into a surface integral via

$$\langle\langle \boldsymbol{P} : \dot{\boldsymbol{F}} \rangle\rangle = \langle\langle \boldsymbol{P} : \mathrm{Grad}\, \dot{\varphi} \rangle\rangle = \langle\langle \mathrm{Div}\,(\dot{\varphi} \cdot \boldsymbol{P}) - \underbrace{\mathrm{Div}\, \boldsymbol{P}}_{=\boldsymbol{0}} \cdot \dot{\varphi} \rangle\rangle = \langle (\boldsymbol{P} \cdot \boldsymbol{N}) \cdot \dot{\varphi} \rangle, \tag{30}$$

where we consecutively used the product rule, the balance of linear momentum at the micro-scale $(24)_1$, and the divergence theorem. By using the definition of the microscopic displacements and the superposition principle of the first Piola-Kirchhoff stress (25) respectively, we can further reformulate the latter term via

[4] See Schröder (2009) for a detailed discussion on how to fulfill this condition by setting $\mathcal{P}^{\mathrm{mech}} = 0$ and $\mathcal{P}^{\mathrm{elec}} = 0$ individually.

$$\langle (\boldsymbol{P} \cdot \boldsymbol{N}) \cdot \dot{\varphi} \rangle = \overline{\boldsymbol{P}} : \underbrace{\langle \dot{\tilde{\varphi}} \otimes \boldsymbol{N} \rangle}_{=\langle\langle \dot{\tilde{\boldsymbol{F}}} \rangle\rangle = \boldsymbol{0}} + \langle (\widetilde{\boldsymbol{P}} \cdot \boldsymbol{N}) \cdot \dot{\tilde{\varphi}} \rangle + \langle (\boldsymbol{P} \cdot \boldsymbol{N}) \otimes \boldsymbol{X} \rangle : \dot{\overline{\boldsymbol{F}}}. \tag{31}$$

Above, we extracted the macroscopic quantities $\overline{\boldsymbol{P}}$ and $\dot{\overline{\boldsymbol{F}}}$ from the volume integrals and noted that the first term on the right-hand side vanishes by definition $(25)_2$. Furthermore, the last term on the right-hand side can be transformed into

$$\langle (\boldsymbol{P} \cdot \boldsymbol{N}) \otimes \boldsymbol{X} \rangle : \dot{\overline{\boldsymbol{F}}} = \langle \boldsymbol{t} \otimes \boldsymbol{X} \rangle : \dot{\overline{\boldsymbol{F}}} = \overline{\boldsymbol{P}} : \dot{\overline{\boldsymbol{F}}} \tag{32}$$

where we have used $(20)_3$. Now, the latter equation can be combined with (30) and (31) so that it gives

$$\underbrace{\langle\langle \boldsymbol{P} : \dot{\boldsymbol{F}} \rangle\rangle - \overline{\boldsymbol{P}} : \dot{\overline{\boldsymbol{F}}}}_{=\mathcal{P}^{\mathrm{mech}}} = \langle (\widetilde{\boldsymbol{P}} \cdot \boldsymbol{N}) \cdot \dot{\tilde{\varphi}} \rangle. \tag{33}$$

With the requirement that $\mathcal{P}^{\mathrm{mech}} = 0$ in (29) we arrive at the expression

$$\langle (\widetilde{\boldsymbol{P}} \cdot \boldsymbol{N}) \cdot \dot{\tilde{\varphi}} \rangle = 0. \tag{34}$$

By again using the superposition principles noted in (25) and $(26)_1$ we finally obtain

$$\langle (\widetilde{\boldsymbol{P}} \cdot \boldsymbol{N}) \cdot \dot{\tilde{\varphi}} \rangle = \langle [\boldsymbol{T} - \overline{\boldsymbol{P}} \cdot \boldsymbol{N}] \cdot [\dot{\varphi} - \dot{\overline{\boldsymbol{F}}} \cdot \boldsymbol{X}] \rangle = 0 \tag{35}$$

The above procedure can analogously be applied to the electric part of the Hill-Mandel condition, see also Keip et al. (2014). There we use the fact that the referential electric field is defined as the material gradient of the scalar electric potential. Based on that, the second volume integral in (29) can be transformed into a surface integral as follows

$$-\langle\langle \boldsymbol{D} : \dot{\boldsymbol{E}} \rangle\rangle = \langle\langle \boldsymbol{D} : \mathrm{Grad}\, \dot{\phi} \rangle\rangle = \langle\langle \mathrm{Div}\, (\boldsymbol{D}\dot{\phi}) - \underbrace{\mathrm{Div}\, \boldsymbol{D}}_{=0} \cdot \dot{\phi} \rangle\rangle = \langle (\boldsymbol{D} \cdot \boldsymbol{N}) \cdot \dot{\phi} \rangle \tag{36}$$

where—in analogy to the mechanical case—we have used the product rule, Gauss' law $(24)_1$, and the divergence theorem. Now, in consideration of specific form of the microscopic potential along with the superposition principle of the electric displacements (25) respectively, this latter term can be manipulated according to

$$\langle (\boldsymbol{D} \cdot \boldsymbol{N}) \dot{\phi} \rangle = \overline{\boldsymbol{D}} \cdot \underbrace{\langle \dot{\tilde{\phi}} \boldsymbol{N} \rangle}_{=\langle\langle \dot{\tilde{\boldsymbol{E}}} \rangle\rangle = \boldsymbol{0}} + \langle (\widetilde{\boldsymbol{D}} \cdot \boldsymbol{N}) \dot{\tilde{\phi}} \rangle - \langle (\boldsymbol{D} \cdot \boldsymbol{N}) \boldsymbol{X} \rangle \cdot \dot{\overline{\boldsymbol{E}}}. \tag{37}$$

Note that in the above transformation, we have extracted the macroscopic variables $\overline{\boldsymbol{D}}$ and $\dot{\overline{\boldsymbol{E}}}$ from the volume integrals and exploited the fact that the first term on the right-hand side vanishes identically by definition $(25)_2$. Moreover, the last expression on the right-hand side can be recast into

$$-\langle (\boldsymbol{D}\cdot\boldsymbol{N})\boldsymbol{X}\rangle\cdot\dot{\overline{\boldsymbol{E}}} = \langle Q\boldsymbol{X}\rangle\cdot\dot{\overline{\boldsymbol{E}}} = -\overline{\boldsymbol{D}}:\dot{\overline{\boldsymbol{E}}} \tag{38}$$

where we have used $(20)_4$. Now, we combine the latter equation with (36) and (37) in order to arrive at

$$\underbrace{\overline{\boldsymbol{D}}:\dot{\overline{\boldsymbol{E}}} - \langle\langle \boldsymbol{D}:\dot{\boldsymbol{E}}\rangle\rangle}_{=\mathcal{P}^{\text{elec}}} = -\langle(\widetilde{\boldsymbol{D}}\cdot\boldsymbol{N})\dot{\widetilde{\phi}}\rangle. \tag{39}$$

According to the requirement that $\mathcal{P}^{\text{elec}} = 0$ in (29), we can equivalently state that

$$-\langle(\widetilde{\boldsymbol{D}}\cdot\boldsymbol{N})\dot{\widetilde{\phi}}\rangle = 0. \tag{40}$$

In consideration of the superposition principles (25) and $(26)_2$ this brings us to

$$-\langle(\widetilde{\boldsymbol{D}}\cdot\boldsymbol{N})\dot{\widetilde{\phi}}\rangle = \langle[Q + \overline{\boldsymbol{D}}\cdot\boldsymbol{N}]\cdot[\dot{\phi} + \dot{\overline{\boldsymbol{E}}}\cdot\boldsymbol{X}]\rangle = 0 \tag{41}$$

Putting together the above results finally gives the reformulated statement of (29)

$$\underbrace{\langle[\boldsymbol{t} - \overline{\boldsymbol{P}}\cdot\boldsymbol{N}]\cdot[\dot{\boldsymbol{\varphi}} - \dot{\overline{\boldsymbol{F}}}\cdot\boldsymbol{X}]\rangle}_{\widehat{\mathcal{P}}^{\text{mech}}} + \underbrace{\langle[Q + \overline{\boldsymbol{D}}\cdot\boldsymbol{N}]\,[\dot{\phi} + \dot{\overline{\boldsymbol{E}}}\cdot\boldsymbol{X}]\rangle}_{\widehat{\mathcal{P}}^{\text{elec}}} = 0. \tag{42}$$

From the conditions $\widehat{\mathcal{P}}^{\text{mech}} = 0$ and $\widehat{\mathcal{P}}^{\text{elec}} = 0$ we can now derive the energetically consistent Dirichlet and Neumann boundary conditions in terms of the mechanical quantities

$$\boldsymbol{\varphi} = \overline{\boldsymbol{F}}\cdot\boldsymbol{X} \quad \text{and} \quad \boldsymbol{t} = \overline{\boldsymbol{P}}\cdot\boldsymbol{N} \quad \text{on } \partial\mathcal{B}, \tag{43}$$

and the electrical quantities

$$\phi = -\overline{\boldsymbol{E}}\cdot\boldsymbol{X} \quad \text{and} \quad Q = -\overline{\boldsymbol{D}}\cdot\boldsymbol{N} \quad \text{on } \partial\mathcal{B}, \tag{44}$$

respectively. Periodic boundary conditions satisfying the two conditions are given on the mechanical side by

$$\boldsymbol{t}(\boldsymbol{X}^+) = -\boldsymbol{t}(\boldsymbol{X}^-) \quad \text{and} \quad \widetilde{\boldsymbol{\varphi}}(\boldsymbol{X}^+) = \widetilde{\boldsymbol{\varphi}}(\boldsymbol{X}^-) \quad \text{on } \boldsymbol{X}^\pm \in \partial\mathcal{B}^\pm, \tag{45}$$

and on the electrical side by

$$Q(\boldsymbol{X}^+) = -Q(\boldsymbol{X}^-) \quad \text{and} \quad \widetilde{\phi}(\boldsymbol{X}^+) = \widetilde{\phi}(\boldsymbol{X}^-) \quad \text{on } \boldsymbol{X}^\pm \in \partial\mathcal{B}^\pm, \tag{46}$$

where $\boldsymbol{X}^\pm \in \partial\mathcal{B}^\pm$ denote points on opposite faces of the RVE.

3.4 Consistent Linearization of Macroscopic Field Equations

For the solution of the nonlinear macroscopic and microscopic BVPs we apply the finite element method (FEM) at both scales. In order to keep the number of macroscopic iterations as small as possible, we have to linearize the macroscopic weak forms consistently. We now give a short and compact derivation of the algorithmic macroscopic tangent.

The linearization of the macroscopic weak forms goes along with the linearization of the macroscopic total Piola-Kirchhoff stresses and the macroscopic electric displacements

$$\Delta\overline{\boldsymbol{P}} = \frac{\partial\overline{\boldsymbol{P}}}{\partial\overline{\boldsymbol{F}}} : \Delta\overline{\boldsymbol{F}} + \frac{\partial\overline{\boldsymbol{P}}}{\partial\overline{\boldsymbol{E}}} \cdot \Delta\overline{\boldsymbol{E}} \quad \text{and} \quad \Delta\overline{\boldsymbol{D}} = \frac{\partial\overline{\boldsymbol{D}}}{\partial\overline{\boldsymbol{F}}} : \Delta\overline{\boldsymbol{F}} + \frac{\partial\overline{\boldsymbol{D}}}{\partial\overline{\boldsymbol{E}}} \cdot \Delta\overline{\boldsymbol{E}}, \tag{47}$$

where $\overline{\boldsymbol{P}}$ and $\overline{\boldsymbol{D}}$ are defined through volume averages $(22)_{3,4}$. The challenge in the derivation of the macroscopic moduli is the computation of the partial derivatives of the volume averages with respect to macroscopic gradient fields. We define

$$\frac{\partial\langle\langle\boldsymbol{P}\rangle\rangle}{\partial\overline{\boldsymbol{F}}} := \overline{\mathbb{A}}, \quad \frac{\partial\langle\langle\boldsymbol{D}\rangle\rangle}{\partial\overline{\boldsymbol{F}}} = \left[-\frac{\partial\langle\langle\boldsymbol{P}\rangle\rangle}{\partial\overline{\boldsymbol{E}}}\right]^T := \overline{\boldsymbol{q}}, \quad \text{and} \quad \frac{\partial\langle\langle\boldsymbol{D}\rangle\rangle}{\partial\overline{\boldsymbol{E}}} := \overline{\boldsymbol{\epsilon}} \tag{48}$$

with the macroscopic mechanical moduli $\overline{\mathbb{A}}$, dielectric moduli $\overline{\boldsymbol{\epsilon}}$, and electro-elastic moduli $\overline{\boldsymbol{q}}$ with $[\overline{\boldsymbol{q}}^T]_{\text{ijk}} := \overline{q}_{\text{kij}}$.

In order to arrive at a compact and generalized derivation of these quantities, we introduce the generalized-stress, generalized-gradient and generalized-tangent fields

$$\overline{\mathbb{S}} := \begin{bmatrix} \overline{\boldsymbol{P}} \\ -\overline{\boldsymbol{D}} \end{bmatrix}, \quad \overline{\mathbb{G}} := \begin{bmatrix} \overline{\boldsymbol{F}} \\ \overline{\boldsymbol{E}} \end{bmatrix} \quad \text{and} \quad \overline{\mathbb{C}} := \begin{bmatrix} \overline{\mathbb{A}} & -\overline{\boldsymbol{q}}^T \\ -\overline{\boldsymbol{q}} & -\overline{\boldsymbol{\epsilon}} \end{bmatrix} \tag{49}$$

with $[\overline{\boldsymbol{q}}^T]_{\text{ijk}} := \overline{q}_{\text{kij}}$. Consequently, we can write the linearization (47) into the simple format

$$\Delta\overline{\mathbb{S}} = \overline{\mathbb{C}} * \Delta\overline{\mathbb{G}}. \tag{50}$$

where the symbol $*$ denotes a generic operator that contracts the given tensor fields in an appropriate manner. Going on forward, we introduce associated generalized-stress, generalized-gradient and generalized-tangent fields at the microscale

$$\mathbb{S} := \begin{bmatrix} \boldsymbol{P} \\ -\boldsymbol{D} \end{bmatrix}, \quad \mathbb{G} := \begin{bmatrix} \boldsymbol{F} \\ \boldsymbol{E} \end{bmatrix} \quad \text{and} \quad \mathbb{C} := \begin{bmatrix} \mathbb{A} & -\overline{\boldsymbol{q}}^T \\ -\boldsymbol{q} & -\overline{\boldsymbol{\epsilon}} \end{bmatrix}. \tag{51}$$

where we have denoted the microscopic elastic, electro-elastic, and dielectric moduli through the partial derivatives

$$\mathbb{A} := \frac{\partial \boldsymbol{P}}{\partial \boldsymbol{F}}, \quad \boldsymbol{q} := \frac{\partial \boldsymbol{D}}{\partial \boldsymbol{F}} = -\left[\frac{\partial \boldsymbol{P}}{\partial \boldsymbol{E}}\right]^T, \quad \text{and} \quad \epsilon := \frac{\partial \boldsymbol{D}}{\partial \boldsymbol{E}}. \tag{52}$$

such that $[\boldsymbol{q}^T]_{\mathrm{ijk}} := q_{\mathrm{kij}}$.

Now, by using the superposition principle (25) together with the chain rule we can rewrite (50) in the form

$$\Delta\overline{\mathbb{S}} = \langle\langle \mathbb{C} + \mathbb{C} * \frac{\partial \widetilde{\mathbb{G}}}{\partial \overline{\mathbb{G}}} \rangle\rangle \Delta\overline{\mathbb{G}}. \tag{53}$$

In the latter equation the only unknown term is the partial derivative of the microscopic fluctuations with respect to macroscopic counterparts. The determination of this term will now be accomplished by means of the finite-element scheme of the microscopic BVP. This scheme is briefly summarized next (for more detailed information please refer to Keip et al. 2014).

In a Galerkin procedure, the weak forms of the balance equations are obtained in the above compact notation as

$$G = -\int_{\mathcal{B}} \mathrm{Div}\mathbb{S} * \delta\widetilde{\mathbb{p}}\, \mathrm{d}V = \int_{\mathcal{B}} \delta\widetilde{\mathbb{G}} * \mathbb{S}\, \mathrm{d}V - \int_{\partial\mathcal{B}} \delta\widetilde{\mathbb{p}} * (\mathbb{S} \cdot \boldsymbol{N})\, \mathrm{d}A, \tag{54}$$

where we have introduced the generalized-primary-variable field

$$\mathbb{p} := \begin{bmatrix} \boldsymbol{u} \\ \phi \end{bmatrix}. \tag{55}$$

Under the assumption of conservative loadings the linearization of (54) gives

$$\Delta G = \int_{\mathcal{B}} \delta\widetilde{\mathbb{G}} * \mathbb{C} * (\Delta\overline{\mathbb{G}} + \Delta\widetilde{\mathbb{G}})\, \mathrm{d}V \tag{56}$$

and at equilibrium it holds that

$$\Delta G = 0. \tag{57}$$

Based on the above equations we can now derive the macroscopic tangent through a finite-element approximation. The strategy is briefly discussed in the following. In order to again arrive at a compact notation we employ vector-matrix notation. In doing so, we introduce suitable vectors and matrices and label them with an underline (for example $\boldsymbol{u} \to \underline{\mathsf{u}}$). By means of this notation, we are able to introduce some basic fields. According to $(51)_{1,2}$, we define a *generalized-gradient* and a *generalized-stress* field in Voigt notation as

$$\begin{aligned} \underline{\mathsf{g}} &= [\mathsf{F}_{11}, \mathsf{F}_{22}, \mathsf{F}_{33}, \mathsf{F}_{12}, \mathsf{F}_{23}, \mathsf{F}_{13}, \mathsf{F}_{21}, \mathsf{F}_{32}, \mathsf{F}_{31}, \mathsf{E}_1, \mathsf{E}_2, \mathsf{E}_3]^\mathsf{T}, \\ \underline{\mathsf{s}} &= [\mathsf{P}_{11}, \mathsf{P}_{22}, \mathsf{P}_{33}, \mathsf{P}_{12}, \mathsf{P}_{23}, \mathsf{P}_{13}, \mathsf{P}_{21}, \mathsf{P}_{32}, \mathsf{P}_{31}, -\mathsf{D}_1, -\mathsf{D}_2, -\mathsf{D}_3]^\mathsf{T}, \end{aligned} \tag{58}$$

respectively. Consequently, we express the incremental constitutive relations at microscale through

$$\Delta\underline{s} = \underline{C}\,\Delta\underline{g} \quad \text{with} \quad \underline{C} := \begin{bmatrix} \underline{A} & -\underline{q}^T \\ -\underline{q} & -\underline{\epsilon} \end{bmatrix}, \tag{59}$$

where $\underline{A}$, $\underline{q}$, and denote the constitutive moduli $(51)_3$ in classical Voigt notation. Furthermore, we also introduce a *generalized-primary-variable* field in accordance to (55) as

$$\underline{p} = [u_1, u_2, u_3, \phi]^T. \tag{60}$$

At FE level, its FE approximations are denoted as

$$\{\underline{p}, \delta\underline{p}, \Delta\underline{p}\} = \underline{N}^e\,\{\underline{d}^e, \delta\underline{d}^e, \Delta\underline{d}^e\}, \tag{61}$$

where δ denotes a variation and Δ denotes a linear increment. The matrix $\underline{N}^e$ comprises the shape functions of the finite element. The vector $\underline{d}^e$ denotes the degrees of freedom of the element (i.e. displacements and electric potential).

The corresponding approximations of the generalized gradient fields can be given by

$$\{\underline{g}, \delta\underline{g}, \Delta\underline{g}\} = \underline{B}^e\,\{\underline{d}^e, \delta\underline{d}^e, \Delta\underline{d}^e\} \tag{62}$$

where $\underline{B}^e$ are so-called B-matrices that contain derivatives of the shape functions. Inserting the above approximations into (57), we arrive at the compact form

$$\sum_{e=1}^{n_{elem}} \delta\underline{\tilde{d}}^{e,T}\,[\,\underbrace{\int_{\mathcal{B}^e} \underline{B}^{e,T}\,\underline{C}\,dV}_{\underline{L}^e}\,\Delta\underline{\bar{g}} + \underbrace{\int_{\mathcal{B}^e} \underline{B}^{e,T}\,\underline{C}\,\underline{B}^e\,dV}_{\underline{K}^e}\,\Delta\underline{\tilde{d}}^e\,] = 0. \tag{63}$$

Here, the matrix $\underline{K}^e$ is the element stiffness matrix and the matrix $\underline{L}^e$ is a so-called localization matrix. By means of typical finite-element assembling we rewrite the latter statement in global format as

$$\delta\underline{\tilde{d}}^T[\,\underline{L}\,\Delta\underline{\bar{g}} + \underline{K}\,\Delta\underline{\tilde{d}}\,] = 0. \tag{64}$$

This makes it possible to express the incremental nodal fluctuations as

$$\Delta\underline{\tilde{d}} = -\underline{K}^{-1}\underline{L}\,\Delta\underline{\bar{g}}. \tag{65}$$

Now, in order to derive the coupled macroscopic tangent, we plug the finite-element approximations of the generalized-gradient fields (62) and the L-matrices into (53). We obtain

$$\underline{\bar{C}} = \langle\langle\,\underline{C}\,\rangle\rangle + \frac{1}{V}\,\underline{L}^T\,\frac{\partial(\Delta\underline{\tilde{d}})}{\partial\underline{\bar{g}}}, \tag{66}$$

in which the partial derivative can be computed based on the solution of the global BVP (65) in time-discrete fashion as

$$\frac{\partial(\Delta\underline{\widetilde{\mathsf{d}}})}{\partial\overline{\mathsf{g}}} = \frac{\partial(\underline{\widetilde{\mathsf{d}}}_{n+1} - \underline{\widetilde{\mathsf{d}}}_n)}{\partial\overline{\mathsf{g}}_{n+1}} = -\underline{\mathsf{K}}^{-1}\,\underline{\mathsf{L}}. \tag{67}$$

This then finally yields the macroscopic tangent moduli

$$\boxed{\overline{\mathsf{C}} = \langle\langle\,\underline{\mathsf{C}}\,\rangle\rangle - \frac{1}{V}\underline{\mathsf{L}}^{\mathsf{T}}\,\underline{\mathsf{K}}^{-1}\,\underline{\mathsf{L}}.} \tag{68}$$

A final note may be on the numerical implementation of the macroscopic tangent (refer also to Schröder 2014). On first view, its calculation seems costly since it contains the inverse of the global microscopic stiffness matrix. However, there is a remedy: For its motivation we recall that formally the solution of a linear system of equations of the type $\underline{\mathsf{A}}\,\underline{\mathsf{x}} = \underline{\mathsf{b}}$ can formally always be obtained by writing $\underline{\mathsf{b}} = \underline{\mathsf{A}}^{-1}\underline{\mathsf{x}}$. This, however, does not mean that the solution of the system necessitates the computation of the inverse $\underline{\mathsf{A}}^{-1}$. In fact, the computation of $\underline{\mathsf{A}}^{-1}$ will for large matrices be extremely expensive so that one would prefer alternative methods. In the same way, one has to imagine the last $\underline{\mathsf{L}}$ in (68) as an array of several right-hand sides $\underline{\mathsf{b}}$ (the number of columns n of this array would in the present context be $\mathsf{n} = 4 + 2 = 6$ in 2D and $\mathsf{n} = 9 + 3 = 12$ in 3D). Then, one could write

$$\underline{\mathsf{K}}^{-1}\underline{\mathsf{L}} = \underline{\mathsf{Y}} \quad \Leftrightarrow \quad \underline{\mathsf{K}}\,\underline{\mathsf{Y}} = \underline{\mathsf{L}} \tag{69}$$

In this equation, the array $\underline{\mathsf{Y}}$ obviously also contains n columns. As in the above context, the solution of (69) does not afford the computation of the inverse. Instead, one can compute the solution corresponding to one of the columns, say $\underline{\mathsf{Y}}_1$, by any other (more sustainable, effective) means and employ this result for the determination of the remaining entries of $\underline{\mathsf{Y}}$. Then, it is straight-forward to rewrite (68) through

$$\underline{\mathsf{L}}^{\mathsf{T}}\,\underline{\mathsf{K}}^{-1}\,\underline{\mathsf{L}} = \underline{\mathsf{L}}^{\mathsf{T}}\,\underline{\mathsf{Y}}, \tag{70}$$

thus making the explicit computation of the inverse completely unnecessary.

As a final remark, it should be noted that in fact one does not even need to solve for $\underline{\mathsf{Y}}_1$ since the microscopic BVP needs to be solved anyway. Thus, once the solution of the microscopic BVP is obtained, the computation of the macrosopic tangent can be done "on the fly". For more information on how to compute the effective tangent we again refer to Schröder (2014) and the references cited therein.

4 Numerical Examples

In this section, we apply the above formulation to the multiscale simulation of electro-active polymer composites. In the first examples, we will quantify the effective actuation of some three-dimensional particle-reinforced elastomers loaded with a homogeneous macroscopic electric field. Thereafter, the method will be employed for the two-scale simulation of an electro-mechanical actuator with composite microstructure.

In the simulations we consider periodic microstructures and model elastomer and ceramic inclusions with the Neo-Hookean-type free-energy function

$$\begin{aligned}\widehat{\Psi}(\boldsymbol{C},\boldsymbol{E}) = &\frac{1}{2}\mu(\mathrm{tr}[\boldsymbol{C}]-3) + \frac{\lambda}{4}(J^2-1) - (\frac{\lambda}{2}+\mu)\ln J \\ &- \frac{1}{2}\epsilon_0(1+\frac{\chi}{J})\, J\, [\boldsymbol{C}^{-1} : (\boldsymbol{E}\otimes\boldsymbol{E})],\end{aligned} \tag{71}$$

where λ and μ are the Lamé parameters and χ is the electric susceptibility. The latter is defined zero in free space and greater than zero in dielectric materials. The material parameters are chosen in such a way that the matrix material mimics an elastomer and the inclusion represents some ceramic substance. The electric susceptibility of the matrix is $\chi^{\mathrm{mat}} = 7$ and the electric susceptibility of the inclusions is $\chi^{\mathrm{incl}} = 700$. The Young modulus of matrix and inclusion are given by $Y^{\mathrm{mat}} = 2\cdot 10^5\ \frac{\mathrm{N}}{\mathrm{m}^2}$ and $Y^{\mathrm{incl}} = 2\cdot 10^8\ \frac{\mathrm{N}}{\mathrm{m}^2}$. In order to account for the incompressibility[5] of the matrix we set the Poisson ratio to $\nu^{\mathrm{mat}} = 0.499$. The Poisson ratio of the inclusion is set to $\nu^{\mathrm{incl}} = 0.3$. For the finite-element discretization at micro-level tetrahedral elements with quadratic shape functions are employed in the 3D case. In the 2D case we employ triangular elements with quadratic shape functions at both scales.

4.1 Determination of the Effective Response of Electroactive Polymers with Spherical and Ellipsoidal Inclusions

In the following we investigate the effective response of electroactive polymer composites with different microstructures. The computations will be performed in three dimensions and consider particle-reinforced microstructures with cubic arrangement of spherical and ellipsoidal inclusions. The considered volume fractions of the inclusions are 4 and 8%. The aspect ratio of the ellipsoids is 3:1.

The effective actuation under applied macroscopic electric field in vertical direction is shown in Fig. 1 and the electric fields observed at microlevel are depicted in Figs. 2 and 3. Clearly, strong electric-field concentrations occur above and below

[5]In general, the electro-elastic response of elastomers is sensitive of the choice of Poisson's ratio. For simulations in the quasi-incompressible limit, i.e. $\nu \to 0.5$ mixed finite-element methods as discussed in Klinkel et al. (2013) and Ask et al. (2013) can be employed.

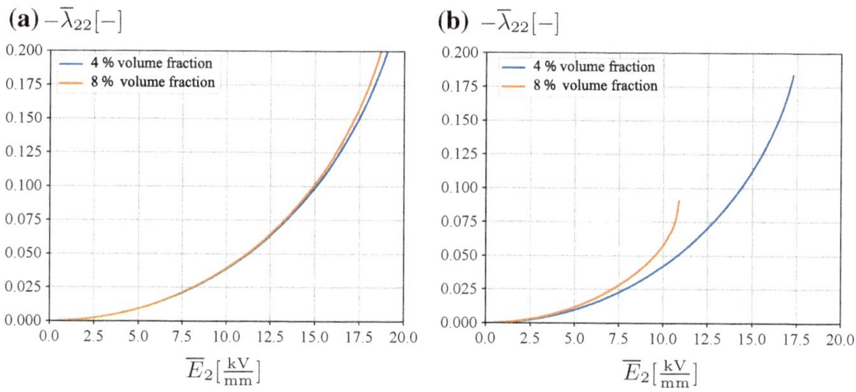

Fig. 1 Actuation of EAP composite under applied electric field: **a** actuation of the composite with spherical inclusion; **b** actuation of the composite with ellipsoidal inclusion. The major axis of the ellipsoidal inclusion points in the direction of the applied electric field. Obviously, the ellipsoidal inclusions leads to stronger coupling, but also to earlier instabilities

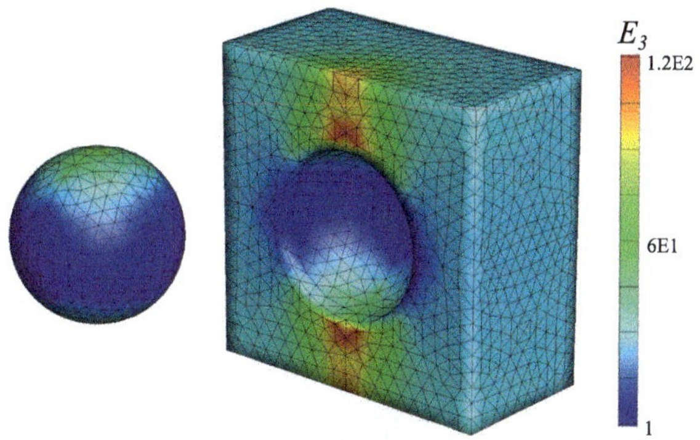

Fig. 2 Distribution of electric field in $\frac{\mathrm{kV}}{\mathrm{mm}}$ on micro-level for the spherical inclusion with 8% volume fraction. The concentrations of the electric field above and below the inclusion is obvious

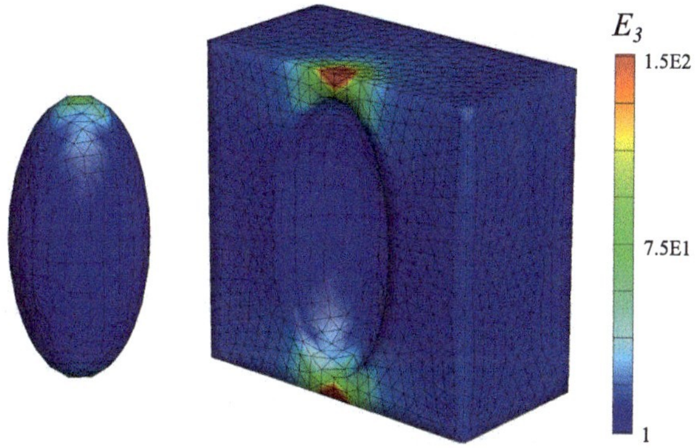

Fig. 3 Distribution of electric field in $\frac{\mathrm{kV}}{\mathrm{mm}}$ on micro-level for the ellipsoidal inclusion with 8% volume fraction (major axis oriented in field direction). The concentrations of the electric field above and below the inclusion is obvious

the inclusions. As expected, in case of ellipsoidal inclusions we oberve stronger coupling, but also earlier instabilities.

4.2 *Multiscale Simulation of Electromechanical Actuator with Composite Microstructure*

As second example, we consider a slender two-dimensional electrostatic actuator that is loaded via opposite electric-potential boundary conditions on top and on bottom. Mechanically the actuator is contrained in horizontal direction at its left and right end. For symmetry reasons, we have only simulated one half of the specimen. The microstructure of the material is given by a two-dimensional RVE with 20% volume fraction of inclusions. The deformation of the macroscopic body and two exemplary microstructures at different loads are shown in Fig. 4.

The sample as well as the microstructures deform as expected. As a result of the applied electric loading, the specimen contracts in vertical direction. Due to the Poisson effect, this leads to a horizontal extension which in turn lets the system deform in vertical direction. As can be seen, the deformations at micro- and macro-level are considerably high leading to a pronounced actuation of the electromechanical device.

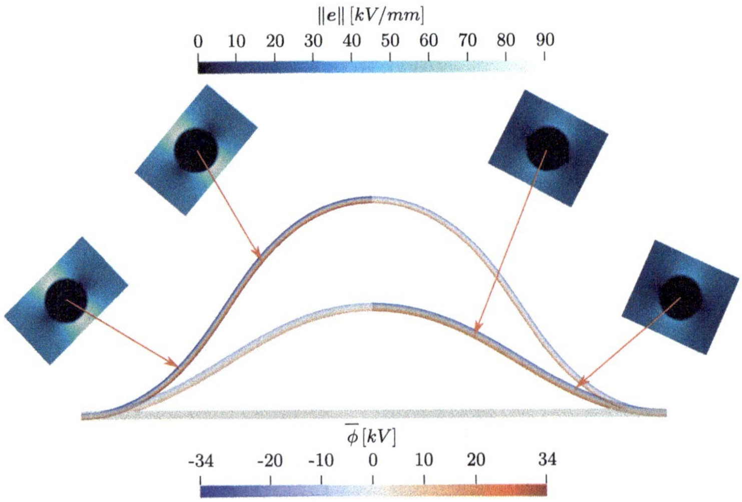

Fig. 4 Deformation of two-dimensional electrostatic actuator with periodic heterogeneous microstructure. The straight, horizontal geometry shows the initial configuration. Above, deformation states of the macro- and two selected micro-structures at two different loading stages are shown. The contour of the macroscopic specimen shows the electric potential resulting from the applied electric loading. The contour of the RVEs shows the norm of the microscopic electric field in current configuration. As can be observed large deformations at both micro- and macro-scale occur. Note that the RVEs have square shape in their initial configuration

5 Summary and Outlook

We have discussed a multiscale method for the simulation of electroelastic boundary value problems at large strains. Theory and implementation follow our previous work Keip et al. (2014), which now has been applied to the general three-dimensional case. As in the cited contribution, we have analyzed the effect of particle reinforcement on the effective actuation of electroactive polymers. We could confirm that the addition of particles with high electric permittvity has an influence on the effective actuation properties. In order to check the numerical feasability of the implementation we have applied the framework to the multiscale simulation of a typical electrostatic actuator. There we could observe both large strains and large rotations of the macroscopic specimen and the attached microstructures.

Acknowledgements This research has been facilitated through financial funding of the German Research Foundation (Research Group 1509 *Ferroic Functional Materials—Multiscale Modeling and Experimental Characterization*, grants no. KE 1849/2-2 & SCHR 570/12-1 and the Cluster of Excellence EXC 310 in *Simulation Technology*). This funding is gratefully acknowledged.

References

Ask, A., Menzel, A., & Ristinmaa, M. (2012). Electrostriction in electro-viscoelastic polymers. *Mechanics of Materials*, *50*, 9–21.

Ask, A., Denzer, R., Menzel, A., & Ristinmaa, M. (2013). Inverse-motion-based form finding for quasi-incompressible finite electroelasticity. *International Journal for Numerical Methods in Engineering*, *94*, 554–572.

Bar-Cohen, Y. (Eds.). (2001). *Electroactive polymer (EAP) actuators as artificial muscles: Reality, potential, and challenges*. SPIE Press.

Bertoldi, K., & Gei, M. (2011). Instabilities in multilayered soft dielectrics. *Journal of the Mechanics and Physics of Solids*, *59*(1), 18–42.

Bhattacharya, K., Li, J., & Xiao, Y. (2001). Electromechanical models for optimal design and effective behavior of electroactive polymers. In Y. Bar-Cohen (Ed.), *Electroactive polymer (EAP) actuators as artificial muscles: Reality, potential, and challenges* (pp. 309–330). SPIE Press.

Bustamante, R., Dorfmann, A., & Ogden, R. W. (2009). On electric body forces and maxwell stresses in nonlinearly electroelastic solids. *International Journal of Engineering Science*, *47*(11), 1131–1141.

Cao, C., & Zhao, X. (2013). Tunable stiffness of electrorheological elastomers by designing mesostructures. *Applied Physics Letters*, *103*(4), 041901.

Carpi, F., & Rossi, D. D. (2005). Improvement of electromechanical actuating performances of a silicone dielectric elastomer by dispersion of titanium dioxide powder. *IEEE Transactions on Dielectrics and Electrical Insulation*, *12*(4), 835–843. ISSN 1070-9878.

Carpi, F., De Rossi, D., Kornbluh, R., Pelrine, R. E., & Sommer-Larsen, P. (Eds.). (2011). *Dielectric elastomers as electromechanical transducers: Fundamentals, materials, devices, models and applications of an emerging electroactive polymer technology*. Elsevier.

Cohen, N. (2014). Multiscale analysis of the electromechanical coupling in dielectric elastomers. *European Journal of Mechanics-A/Solids*, *48*, 48–59.

Cohen, N., Menzel, A., & DeBotton, G. (2016). Towards a physics-based multiscale modelling of the electro-mechanical coupling in electro-active polymers. *Proceedings of the Royal Society A*, *472*, 20150462. (The Royal Society).

DeBotton, G., Tevet-Deree, L., & Socolsky, E. A. (2007). Electroactive heterogeneous polymers: analysis and applications to laminated composites. *Mechanics of Advanced Materials and Structures*, *14*(1), 13–22.
Dorfmann, A., & Ogden, R. W. (2005). Nonlinear electroelasticity. *Acta Materialia*, *174*(3–4), 167–183. ISSN 0001-5970.
Dorfmann, A., & Ogden, R. W. (2006). Nonlinear electroelastic deformations. *Journal of Elasticity*, *82*(2), 99–127. ISSN 0374-3535.
Eringen, A. C. (1963). On the foundations of electroelastostatics. *International Journal of Engineering Science*, *1*(1), 127–153.
Eringen, A. C., & Maugin, G. A. (1990). *Electrodynamics of continua*. New York: Springer.
Goshkoderia, A., & Rudykh, S. (2017). Electromechanical macroscopic instabilities in soft dielectric elastomer composites with periodic microstructures. *European Journal of Mechanics-A/Solids*, *65*, 243–256.
Goulbourne, N. C., Mockensturm, E. M., & Frecker, M. I. (2007). Electro-elastomers: Large deformation analysis of silicone membranes. *International Journal of Solids and Structures*, *44*(9), 2609–2626.
Hill, R. (1963). Elastic properties of reinforced solids—some theoretical principles. *Journal of the Mechanics and Physics of Solids*, *11*, 357–372.
Huang, C., Zhang, Q. M., & Su, J. (2003). High-dielectric-constant all-polymer percolative composites. *Applied Physics Letters*, *82*(20), 3502–3504.
Huang, C., Zhang, Q. M., DeBotton, G., & Bhattacharya, K. (2004). All-organic dielectric-percolative three-component composite materials with high electromechanical response. *Applied Physics Letters*, *84*, 4391–4393.
Huang, C., Zhang, Q. M., Li, J. Y., & Rabeony, M. (2005). Colossal dielectric and electromechanical responses in self-assembled polymeric nanocomposites. *Applied Physics Letters*, *87*(18), 182901–182901–3. ISSN 0003-6951.
Javili, A., Chatzigeorgiou, G., & Steinmann, P. (2013). Computational homogenization in magneto-mechanics. *International Journal of Solids and Structures*, *50*(25–26), 4197–4216. ISSN 0020-7683.
Keip, M.-A. (2012). *Modeling of electro-mechanically coupled materials on multiple scales*. Ph.D. thesis, Institute of Mechanics, Department Civil Engineering, University of Duisburg-Essen.
Keip, M.-A., & Rambausek, M. (2016). A multiscale approach to the computational characterization of magnetorheological elastomers. *International Journal for Numerical Methods in Engineering*, *107*, 338–360.
Keip, M.-A., & Rambausek, M. (2017). Computational and analytical investigations of shape effects in the experimental characterization of magnetorheological elastomers. *International Journal of Solids and Structures*, *121*, 1–20. https://doi.org/10.1016/.ijsolstr.2017.04.012.
Keip, M.-A., Steinmann, P., & Schröder, J. (2014). Two-scale computational homogenization of electro-elasticity at finite strains. *Computer Methods in Applied Mechanics and Engineering*, *278*, 62–79.
Keip, M.-A., Schrade, D., Thai, H., Schröder, J., Svendsen, B., Müller, R., et al. (2015). Coordinate-invariant phase field modeling of ferro-electrics, part ii: Application to composites and polycrystals. *GAMM-Mitteilungen*, *38*(1), 115–131.
Kim, K. J., & Tadokoro, S. (2007). *Electroactive polymers for robotics applications: Artificial muscles and sensors*. New York: Springer.
Klassen, M., Xu, B.-X., Klinkel, S., & Müller, R. (2012). Material modeling and microstructural optimization of dielectric elastomer actuators. *Technische Mechanik*, *32*(1), 38–52.
Klinkel, S., Zwecker, S., & Müller, R. (2013). A solid shell finite element formulation for dielectric elastomers. *Journal of Applied Mechanics*, *80*, 021026.
Kouznetsova, V., Geers, M. G. D., & Brekelmans, W. A. M. (2002). Multi-scale constitutive modelling of heterogeneous materials with a gradient-enhanced computational homogenization scheme. *International Journal for Numerical Methods in Engineering*, *54*(8), 1235–1260.

Kovacs, G., Lochmatter, P., & Wissler, M. (2007). An arm wrestling robot driven by dielectric elastomer actuators. *Smart Materials and Structures*, *16*(2), S306.
Kovetz, A. (2000). *Electromagnetic theory*. Oxford: Oxford University Press.
Labusch, M., Etier, M., Lupascu, D. C., Schröder, J., & Keip, M.-A. (2014). Product properties of a two-phase magneto-electric composite: Synthesis and numerical modeling. *Computational Mechanics*, *54*(1), 71–83.
Lefèvre, V., & Lopez-Pamies, O. (2017). Nonlinear electroelastic deformations of dielectric elastomer composites: Li-non-gaussian elastic dielectrics. *Journal of the Mechanics and Physics of Solids*, *99*, 438–470.
Li, J. Y., Huang, C., & Zhang, Q. M. (2004). Enhanced electromechanical properties in all-polymer percolative composites. *Applied Physics Letters*, *84*(16), 3124–3126. ISSN 0003-6951.
Li, W., & Landis, C. M. (2012). Deformation and instabilities in dielectric elastomer composites. *Smart Materials and Structures*, *21*(9), 094006.
Lopez-Pamies, O. (2014). Elastic dielectric composites: Theory and application to particle-filled ideal dielectrics. *Journal of the Mechanics and Physics of Solids*, *64*, 61–82. ISSN 0022-5096.
Lopez-Pamies, O., Goudarzi, T., Meddeb, A. B., & Ounaies, Z. (2014). Extreme enhancement and reduction of the dielectric response of polymer nanoparticulate composites via interphasial charges. *Applied Physics Letters*, *104*(24), 242904.
Markovic, D., Niekamp, R., Ibrahimbegovic, A., Matthies, H. G., & Taylor, R. L. (2005). Multi-scale modeling of heterogeneous structures with inelastic constitutive behavior. *International Journal for Computer-Aided Engineering and Software*, *22*(5/6), 664–683.
Maugin, G. (1988). *Continuum mechanics of electromagnetic solids* (Vol. 33). Amsterdam: North-Holland.
McMeeking, R. M., & Landis, C. M. (2005). Electrostatic forces and stored energy for deformable dielectric materials. *Journal of Applied Mechanics*, *72*(4), 581–590.
Michel, J. C., Moulinec, H., & Suquet, P. (1999). Effective properties of composite materials with periodic microstructure: A computational approach. *Computer Methods in Applied Mechanics and Engineering*, *172*, 109–143.
Miehe, C., & Koch, A. (2002). Computational micro-to-macro transitions of discretized microstructures undergoing small strains. *Archive of Applied Mechanics*, *72*(4), 300–317.
Miehe, C., Schotte, J., & Schröder, J. (1999a). Computational micro-macro transitions and overall moduli in the analysis of polycrystals at large strains. *Computational Materials Science*, *16*(1–4), 372–382.
Miehe, C., Schröder, J., & Schotte, J. (1999b). Computational homogenization analysis in finite plasticity. Simulation of texture development in polycrystalline materials. *Computer Methods in Applied Mechanics and Engineering*, *171*, 387–418.
Miehe, C., Vallicotti, D., & Teichtmeister, S. (2015a). Homogenization and multiscale stability analysis in finite magneto-electro-elasticity. *GAMM-Mitteilungen*, *38*(2), 313–343.
Miehe, C., Vallicotti, D., & Zäh, D. (2015b). Computational structural and material stability analysis in finite electro-elasto-statics of electro-active materials. *International Journal for Numerical Methods in Engineering*, *102*(10), 1605–1637.
Miehe, C., Vallicotti, D., & Teichtmeister, S. (2016). Homogenization and multiscale stability analysis in finite magneto-electro-elasticity. application to soft matteree, me and mee composites. *Computer Methods in Applied Mechanics and Engineering*, *300*, 294–346.
Müller, R., Xu, B.-X., Gross, D., Lyschik, M., Schrade, D., & Klinkel, S. (2010). Deformable dielectrics-optimization of heterogeneities. *International Journal of Engineering Science*, *48*(7), 647–657.
Özdemir, I., Brekelmans, W. A. M., & Geers, M. G. D. (2008a). Computational homogenization for heat conduction in heterogeneous solids. *International Journal for Numerical Methods in Engineering*, *73*(2), 185–204.
Özdemir, I., Brekelmans, W. A. M., & Geers, M. G. D. (2008b). FE^2 computational homogenization for the thermo-mechanical analysis of heterogeneous solids. *Computer Methods in Applied Mechanics and Engineering*, *198*(34), 602–613.

Pao, Y. H. (1978). Electromagnetic forces in deformable continua. In S. Nemat-Nasser (Ed.), *Mechanics today* (Vol. 4, pp. 209–306). Oxford: Pergamon Press.

Pelrine, R., Kornbluh, R., Pei, Q., & Joseph, J. (2000). High-speed electrically actuated elastomers with strain greater than 100%. *Science*, *287*(5454), 836–839.

Pelteret, J.-P., Davydov, D., McBride, A., Vu, D. K., & Steinmann, P. (2016). Computational electro-elasticity and magneto-elasticity for quasi-incompressible media immersed in free space. *International Journal for Numerical Methods in Engineering*, *108*(11), 1307–1342.

Plante, J.-S., & Dubowsky, S. (2006). Large-scale failure modes of dielectric elastomer actuators. *International Journal of Solids and Structures*, *43* (25–26), 7727–7751. ISSN 0020-7683.

Ponte Castañeda, P., & Siboni, M. H. (2012). A finite-strain constitutive theory for electro-active polymer composites via homogenization. *International Journal of Non-Linear Mechanics*, *47*(2), 293–306.

Rudykh, S., Lewinstein, A., Uner, G., & DeBotton, G. (2013). Analysis of microstructural induced enhancement of electromechanical coupling in soft dielectrics. *Applied Physics Letters*, *102*(15), 151905.

Rudykh, S., Bhattacharya, K., & DeBotton, G. (2014). Multiscale instabilities in soft heterogeneous dielectric elastomers. *Proceedings of the Royal Society A*, *470*(2162), 20130618.

Schröder, J. (2000). *Homogenisierungsmethoden der nichtlinearen Kontinuumsmechanik unter Beachtung von Instabilitäten*. Habilitation, Bericht aus der Forschungsreihe des Instituts für Mechanik (Bauwesen), Lehrstuhl I, Universität Stuttgart.

Schröder, J. (2009). Derivation of the localization and homogenization conditions for electro-mechanically coupled problems. *Computational Materials Science*, *46*(3), 595–599.

Schröder, J. (2014). A numerical two-scale homogenization scheme: The FE^2-method. In J. Schröder & K. Hackl (Eds.), *Plasticity and beyond, CISM International Centre for Mechanical Sciences* (Vol. 550, pp. 1–64). Springer. ISBN 978-3-7091-1624-1.

Schröder, J., & Keip, M.-A. (2010). A framework for the two-scale homogenization of electro-mechanically coupled boundary value problems. In M. Kuszma & K. Wilmanski (Eds.), *Computer methods in mechanics* (Vol. 1, pp. 311–329). Berlin, Heidelberg: Springer. ISBN 978-3-642-05241-5.

Schröder, J., & Keip, M.-A. (2012). Two-scale homogenization of electromechanically coupled boundary value problems. *Computational mechanics*, *50*, 229–244. ISSN 0178-7675.

Schröder, J., Labusch, M., Keip, M.-A., Kiefer, B., Brands, D., & Lupascu, D. C. (2015). Computation of non-linear magneto-electric product properties of 0–3 composites. *GAMM-Mitteilungen*, *38*(1), 8–24.

Schröder, J., Labusch, M., & Keip, M.-A. (2016). Algorithmic two-scale transition for magneto-electro-mechanically coupled problems: Fe2-scheme: Localization and homogenization. *Computer Methods in Applied Mechanics and Engineering*, *302*, 253–280.

Siboni, H. M., & Ponte, P. (2013). Castañeda. Dielectric elastomer composites: Small-deformation theory and applications. *Philosophical Magazine*, *93*(21), 2769–2801.

Siboni, M. H., & Ponte Castañeda, P. (2014). Finite-strain response and stability analysis, Fiber-constrained, dielectric-elastomer composites. *Journal of the Mechanics and Physics of Solids*, *68*, 211–238.

Siboni, M. H., Avazmohammadi, R., & Ponte, P. (2015). Castañeda. Electromechanical instabilities in fiber-constrained, dielectric-elastomer composites subjected to all-around dead-loading. *Mathematics and Mechanics of Solids*, *20*(6), 729–759.

Skatulla, S., Arockiarajan, A., & Sansour, C. (2009). A nonlinear generalized continuum approach for electro-elasticity including scale effects. *Journal of the Mechanics and Physics of Solids*, *57*(1), 137–160.

Skatulla, S., Sansour, C., & Arockiarajan, A. (2012). A multiplicative approach for nonlinear electro-elasticity. *Computer Methods in Applied Mechanics and Engineering*, *245–246*, 243–255.

Smit, R. J. M., Brekelmans, W. A. M., & Meijer, H. E. H. (1998). Prediction of the mechanical behavior of nonlinear heterogeneous systems by multi-level finite element modeling. *Computer Methods in Applied Mechanics and Engineering*, *155*, 181–192.

Somer, D. D., de Souza Neto, E. A., Dettmer, W. G., & Peric, D. (2009). A sub-stepping scheme for multi-scale analysis of solids. *Computer Methods in Applied Mechanics and Engineering*, *198*(9–12), 1006–1016.

Sridhar, A., Keip, M.-A., & Miehe, C. (2016). Homogenization in micro-magneto-mechanics. *Computational Mechanics*, *58*(1), 151–169.

Steinmann, P. (2011). Computational nonlinear electro-elasticity–getting started. In R. W. Ogden & D. J. Steigmann (Eds.), *Mechanics and electrodynamics of magneto-and electro-elastic materials, CISM International Centre for Mechanical Sciences* (Vol. 527, pp. 181–230). Springer.

Suo, Z., Zhao, X., & Greene, W. H. (2008). A nonlinear field theory of deformable dielectrics. *Journal of the Mechanics and Physics of Solids*, *56*(2), 467–486.

Terada, K., & Kikuchi, N. (2001). A class of general algorithms for multi-scale analyses of heterogeneous media. *Computer Methods in Applied Mechanics and Engineering*, *190*(40–41), 5427–5464.

Terada, K., Saiki, I., Matsui, K., & Yamakawa, Y. (2003). Two-scale kinematics and linearization for simultaneous two-scale analysis of periodic heterogeneous solids at finite strain. *Computer Methods in Applied Mechanics and Engineering*, *192*(31–32), 3531–3563.

Tian, L., Tevet-Deree, L., DeBotton, G., & Bhattacharya, K. (2012). Dielectric elastomer composites. *Journal of the Mechanics and Physics of Solids*, *60*(1), 181–198.

Toupin, R. A. (1956). The elastic dielectric. *Journal of Rational Mechanics and Analysis*, *5*(6), 849–915.

Vu, D. K., & Steinmann, P. (2007). Nonlinear electro- and magneto-elastostatics: material and spatial settings. *International Journal of Solids and Structures*, *44*(24), 7891–7905.

Vu, D. K., & Steinmann, P. (2010). A 2-d coupled bem-fem simulation of electro-elastostatics at large strain. *Computer Methods in Applied Mechanics and Engineering*, *199*(17), 1124–1133.

Vu, D. K., Steinmann, P., & Possart, G. (2007). Numerical modelling of non-linear electroelasticity. *International Journal for Numerical Methods in Engineering*, *70*(6), 685–704. ISSN 1097-0207.

Xu, B.-X., Mueller, R., Klassen, M., & Gross, D. (2010). On electromechanical stability analysis of dielectric elastomer actuators. *Applied Physics Letters*, *97*(16), 162908.

Xu, B.-X., Mueller, R., Theis, A., Klassen, M., & Gross, D. (2012). Dynamic analysis of dielectric elastomer actuators. *Applied Physics Letters*, *100*(11), 112903.

Zäh, D., & Miehe, C. (2013). Computational homogenization in dissipative electro-mechanics of functional materials. *Computer Methods in Applied Mechanics and Engineering*, *267*, 487–510.

Zäh, D., & Miehe, C. (2015). Multiplicative electro-elasticity of electroactive polymers accounting for micromechanically-based network models. *Computer Methods in Applied Mechanics and Engineering*, *286*, 394–421.

Zhang, Q. M., Li, H., Poh, M., Xia, F., Cheng, Z.-Y., Xu, H., & Huang, C. (2002). An all-organic composite actuator material with a high dielectric constant. *Nature*, *419*(6904), 284–287.

Zhang, S., Huang, C., Klein, R. J., Xia, F., Zhang, Q. M., & Cheng, Z.-Y. (2007). High performance electroactive polymers and nano-composites for artificial muscles. *Journal of Intelligent Material Systems and Structures*, *18*(2), 133–145.

Zhao, X., & Suo, Z. (2007). Method to analyze electromechanical stability of dielectric elastomers. *Applied Physics Letters*, *91*(6), 061921.

Zhao, X., & Wang, Q. (2014). Harnessing large deformation and instabilities of soft dielectrics: Theory, experiment, and application. *Applied Physics Reviews*, *1*(2), 021304.

MIX
Papier aus verantwortungsvollen Quellen
Paper from responsible sources
FSC® C105338

Printed by Books on Demand, Germany